PHYSICAL PROPERTIES OF POLYMERIC GELS

PHYSICAL PROPERTIES OF POLYMERIC GELS

Edited by

J. P. Cohen Addad

Université Joseph-Fourier Grenoble 1

JOHN WILEY & SONS

Chichester · New York · Brisbane · Toronto · Singapore

National 01243 779777
International (+44) 1243 779777

Other Wiley Editorial Offices

John Wiley & Sons, Inc., 605 Third Avenue,
New York, NY 10158-0012, USA

Jacaranda Wiley Ltd, 33 Park Road, Milton,
Queensland 4064, Australia

John Wiley & Sons (Canada) Pte Ltd, 22 Worcester Road,
Rexdale, Ontario M9W 1L1, Canada

John Wiley & Sons (SEA) Pte Ltd, 37 Jalan Pemimpin #05–04,
Block B, Union Industrial Building, Singapore 2057

Library of Congress Cataloging-in-Publication Data

Physical properties of polymeric gels / edited by J. P. Cohen Addad.
p. cm.
Includes bibliographical references (p. –) and index.
ISBN 0-471-93971-4 (alk. paper)
1. Polymer colloids. 2. Organic compounds. I. Cohen Addad, J. P.
QD549.2.P64P48 1996 95-10955
547.7'0454513—dc20 CIP

British Library Cataloguing in Publication Data

A catalogue record for this book is available from the British Library

ISBN 0 471 93971 4

Typeset in 10/12pt Times by Thomson Press (India) Ltd, New Delhi

CONTENTS

LIST OF CONTRIBUTORS

Dr M. Adam
Laboratoire Léon Brillouin
CEA–CNRS
CE–Saclay
91191 Gif-sur-Yvette Cedex
France

Dr J. Bastide
Institut Charles Sadron
6, rue Boussingault
67083 Strasbourg, Cedex
France

Dr S. J. Candau
Laboratoire d'Ultrasons et de
Dynamique des fluides Complexes
URA CNRS Université Louis Pasteur
4, rue Blaise Pascal
676070 Strasbourg, Cedex
France

Professor J. P. Cohen Addad
Laboratoire de Spectrométrie Physique
Université Joseph-Fourier, Grenoble 1
BP 87, 38402 St Martin d'Hères Cedex
France

Professor J. F. Joanny
Institut Charles Sadron
6 rue Boussingault
67083 Strasbourg Cedex
France

Dr D. Lairez
Laboratoire Léon Brillouin
CEA–CNRS
CE–Saclay
91191 Gif-sur-Yvette Cedex
France

Professor Wilhelm Oppermann
Institute for Textile and Fibrechemistry
University of Stuttgart
Pfaffenwaldring 55
D-70550 Stuttgart
Germany

Dr Uwe P. Schröder,
Institute for Textile and Fibrechemistry
University of Stuttgart
Pfaffenwaldring 55
D-70550 Stuttgart
Germany

FOREWORD

Gels are a spectacular form of soft matter—responding to very weak external perturbations. They swell or contract under minute changes of certain control parameters: adding salt, changing the pH, changing temperature; even shifting from an isotopic solvent to a nematic solvent. Most of these remarkable features are considered here—especially in the case of polyelectrolytes.

For many of us, the persistent dream about gels has been to construct some form of artificial muscle. This started long ago with A. Katchalsky. The obvious difficulty is related to *time constants*. For instance, if we want to create a volume charge by injecting H^+ ions, the diffusion of the ions over the macroscopic dimensions of the gel is very slow.

I have naively believed that this defect could be avoided if the gel was made of two interpenetrating networks: one elastic component (carrying carboxylic groups which are sensitive to pH) and one polymer conductor network. In our days, polymer conductors which are soluble in certain organic solvents—thanks to suitable size groups—begin to be available: so this whole idea is not entirely science fiction. The conducting network allows to set up a distributing electrode-emitting or absorbing protons inside all regions of the gel, and reducing certain time constants.

But even this sophisticated (and futuristic) approach is not sufficient: in a very recent Faraday discussion (Paris, September 1995) Prof. Kokhlov reminded us that when the gel contracts, it has to squeeze out the solvent: this, again, occurs by a diffusion process, which we have studied long ago, and which is intrinsically slow for macroscopic samples.

Thus, there is still room for a lot of imagination. And, as always, this imagination must be based on a precise understanding of our materials: how gels are formed, what inhomogeneities (extrinsic and intrinsic) exist inside them, what is their resistance to tear, etc. The present book, written by leading experts, gives an excellent vision of many such problems. I am convinced that it will be extremely useful.

P. G. de Gennes

PREFACE

The gel is a state of matter, neither liquid nor solid, or conversely both liquid and solid. It is much easier to describe a gel than to define it because its precise definition must refer to molecular structure and to the connectivity concept. Any gel consists of at least two components associated with a molecular network which is swollen by a liquid, whether or not this is a polymer. As a result, it possesses both the cohesive properties of solids and the deffusive transport properties of liquids. Gels tend to the soft elastically and highly active osmotically. This duality confers on gels their appealing absorptive and rheological properties, which are the bases of many important applications.

This book deals with organic polymeric gels. Network structures can be obtained as a result of chemical or physical processes. Covalent crosslinking leads to the formation of irreversible gels. In such gels the number of tie-points essentially does not change upon a variation of external conditions such as temperature, concentration or stress. By contrast, crosslinks resulting from weaker interactions (hydrogen bond, ion complexation, van der Walls' forces) can be broken, and their number and strength depend on thermodynamic and mechanical parameters; corresponding gels are reversible. When formed by physical processes, the tie-points often result from a complex organization of macromolecules (e.g. formation of small crystallites, mesophases, egg-box structures in ion complexations). Although the strength and the lifetime of crosslinks vary enormously with the nature of gel, many properties of reversible and covalent gels can be treated on an equal footing. Gels are frozen systems whose properties often reflect the conditions of preparation. As with glasses, there are major difficulties in understanding these systems. Although much progress has been made, many questions still remain open.

This book is presented to provide an overview of physical properties of polymeric gels with particular emphasis on the relation between the microscopic structure and the macroscopic behavior. Experimental techniques (neutron scattering, NMR, quasi-elastic light scattering) have been recently applied to probe systematically the structure of networks, enabling researchers to test and develop new theoretical approaches.

This book deals mainly with experimental methods which have been introduced to study irreversible as well as reversible gels; it also discusses concepts underlying the description of those systems.

The volume has five chapters. Each chapter includes a theoretical section which presents the theory underlying a particular topic and the status of the present understanding. Chapter 1 describes properties of semi-dilute solutions which can be considered as tenuous physical gels.

Chapter 2 is devoted to polyelectrolyte gels; typical properties such as swelling and elasticity are well illustrated by considering several experimental examples.

Chapter 3 deals with statistical structures of polymeric gels revealed from the transverse magnetic relaxation of protons attached to network chain segments. A general relationship between NMR and swelling or stretching properties is established.

Approaches to the sol–gel transition which are presented, in great detail, in Chapter 4, are both experimental and theoretical. Last concepts are evoked, while recent rheological and quasi-elastic light scattering measurements are reported and quantitatively discussed.

Microscopic structures of well-formed networks as probed from scattering techniques are described quantitatively in Chapter 5; polymeric gels are observed far from the gel-point, considering swelling or stretching effects to reveal structural aspects specific to neutral as well as charged polymeric gels.

1 SEMI-DILUTE POLYMER SOLUTIONS

J. F. Joanny
Institut Charles Sadron, Strasbourg, France

1.1 Introduction

In extremely dilute polymer solutions, different macromolecules do not overlap and their physical properties may be considered as single chain properties. As the concentration is increased, the volume fraction occupied by the chains in the solution becomes larger and interchain interactions are more and more relevant. As long as the volume fraction occupied by the chains is much smaller than one, the solution may be considered as a gas of chains interacting essentially as hard spheres (in a good solvent). When the volume fraction occupied by the chains is of

The Physical Properties of Polymeric Gels. Edited by J. P. Cohen Addad.

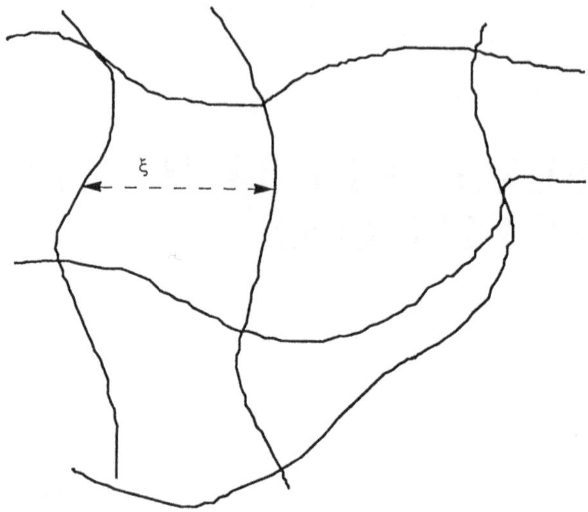

Figure 1. Qualitative picture for semi-dilute polymer solutions

the order of one, the chains start to overlap and one cannot consider them individually; their behavior becomes cooperative. The so-called overlap concentration ϕ^* (expressed as a volume fraction of the monomers) where the chains start to overlap is, however, still very small for polymer chains of high degree of polymerization; there exists a broad range of monomer volume fraction ϕ so that $\phi^* \ll \phi \ll 1$, called the semi-dilute regime where the chains overlap and make a dilute network in the solvent (Figure 1). This network shares many properties with the polymeric gels discussed in this book, but with the major difference that it constantly fluctuates and thus that the entanglements between different chains, that to some extent play the role of crosslinks, have a short lifetime. The theoretical study of semi-dilute polymer solutions has now reached a high degree of sophistication (de Gennes, 1985; Doi and Edwards, 1986; des Cloizeaux, and Jannink, 1987) and seems far simpler than that of equivalent polymeric gels (of the same polymer at the same concentration) because of the existence in the gel of the frozen constraints due to the crosslinks. For many purposes, an equivalent semi-dilute polymer solution can thus serve as a guide and a reference in the understanding of the physical properties of polymeric gels.

We give here a brief review of our theoretical understanding of semi-dilute polymer solutions. Historically the first thermodynamic study of these systems was the lattice mean field model constructed independently by Flory (1973) and Huggins (1942), this model as well as its extensions to describe spatial correlations in the solution are discussed in the following section. The mean field theory, however, implicitly assumes that polymer chains have a Gaussian structure at short length scales. This is true in a θ solvent or a poor solvent, and the, Flory–Huggins theory well describes the experimental results in these situations. In a good solvent, however, the chains are swollen at short length scales and one cannot neglect the correlations introduced by the excluded volume. A more refined theory has been initiated by des Cloizeaux using an analogy with critical phenomena and the renormalization group approach. This is the basis of the scaling theory which gives a good description of the properties of semi-dilute polymer solutions in good solvents. The scaling approach is summarized in section 1.3. The last section is devoted to dynamic properties. We consider first the cooperative behavior and then the individual reptative motion of the chains which has no analog in irreversible chemical gels but which may be of some relevance for physical gels.

1.2 Flory–Huggins theory of semi-dilute solutions

1.2.1 FREE ENERGY, OSMOTIC PRESSURE

Flory and Huggins have proposed a lattice model of semi-dilute polymer solutions where each site of the lattice is occupied either by a monomer or by a solvent molecule. The lattice constant a is equal to both the size of a monomer and that of a solvent molecule. Each chain has N monomers and the fraction of lattice sites occupied by the monomers is equal to the volume fraction ϕ. The free energy of the solution is written as the sum of an entropic contribution and an enthalpic contribution. The entropy is obtained by counting the number of ways to put the chains on the lattice. The entropy of mixing per site is

$$-S = \frac{\phi}{N} \log \phi + (1 - \phi) \log(1 - \phi)$$

We have chosen here and in the following temperature units where the Boltzmann constant is equal to 1.

This entropy may be interpreted as the sum of the translational entropies of a gas of chains of concentration ϕ/N and a gas of solvent molecules of concentration $1 - \phi$.

The enthalpy of mixing is calculated in a mean field manner by attributing interaction energies ε_{SS}. ε_{SP}, ε_{SP} to neighboring solvent–solvent, solvent– monomer and monomer–monomer molecules respectively. One obtains per site

$$H = T\chi\phi(1 - \phi)$$

The Flory parameter χ is defined as

$$\chi = \frac{z}{2T}(-\varepsilon_{SS} - \varepsilon_{PP} + 2\varepsilon_{SP}) \qquad (1)$$

where z is the coordination number of the lattice

The free energy per site is thus

$$\frac{F}{T}=\frac{\phi}{N}\log\phi+(1-\phi)\log(1-\phi)+\chi\phi(1-\phi) \tag{2}$$

In the semi-dilute regime ϕ is much smaller than one and the free energy may be expanded in powers of ϕ. The final form is written as a free energy per unit volume as a function of the concentration $c=\phi/a^3$

$$\frac{F}{T}=\frac{c}{N}\log ca^3+\frac{1}{2}vc^2+\frac{1}{6}w^2c^3 \tag{3}$$

where $v=a^3(1-2\chi)$ is the Edwards excluded followed parameter that vanishes at the θ temperature and $w^2(=a^6)$ is the third virial coefficient between monomers in the solvent. In Eq. (3), we have also ignored a contribution to the free energy proportional to the concentration ($c(\chi-1)$). In all the situations considered here this term is not important and can be eliminated by a redefinition of the origin of the chemical potential of the monomers.

The Flory–Huggins free energy is very similar to the free energy of a solution of small molecules, the only polymeric effect shows up in the entropic term which is smaller by a factor equal to the number of monomers N. In polymer solutions the entropic effects are thus much smaller than in small molecule solutions.

The osmotic pressure of the solution measures the change of the chemical potential of the solvent upon dissolution of the polymer $\pi=-(\mu_S-\mu_S^0)/a^3$. It is related to the free energy density by

$$\pi=c^2\frac{(\partial F/c)}{\partial c}$$

and reads

$$\frac{\pi}{T}=\frac{c}{N}+\frac{1}{2}vc^2+\frac{1}{3}w^2c^3 \tag{4}$$

In a semi-dilute solution the ideal gas contribution is small and in a good solvent ($v>N^{-1/2}$) the osmotic pressure is proportional to c^2.

1.2.2 CONCENTRATION CORRELATIONS

The structure factor of the solution is defined as $S(q)=\langle\delta c_q\delta c_q\rangle$ where

$$\delta c_q=\int \mathrm{d}\mathbf{r}\,\delta c(\mathbf{r})e^{-i\mathbf{qr}}$$

is the Fourier transform at a wave vector $\mathbf{q}$ of the local concentration fluctuation. This is the quantity measured in scattering experiments. Within the framework of the classical mean field theory, it can be obtained via the so-called random phase approximation (RPA).

An external potential $\delta\psi_q$ periodic in space with a wave vector $\mathbf{q}$, acting on all the monomers of an isolated Gaussian chain, creates a concentration fluctuation proportional to $\delta\psi_q$:

$$\delta c_q = -\chi_0(q)\delta\psi_q$$

The response function $\chi_0(q)$ is very generally related to the fluctuations of concentration characterized by the structure factor $S_0(q)$ of the Gaussian polymer chain by the so-called linear response theory.

$$S_0(q) = T\chi_0(q) \tag{5}$$

The structure factor $S_0(q)$ of an isolated Gaussian chain is given by the Debye function that we approximate at intermediate wave vectors by

$$S_0(q) = \frac{12c}{q^2a^2} \tag{6}$$

In a semi-dilute solution, the local fluctuations of concentration created by an external potential induce an internal potential due to the interactions with neighbouring monomers $T(v + w^2c)\delta c_q$. The random phase approximation considers that each chain in the semi-dilute solution is perturbed by a total potential which is the sum of the external potential and this mean field induced potential. The concentration fluctuation δc_q is then calculated self-consistently as

$$\delta c_q = -\chi_0(q)[\delta\psi_q + T(v + w^2c)\delta c_q] \tag{7}$$

The response function of the solution defined by $\delta c_q = -\chi(q)\delta\psi_q$ is then

$$\chi(q) = \frac{\chi_0(q)}{1 + T\chi_0(q)(v + w^2c)} \tag{8}$$

Using the linear response Eq. (5) and Eq. (6) we obtain the structure factor

$$S(q) = \frac{c}{vc + w^2c^2 + q^2a^2/12} \tag{9}$$

The structure factor has thus a Lorentzian form as in the Ornstein–Zernicke approximation for simple fluids. Notice, however, that the q^2 term in the denominator does not stem from the short-range interaction between monomers (considered here as local) but from the elasticity of the polymer chains.

The value of the response function at zero wave vector is the inverse of the second derivative of the free energy and is consistent with Eq. (4) in the limit of high molecular weights; in this mean field theory, in a good solvent it scales as c^0.

The width in q of the structure factor defines the correlation length of the solution as

$$\xi = \frac{a}{[12(vc + w^2c^2)]^{1/2}} \tag{10}$$

In a good solvent, the Flory–Huggins theory predicts a variation of the correlation length with the concentration as $\xi - c^{-1/2}$.

1.2.3 SCREENING OF EXCLUDED VOLUME INTERACTIONS

In a good solvent, isolated polymer chains are swollen and their radius is larger than the Gaussian radius $R_0 = N^{1/2}a$. However, it was shown by Flory (1949) that in more concentrated solutions and melts, the excluded volume interaction is screened and that the polymer chains remain Gaussian. The screening may be studied quantitatively in a mean field theory with the help of the random phase approximation. We consider here a semi-dilute solution containing the polymer at a concentration c and a very small concentration c_{m} of free monomers of the same chemical nature as the polymer. In order to calculate the effective interactions between monomers in the polymer, we study the interactions between the free monomers mediated by the chains.

In a good solvent where we may ignore the three body interactions ($vc \gg w^2c^2$), the interaction free energy of the solution per unit volume reads

$$F_{\mathrm{int}} = \frac{T}{2}\, v(c + c_{\mathrm{m}})^2 \tag{11}$$

We now introduce an external potential $\delta\psi_q$ acting only on the free monomers that produces fluctuations of the concentrations δc_q and $\delta c_{\mathrm{m}q}$. The RPA equations relating these concentration fluctuations to the external potential are obtained by generalizing Eq. (7)

$$\begin{aligned} \delta c_q &= -\chi_0(q)[Tv(\delta c_q + \delta c_{\mathrm{m}q})] \\ \delta c_{\mathrm{m}q} &= -\chi_{\mathrm{m}}(q)[\delta\psi_q + Tv(\delta c_q + \delta c_{\mathrm{m}q})] \end{aligned} \tag{12}$$

where $\chi_{\mathrm{m}}(q)$ is the linear response function of free monomers, $\chi_{\mathrm{m}}(q) = c_{\mathrm{m}}/T$.

Eliminating the polymer concentration from Eq. (12) we obtain

$$\delta c_{\mathrm{m}q} = -\chi_{\mathrm{m}}(q)[\delta\psi_q + Tv'(q)\delta c_{\mathrm{m}q}] \tag{13}$$

$v'(q)$ is precisely the effective interaction between monomers at a wave vector $\mathbf{q}$ and is given by

$$v'(q) = v\left[1 - \frac{1}{1 + \gamma/Tv\chi_0(q)}\right] \tag{14}$$

The response function of a Gaussian polymer chain $\chi_0(q)$ is given by Eq. (5). The effective interaction between monomers is obtained by taking the inverse Fourier transform of Eq. (14)

$$v'(r) = v\left[\delta(\mathbf{r}) - \frac{1}{4\pi r\xi^2}\exp(-r/\xi)\right] \tag{15}$$

where the correlation length ξ is given by Eq. (10) with $w = 0$ and $\delta(\mathbf{r})$ is the Dirac delta function.

The effective interaction is thus the sum of two terms, a short-range strongly repulsive contribution and a weakly attractive contribution. At distances larger than ξ the two contributions compensate, there is thus no excluded volume interaction between monomers with an average distance larger than ξ. This is the screening effect: over length scales larger than ξ, the chains retain Gaussian statistics.

1.3 Scaling theory of semi-dilute solutions

The Flory–Huggins theory is a mean field model which assumes that the polymer chains are Gaussian at all length scales (the monomers are distributed at random on the lattice). However, the repulsive excluded volume interactions between monomers in a good solvent swell the chains at short length scales (smaller than the correlation length ξ) and introduce correlations that are not properly taken into account by the Flory–Huggins theory. A more detailed description incorporating excluded volume effects is now given using scaling arguments. We first briefly recall the behavior of isolated chains in a good solvent and then discuss semi-dilute solutions.

1.3.1 ISOLATED POLYMER CHAINS IN A GOOD SOLVENT

The basic idea underlying scaling arguments is that in the limit of infinite molecular weights the structure of a polymer chain is independent of the length scale and that a polymer chain is a self-similar fractal object (Mandelbrojt, 1982). The fractal dimension is defined by the variation of the number of monomers N with the chain radius R

$$N \sim R^{d_f} \qquad (16)$$

Alternatively, one may characterize the size of the chains by the swelling exponent defined as $\nu = 1/d_f$. The important property is that in a given universality class (for all linear polymers in a good solvent) the fractal dimension or equivalently the swelling exponent are universal, they do not depend on the actual chemical nature of the monomers provided that they are in a good solvent.

The first accurate determination of ν in a good solvent was made by Flory who estimated the chain radius by balancing the loss of entropy due to the swelling with the repulsive interactions between monomers. In a three-dimensional space, this gives

$$\nu = 3/5 \qquad d_f = 5/3 \qquad (17)$$

More precise estimates have been obtained through renormalization group calculations (Leguillou and Zinnjustin, 1985), series expansions (Muthukumar and Nickel, 1984) or numerical simulations (Kremer and Binder, 1988); the commonly accepted value is $\nu = 0.588$. In the following we will often use the Flory value which is a sufficiently good approximation for most purposes.

In a θ solvent, the excluded volume v vanishes and the chains remain Gaussian $\nu = 2$, the three body interactions (measured by w in Eq. 4) should in principle be taken into account, but they only lead to very small logarithmic corrections to the Gaussian behavior (Duplantier, 1980, 1982) that cannot be observed in practice.

The internal structure of the chain may be probed by measuring the structure factor $S(q)$ in the intermediate range ($qR \gg 1$, $qa \ll 1$). The structure factor is the total number of monomers in a volume of size q^{-1} around a given monomer. For a fractal object it thus decays with the wave vector as

$$S(q) \sim \frac{c}{q^{d_f}} \tag{18}$$

In a good solvent, d_f is approximatively given by Eq. (12); in a θ solvent (for a Gaussian chain) this scaling law is in agreement with the mean field result of Eq. (6) and decays with the wave vector as q^{-2}.

1.3.2 SCALING LAWS FOR SEMI-DILUTE SOLUTIONS IN A GOOD SOLVENT

In a good solvent the overlap concentration may be estimated as $c^* \sim N/R^3$ and scales thus as $N^{-4/5}$. Above this concentration, in the semi-dilute regime, different chains interpenetrate and make a fluctuating network. We retain for the description of the physical properties of semi-dilute solutions the qualitative picture provided by the mean field theory: there exists a screening length ξ (of the order of the mesh size of the network) such that monomers with an average distance smaller than ξ have an excluded volume interaction and that the interaction is screened for monomers with an average distance larger than ξ.

It is then convenient to group the monomers into subunits of size ξ called blobs, each containing g monomers. Inside the blobs the chains have an excluded volume statistics and the blobs show the excluded volume fractal dimension $d_f = 5/3$. Over larger distances, the excluded volume interaction is screened and a semi-dilute solution may be considered as an ideal melt of close-packed blobs. This leads to the following relations between ξ and g:

$$\begin{aligned} g &= (\xi/a)^{d_f} \\ g &= c\xi^3 \end{aligned} \tag{19}$$

The screening length varies thus with the volume fraction $\phi = ca^3$ as

$$\xi \sim a\phi^{-1/(3-d_f)} \sim a\phi^{-3/4} \tag{20}$$

The exponent is slightly higher than the mean field value (Eq. (10)) because of the swelling of the blobs induced by the excluded volume. At the overlap concentration ϕ^*, the correlation length becomes of the order of the radius of an isolated chain in a good solvent $R \sim N^{3/5}a$.

The blob picture also allows the determination of the concentration correlations of the semi-dilute solutions. At large wave vector the internal structure of the blob is probed and one sees the single chain structure factor given by Eq. (18). At lower wave vector the blobs are uncorrelated and the structure factor is given by a Guinier formula for the blobs

$$S(q) = \frac{cg}{1 + q^2\xi^2} \tag{21}$$

This has the same Lorentzian form as the structure factor obtained in the classical theory (Eq. (9)), but the concentration power laws are modified; the zero wave vector value decreases with the volume fraction as $S(0) \sim \phi^{-1/4}$.

At length scales larger than the blob size, a labelled chain in the solution has Gaussian statistics and may be viewed an ideal chain of blobs, its radius scales thus as

$$R \sim \left(\frac{N}{g}\right)^{1/2} \xi \sim N^{1/2}\phi^{-1/8}a \tag{22}$$

In a melt, $\phi = 1$ and the chain is Gaussian at all length scales; at the overlap concentration $\phi = N^{-4/5}$, the radius crosses over smoothly to the radius of an isolated chain in a good solvent.

The scaling law for the osmotic pressure is obtained by dimensional analysis by noting that a blob has a free energy of order T

$$\frac{\pi}{T} \sim \xi^{-3} \sim \phi^{9/4} \tag{23}$$

The excluded volume interaction reduces the number of binary contacts between monomers in the solution, the osmotic pressure is thus smaller than the mean field value given by Eq. (4).

All these scaling laws are in very good agreement with several neutron scattering experiments (Daoud *et al*, 1975; de Cloizeaux and Jannink 1987).

The scaling theory of semi-dilute solutions was first proposed by des Cloizeaux by analogy with critical phenomena (des Cloizeaux, 1975). As for critical phenomena, a more detailed theory can be made that does not calculate only universal critical exponents but also universal scaling functions. We take here as an example the osmotic pressure following the lines of des Cloizeaux (1975).

We first define the overlap concentration more precisely as $c^* = N3^{3/2}/R^3$, where R is the end-to-end distance of the chain, and we measure the concentration by $x = c/c^*$. In a dilute solution, $x \ll 1$ and the osmotic pressure is that of an ideal solution $\pi/T = c/N$. At any finite concentration, dimensional analysis suggests that the osmotic pressure in a scaling form is written as

$$\frac{\pi}{T} = \frac{c}{N} h(x) \tag{24}$$

Renormalization group calculations show that the scaling function h is universal (in the asymptotic limit of very high molecular weights). This crossover function must also give the know scaling laws in dilute and semi-dilute solutions.

In a dilute solution ($x \ll 1$) the scaling function is analytic and may be expanded in powers of x:

$$h(x) = 1 + \tfrac{1}{2}(2\pi)^{3/2} g x \tag{25}$$

The first term is the ideal solution law, the second term is the first term of a virial expansion. Different chains interact as hard spheres and the virial coefficient between chains defined by writing the osmotic pressure as

$$\frac{\pi}{T} = \frac{c}{N} + \frac{1}{2} A_2 \left(\frac{c}{N}\right)^2$$

is proportional to the volume R^3 of the chains. Here A_2 is given by Eq. (25), where the dimensionless virial coefficient g has the universal value 0.233 calculated by renormalization group methods.

In semi-dilute solutions $x \gg 1$ the scaling function must behave as a power law

$$h(x) = h_\infty x^p \tag{26}$$

In agreement with Eq. (20), the exponent p is equal to $3/(3 - d_f) \sim 9/4$. The molecular weight dependence of the osmotic pressure cancels in a semi-dilute solution, the osmotic pressure is a local quantity which can only depend on the local concentration and not on the molecular weight of the polymers. The universal amplitude h_∞ has been calculated by renormalization group methods and is equal to 2.46.

A calculation of the function h over the full range of concentration has been made by several authors (Knoll, Schäffer and Witten, 1981; Schäffer, 1982); Ohta and Nakanishi, 1983); we refer to the book of desCloizeaux and Janink (1987) for a more detailed review. Systematic measurements of osmotic pressure and radius of gyration have allowed for a satisfactory comparison between theory and experiment for the scaling function h (desCloizeaux and Noda, 1982).

Finally, let us mention that other scaling functions can be defined for various physical quantities such as the radius of gyration of the chains or the correlation length ξ.

1.3.3 SEMI-DILUTE SOLUTIONS IN θ AND MARGINAL SOLVENTS

At the Flory compensation temperature θ, the excluded volume parameter v vanishes and the polymer chains remain in a first approximation ideal at any concentration. The scaling behavior for the correlation length ξ and the osmotic pressure are given by the mean field theory (Eqs (4) and (10)).

$$\xi \sim \phi^{-1}, \qquad \pi/T \sim \xi^{-3} \sim \phi^3 \tag{26a}$$

When the excluded volume increases from zero to a finite value these laws must cross over to the good solvent scaling laws; this cross-over may be described in terms of so-called thermal blobs. If the dimensionless excluded volume parameter $\tau = v/a^3$ that measures the distance from the θ temperature is much smaller than one, the interaction between monomers is smaller than T, and at short length scales the polymer chains are not swollen. We now group the monomers into thermal blobs, each containing g_t monomers; the interaction free energy between thermal blobs increases with the number of monomers in the blob and we choose g_t such that this interaction is exactly T. Inside the thermal blobs, the excluded volume interaction is small and the thermal blobs have Gaussian statistics, their size is thus $\xi_t = g_t^{1/2}a$; at larger length scales, the excluded volume interaction is relevant and the polymer has good solvent statistics; however, one must renormalize the monomers to the thermal blobs.

The interaction free energy between blobs is $F_b/T \sim vg_t^2/\xi_t^3$ giving the size of the thermal blobs

$$g_t \sim \tau^{-2}, \qquad \xi_t \sim \tau^{-1}a \tag{27}$$

In a semi-dilute solution in a good solvent, the correlation length ξ is larger than the thermal blob size ξ_t and one may use the scaling laws of section 1.3 with a monomer renormalized to the thermal blob; the correlation length scales with temperature and concetration as

$$\xi \sim a\phi^{-3/4}\tau^{-1/4}$$

In a semi-dilute solution in a θ solvent the concentration correlation length becomes smaller than the thermal blob size and the correlation length of the solution is given by Eq. (26a). The cross-over occurs when $\xi = \xi_t$ or on a line in a temperature-concentration plane such that

$$\tau = \phi \tag{28}$$

There are thus two ways to induce the crossover from a good solvent to a θ solvent, either by increasing concentration and thus decreasing ξ until it becomes smaller than ξ_t, or by decreasing temperature and thus increasing ξ_t until it becomes larger than ξ.

All the scaling results are often summarized on a temperature-concentration phase diagram first introduced first by Daoud and Jannink (1976) where the various scaling regimes are displayed and are separated by the crossover lines, this diagram is shown on Figure 2. The region in this diagram where $\xi \sim \xi_t$ ($\phi \sim \tau$) is sometimes called the marginal solvent regime, it becomes broader when the polymer has a finite persistence length (Schaeffer, Joanny and Pincus, 1980).

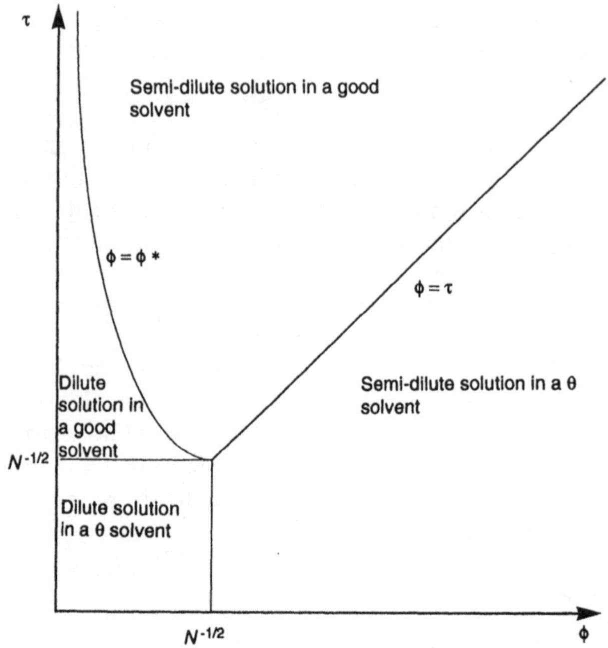

Figure 2. The Daoud–Jannink phase diagram displaying the various possible behaviors of semi-dilute polymer solutions

1.4 Hydrodynamic properties of semi-dilute solutions

We now discuss very briefly the hydrodynamic properties of semi-dilute polymer solutions, limiting the discussion to good solvents which are better understood and for which a scaling description seems appropriate. We first recall the hydrodynamic properties of dilute solutions where the motion of different chains can be studied independently. This also illustrates the role of the hydrodynamic interactions between monomers. In a semi-dilute solution we first focus on the dynamics of the concentration fluctuations and thus on the collective behavior of the solution. The last section is devoted to the individual motion of labelled chains.

The presentation is here limited to scaling laws, we refer to the book of Doi and Edwards (1986) for a more detailed description.

1.4.1 HYDRODYNAMIC INTERACTIONS AND SCREENING

The essential microscopic quantity governing the hydrodynamic properties of polymer solutions is the monomer friction ζ describing the motion of an isolated monomer in the solvent; if we idealize the monomer, as a sphere of radius a, ζ is related to the solvent viscosity η_0 by Stokes' law ($\zeta = 6\pi\eta_0 a$). The problem is to relate the macroscopic measurable quantities such as the chain friction or diffusion constant to this microscopic parameter.

The simplest hydrodynamic model in dilute solution assumes that the solvent drains freely into the polymer chain. In this model introduced by Rouse (1953), the friction force of the solvent on a given monomer depends solely on the velocity of that monomer independently of the other monomers. The total friction on a chain ζ_{ch} is then proportional to the number of monomers

$$\zeta_{\mathrm{ch}} = N\zeta \tag{29}$$

The diffusion constant of a chain is related to the chain friction by the Einstein relation $D_0 = T/\zeta_{\mathrm{ch}}$ and is thus inversely proportional to the number of monomers.

The Rouse free-draining model is based on a local response of the monomers; however, it is well known that the motion of each monomer creates a backflow velocity field in the solvent that decays very slowly with the distance (as $1/r$). This is the origin of the hydrodynamic interaction between the monomers, each monomer moves in the flow field created by all the others. The hydrodynamic interactions are taken into account in a modified version of the Rouse model, the Zimm model (Zimm, 1956). The result is that because of the hydrodynamic interactions, the flow field does not penetrate the polymer chain and the polymer moves as a whole like an equivalent sphere with a hydrodynamic radius R_{H} proportional to its radius of gyration R_{G}. The ratio $R_{\mathrm{G}}/R_{\mathrm{H}}$ is in the limit of infinite molecular weights a universal number equal to 1.50 for a Gaussian chain and slightly larger in a good solvent. The chain friction and the diffusion constant are then given by Stokes' law

$$\zeta_{\mathrm{ch}} = 6\pi\eta_0 R_{\mathrm{H}}, \qquad D_0 = T/6\pi\eta_0 R_{\mathrm{H}} \tag{30}$$

The Zimm model has been extended to semi-dilute solutions by Edwards and Freed (1974) who have shown that similarly to the static interactions, the hydrodynamic interactions between monomers are screened at length scales larger than the hydrodynamic screening length λ: because of the interferences between the velocity fields created by different monomers, the backflow of one monomer decreases exponentially after a length λ. The Zimm model and the screening of the hydrodynamic interactions are the basis of our understanding of the hydrodynamic properties of semi-dilute solutions presented below in a scaling form.

1.4.2 CONCENTRATION FLUCTUATIONS, COOPERATIVE DIFFUSION

The relaxation of the concentration fluctuations is directly measured in a quasielastic scattering experiment of light or neutrons. We consider here a sinusoidal concentration fluctuation along the x-direction $\delta c(x, t) = \delta c_q(t) \sin qx$ and study the relaxation of the amplitude $\delta c_q(t)$ towards zero.

At low wave vectors the relaxation is diffusive, the amplitude decays exponentially with time and the relaxation time τ_q can be written as $\tau_q = 1/D_c q^2$. This defines the cooperative diffusion constant D_c. In a dilute solution, the cooperative diffusion constant is the Zimm diffusion coefficient of single chains (Eq. (30)).

In a semi-dilute-solution, cooperative diffusion is a local mechanism, the concentration fluctuations relax via the breathing modes of the temporary network formed by the chains and do not involve the chain ends. We thus expect the cooperative diffusion constant to be independent of the number of monomers in a chain N. If we define the hydrodynamic correlation length by

$$D_c = \frac{T}{6\pi\eta_0\xi_H} \tag{31}$$

ξ_H depends only on the concentration c. We also expect ξ_H to be of the order of the screening length of the hydrodynamic interactions λ.

The concentration dependence of ξ_H may be obtained through a scaling law (de Gennes, 1976): in a dilute solution it should cross-over to the hydrodynamic radius R_H. We thus assume a scaling form

$$\xi_H \doteq R_H f(c/c^*) \tag{32}$$

the dimensionless function f is also expected to be universal. When $x = c/c^*$ goes to zero, in a dilute solution, f is equal to 1. At a finite but small value of x the scaling function may be expanded in a Taylor series. One obtains the so-called virial expansion of the diffusion constant which is very similar to Einstein's law for the diffusion constant of hard spheres: in dilute solutions, polymer chains can essentially be considered as hard spheres for hydrodynamic properties.

When x is large, in the semi-dilute regime, f decreases as a power law $f(x) \sim x^{-a}$ with an exponent a such that ξ_H is independent of molecular weight.

$$a = 1/(3 - d_f) = \tfrac{3}{4} \tag{33}$$

The hydrodynamic correlation length has thus the same scaling behavior as the static correlation length ξ. The ratio between these two lengths is a universal quantity that has been estimated either using mode coupling theories (Muthukumar and Edwards, 1982) or renormalization group methods (Oono, 1985).

A calculation of the scaling function f has been proposed by Oono and his coworkers (1986) over the full cross-over range.

Equations (33) and (32) show that up to non-important numerical constants, the cooperative diffusion constant is the Stokes diffusion constant of the blobs that can thus be considered as hard spheres for hydrodynamic properties, the screening length of the hydrodynamic interactions being of the order of the blob size. The hydrodynamic friction on a blob is given by the Einstein relation,

$\zeta_b = 6\pi\eta_0\xi_H$. The effective friction on a monomer, measured in sedimentation experiments, is obtained by dividing this friction constant by the number of monomers in a blob $g = c\xi^3$:

$$\tilde{\zeta} = 6\pi\eta_0/c\xi^2 \sim \phi^{1/2} \tag{34}$$

The friction constant on a monomer is thus renormalized by hydrodynamic interactions and becomes concentration dependent.

When the wave vector $\mathbf{q}$ of the scattering experiment is larger than ξ^{-1}, the cooperative diffusion constant becomes wave vector dependent. The scattering experiment probes a length scale q^{-1} smaller than the blob size and measures single chain properties independent of the concentration. The cooperative diffusion constant is then given by a Stokes–Einstein relation for the size q^{-1}

$$D_c \sim \frac{Tq}{6\pi\eta_0} \tag{35}$$

All these results are in good agreement with experimental data obtained either by quasielastic light scattering or by neutron scattering. Although the dynamic scaling theory is less developed than the corresponding static theory, it seems to describe reasonably well most of the experiments on semi-dilute polymer solutions in a good solvent.

1.4.3 REPTATIVE MOTION OF LABELLED CHAINS

The collective behavior of semi-dilute polymer solutions is governed by local concentration fluctuations and does not depend on the finite molecular weight of the chains. When motions over a length scale much larger than the blob size are considered, the effect of the chain ends becomes important and the molecular weight becomes a relevant variable for hydrodynamic phenomena: the hydrodynamic properties are dominated by entanglements. Our understanding of entangled polymeric liquids is mainly based on the reptation or tube model introduced first for polymer melts by deGennes (1971) and Edwards (1967). We will not give here a detailed review of this model because it is of little relevance for many polymeric gels. If the crosslinks have an infinite lifetime, for chemical gels, no motion over length scales larger than the distance between crosslinks is allowed, and the only hydrodynamic modes are the breathing modes governing the collective properties. If the crosslinks have a finite lifetime, for physical gels, a reptative motion is possible, but this has not been investigated in detail and is only partially understood.

The basic idea of the reptation model is that due to the entanglements with the neighboring chains, the curvilinear motion of a labelled chain along itself is much easier than the transverse motion. The effect of the entanglements is then modelled by a tube, inside which the chain is constrained to move. This leads to a relaxation time T_R proportional to N^3 and a self-diffusion constant D_s for labelled chains proportional to N^{-2}. The self-diffusion constant characterizes the motion of a tagged chain in the solution and is different from the cooperative diffusion constant which characterizes the decay of concentration fluctuations.

In a semi-dilute solution the tube size is the blob size ξ and the relaxation time and the diffusion constant may be obtained through scaling arguments

$$T_R \sim N^3\phi^{3/2}, \qquad D_s \sim N^{-2}\phi^{-7/4} \tag{36}$$

Notice that the self-diffusion constant D_s decreases with the concentration in contrast with the cooperative diffusion constant D_c that increases with concentration (Eqs (31, 33)).

1.5 Conclusions

The scaling theory for the static properties of semi-dilute polymer solutions gives a good agreement with many experimental results, not only for universal exponents but also for some universal amplitudes and scaling functions. For dynamic properties, the theory is less precise and the description of the experiments seems less accurate perhaps because the asymptotic limit where the theory is valid requires even larger molecular weights. This is particularly true in a θ solvent where the reptation theory is not well developed (Colby and Rubinstein, 1990) and where the experimental results on the dynamics of concentration fluctuations indicate a more complex behavior than the good solvent behavior described here.

We have confined our review to good or θ solvents which lead to stable solutions. In a poor solvent, at temperatures below the Flory compensation temperature, the polymer phase separates from the solvent. The phase transition is rather well described by the classical Flory–Huggins free energy. Some subtle corrections have been recently calculated in particular by Duplantier (1980, 1982) and are reviewed in des Cloizeaux and Jannink (1987). The phase separation between chemically different polymers in a common solvent has also been studied by extending the scaling theory to mixtures of polymers (Joanny, Leibler and Ball, 1984). In a common good solvent, the demixing transition is driven not by the leading scaling terms that we have discussed here but by the corrections to the scaling behavior; phase separation occurs well in the semi-dilute range.

More generally we have only considered neutral linear polymer chains. Some of the scaling ideas used for these systems have been extended to polymers with a more complex architecture such as branched polymers, which are of particular interest close to the gelation transition (Daoud and Joanny, 1981; Daoud and Leibler, 1988). In the special case of charged polymers, it is not clear whether the scaling approach gives a good description (Odjik, 1979; Joanny and Leibler, 1990). One must take into account the long-range character of the electrostatic interaction, but also the fact that polyelectrolyte solutions are at least three component systems including the solvent, the polymer, the counterions insuring the electrical neutrality and eventually some salt. The static properties of semi-dilute polyelectrolyte solutions are only partially understood and our understanding of their dynamic properties is rather poor (Witten and Pincus, 1987; Ajdari, Leibler and Joanny, 1991).

Finally, the scaling concepts for semi-dilute solutions are applicable if the volume fraction ϕ occupied by the polymer is small (although much larger than the overlap volume fraction), thus they cannot be applied to concentrated solutions. The thermodynamic properties of concentrated solutions have been discussed recently either by extending the Flory–Huggins lattice model (Bawendi and Freed, 1988) via a more precise estimation of both entropy and enthalpy or by numerical simulations (Deutsch and Dickman, 1990).

Acknowledgements

I am grateful to J. B. Bastide (ICS, Strasbourg) and A. Silberberg (Weizmann Institute, Rehovot) for a critical reading of the manuscript.

References

Ajdari, A., Leibler, L., Joanny, J F. (1991) *J. Chem. Phys.* **95**, 4580
Bawendi, M., Reed, K. (1988) *J. Chem. Phys.* **88**, 2741
desCloizeaux, J. (1975) *J. Physique* **36**, 281
desCloizeaux, J., Jannink, G. (1987) *Les Polymères en solution: leur modélisation et leur structure.* Editions de Physique Orsay
desCloizeaux, Noda I. (1982) *Macromolecules* **15**, 1505
Colby, R., Rubinstein, M. R. (1990) *Macromolecules* **23**, 2753
Daoud, M. *et al.* (1987) *Macromolecules* **8**, 804
Daoud, M., Janink, G. (1976) *J. Phys.* **37**, 973
Daoud, M., Joanny, J. F. (1981) *J. Phys.* **42**, 1359
Daoud, M., Leibler, L. (1988) *Macromolecules* **21**, 1497
Deutsch, H. D., Dickman, R. D. (1990) *J. Chem. Phys.* **93**, 8983
Doi, M., Edwards, S. F. (1986) *The Theory of Polymer Dynamics*, Oxford Science publications, Oxford
Duplantier, B. (1980, 1982) *J. Phys.* **41**, L409; **43**, 991
Edwards, S. (1967) *Proc. Phys. Soc. London* **92**, 9
Edwards, S., Freed, K. (1974) *J. Chem. Phys.* **61**, 1189, 3626
Flory, P. J. (1949) *J. Chem. Phys.* **17**, 303
deGennes, P. G. (1971) *J. Chem. Phys.* **55**, 572
Flory, P. J. (1973) *Principles of Polymer Chemistry*, Cornell University Press, Ithaca
de Gennes, P. G. (1976) *Macromolecules* **9**, 587, 594
de Gennes, P. G. (1985) *Scaling Concepts in Polymer Physics*, Cornell University Press, Ithaca
Huggins, M. L. (1942) *J. Phys. Chem.* **46**, 161
Joanny, J. F., Leibler, L. (1990) *J. Phys.* **51**, 545
Joanny, J. F., Leibler, L., Ball, R. (1984) *J. Chem. Phys.* **81**, 4640
Knoll, A., Schaffer, L., Witten, T. (1981) *J. Physique* **42**, 767
Kremer, K. Binder, K. (1988) *Computer Physics Reports* **7**, 260

Leguillou, J. C., Zinnjustin, J. (1985) *J. Phys. Lett.* **46**, L137
Mandelbrojt, B. (1982) The *Fractal. Geometry of Nature*, Freeman, San Francisco
Muthukumar, M, Edwards, S. (1982) *Polymer* **23**, 345
Muthukumar, M., Nickel, B. G. (1984) *J. Chem. Phys.* **80**, 5839
Odijk, T. (1979) *Macromolecules*, **12**, 686
Ohta, T., Nakanishi, A. (1983) *J. Phys.* **A16**, 4155
Ono, Y. (1985) in *Advances in Chemical Physics*, eds Prigogine, I. and Rice, S. Wiley, New York
Oono, Y. Baldwin, P. (1986) *Phys. Rev.* **A33**, 3391
Rouse, P. (1953) *J. Chem. Phys.* **21**, 1272
Schäffer, L. (1982) *Macromolecules* **15**, 652
Schaeffer, D. S., Joanny, J. F., Pincus, P. (1980) *Macromolecules* **13**, 1280
Witten, T., Pincus, P. (1987) *Europhysics Letters* **3**, 315
Zimm, B. (1956) *J. Chem. Phys.* **24**, 269

2 PROPERTIES OF POLYELECTROLYTE GELS

U. P. Schröder and W. Oppermann
University of Stuttgart, Stuttgart, Germany

2.1 Introduction

Polyelectrolyte gels are networks made of ionic macromolecules swollen in water or aqueous fluids. It is the ionic nature of the chains which gives rise to the unique properties of these materials. The most striking characteristic is the ability to absorb up to several hundred times their own weight of water, while retaining coherence and elasticity. This property makes them a useful material known as superabsorbent polymer.

Because of the ionic charges present, the swelling behaviour is markedly dependent on the ionic strength of the solution and on the nature of the low molecular weight electrolytes, in particular on the valency of the counterions. However, the solvent water itself causes some problems as far as a basic physical

The Physical Properties of Polymeric Gels. Edited by J. P. Cohen Addad.

treatment of such a system is concerned: besides the long-range coulombic interactions, short-range interactions such as hydrogen bonding, hydrophobic interactions, etc. have to be taken into account. (Hooper *et al*, 1990; Prange, Hooper and Prausnitz, 1989).

In this chapter we will focus on the elastic properties and on the swelling behaviour of polyelectrolyte gels with emphasis on a general treatment.

When non-ionic networks swollen in organic solvents are considered, the common molecular statistics approach to swelling equilibrium is based on the assumption that the total change in Gibbs free energy ΔF is simply the sum of the free energy of mixing, ΔF_M, and the free energy of elastic dilation of the network, ΔF_{El} (Flory, 1953). ΔF_M is obtained from the liquid-lattice theory of polymer solutions developed by Flory and Huggins, while ΔF_{El} can be deduced from the statistical theory of rubber elasticity (Treloar, 1975):

$$\Delta F_M = RT(n_1 \ln \phi_1 + n_2 \ln \phi_2 + n_1 \chi \phi_2) \tag{1}$$

$$\Delta F_{El} = RTV_0 \nu^* [\tfrac{3}{2} A\eta(\phi_2^{-2/3} - 1)] \tag{2}$$

where n_1 and n_2 are the molar amounts of solvent and polymer in the gel, ϕ_1 and ϕ_2 the corresponding volume fractions, χ the Flory–Huggins interaction parameter, ν^* the number of elastically effective network chains per unit dry volume and V_0 the volume of the dry polymer network. Here R denotes the gas constant, T the absolute temperature and $\eta = \langle r^2 \rangle / \langle r^2 \rangle_0$, with $\langle r^2 \rangle$ being the mean-square end-to-end distance of chains in the unswollen network and $\langle r^2 \rangle_0$ the corresponding quantity for free, uncrosslinked chains under the same conditions; A is a numerical factor of order unity.

The term $n_2 \ln \phi_2$ in Eq. (1) will be neglected in the following because, with networks, n_2 is extremely small.

Swelling equilibrium is established when the free energy has reached a minimum or, equivalently, when there is no difference between the chemical potentials of the solvent in the two coexisting phases (Flory, 1953; Treloar, 1975).

$$0 = \Delta\mu_1 = \Delta\mu_{1,M} + \Delta\mu_{1,El} = \left(\frac{\partial \Delta F_M}{\partial n_1}\right)_{T,p} + \left(\frac{\partial \Delta F_{El}}{\partial n_1}\right)_{T,p} \tag{3}$$

$$\Delta\mu_{1,M} = RT(\ln \phi_1 + \phi_2 + \chi \phi_2^2) \tag{4}$$

$$\Delta\mu_{1,El} = \nu^* RT V_{01} \eta A \phi_2^{1/3} \tag{5}$$

with V_{01} as the molar volume of the solvent. The first two terms in the Flory–Huggins equation are due to the rise of entropy which occurs when solvent molecules are introduced into a polymer matrix. The rise of entropy is of a statistical nature and has its origin in the increase of the number of possible conformations a polymer chain can have when the system is diluted. The third term containing the χ-parameter covers all excess contributions, particularly enthalpic contributions, and reflects the quality of the solvent.

With polyelectrolyte systems there is an additional and even dominating mixing contribution due to the presence of counterions. It is common practice to treat this contribution as an additive term to the chemical potential. In order to calculate $\Delta\mu_{1,\mathrm{Ion}}$ quantitatively one has to know the effective concentration of released counterions in the network. This is a major challenge in the treatment of polyelectrolytes, and some comments on the problems connected therewith are given in section 2.2.

When the polyelectrolyte gel is not swollen in pure water but in a salt solution, in addition to Eq. (3) further equilibrium equations exist which deal with the distribution of added salt inside and outside the gel (Tanford, 1967). A qualitative treatment of these coupled equilibria and a prediction of the dependence of the degree of swelling on salt concentration is attempted in section 2.3.

The fact that polyelectrolyte gels show extraordinarily large degrees of swelling in solutions of low ionic strength makes it necessary to use a more realistic equation for the elastic contribution to the chemical potential, $\Delta\mu_{1,\mathrm{El}}$, instead of the Gaussian approach leading to Eqs 2 and 5 respectively, because the limited chain extensibility has to be incorporated (Opperman, Rose and Rehage, 1985). This was done in section 2.3 as far as the swelling properties are concerned, and in section 2.4 with regard to the elasticity of the swollen gel.

2.2 Some remarks on general properties of polyelectrolytes

Even in a very diluted state, polyelectrolyte solutions show strong deviations from thermodynamically ideal solutions (Armstrong and Strauss, 1969). This non-ideality is caused by the strong electrostatic potential which exists in the surroundings of a highly charged polyion. The physical treatment of such systems is particularly difficult because the shape of a polyion is not fixed and can change as a result of changes in the electrostatic interactions.

Many theoretical treatments of polyelectrolytes start out from a charged rod as a geometrical model of the real situation and solve the appropriate Poisson–Boltzmann equation (Fuoss, Katchalsky and Lifson, 1951; Alexandrowicz and Katchalsky, 1963). One of the important results obtained from such considerations is the notion that counterions are bound to the polyion when the reduced charge density ξ exceeds a certain threshold value, which is unity for monovalent counterions and $\frac{1}{2}$ for divalent counterions.

$$\xi = \frac{e_0^2}{4\pi\varepsilon\varepsilon_0 kTb} \tag{6}$$

where e_0 is the proton charge and b is the distance between (monovalent) charged groups on the polyion. The other quantities have their usual meaning.

For water as solvent at 25 °C, $\xi = 1$ corresponds to a critical value of b of 0.71 nm.

The role of the bound counterions is to reduce the effective charge density on the polyion, hence to increase the effective charge spacing b, thus bringing ξ to the threshold value. The customary expression for this phenomenon, which was first mentioned by Fuoss, Katchalsky and Lifson (1951) and Oosawa, Imai and Kagawa (1954) and Oosawa (1971) and later examined systematically by Manning (1969, 1978), is 'counterion condensation'. The counterions may exist in either of two states: bound to the polyion or free in solution.

With regard to the swelling equilibrium of a polyelectrolyte network, it is only the free counterions which contribute to the reduction of the chemical potential of the water or to the occurrence of an osmotic pressure. According to the above-mentioned theories, all counterions are free when the average charge spacing is larger than $b_{\text{crit}} = 0.71$ nm. When, on the other hand, b is below this value, only the fraction $b/b_{\text{crit}} = 1/\xi$ is free.

The treatment described so far is confined to an infinitely diluted solution of charged rods. The polyions in polyelectrolyte gels differ from this model in two important points: even at very high degrees of swelling, the concentration in terms of ionic strength is quite high, and the macromolecules are highly flexible and far from the rod-like, fully extended conformation.

Although several attempts have been made to investigate the effect of a curvature of the polyion on the electrostatic potential and on counterion condensation, there is no clear or quantitative opinion on this point except for the general view that more bending causes more condensation (Schmitz, 1993).

2.3 Swelling phenomena in polyelectrolyte gels

Polyelectrolyte gels are often swollen in aqueous solutions of low molecular weight salts. In that case, the equilibrium condition, Eq. (3), has to be expanded and similar equations apply to all diffusible components (Tanford, 1967).

The chemical potential of the solvent water outside the gel in the liquid phase, $\mu_1(l)$, then depends on the salt concentration:

$$\mu_1(l) = f(c_s(l)) \tag{7}$$

The chemical potential of water in the gel phase, $\mu_1(g)$, depends on the salt concentration there, $c_S(g)$, as well as on polymer concentration:

$$\mu_1(g) = f(c_s(g), \phi_2) \tag{8}$$

The equilibrium condition can then be written as

$$0 = \Delta\mu_1 = \Delta\mu_{1,\text{M}} + \Delta\mu_{1,\text{El}} + \Delta\mu_{1,\text{Ion}} \tag{9}$$

where the first two terms, $\Delta\mu_{1,\text{M}}$ and $\Delta\mu_{1,\text{El}}$, were already discussed in the introduction (Eqs (4) and (5)). The third term takes care of the fact that $c_S(g)$ and $c_S(l)$ are usually not identical. When one replaces activities by concentrations, $\Delta\mu_{1,\text{Ion}}$ can be expressed as (Ricka and Tanaka, 1984; Harsh and Gehrke, 1990;

Brannon-Peppas and Peppas, 1990):

$$\Delta\mu_{1,\text{Ion}} = -RTV_{01}\left(\sum c_i(g) - \sum c_i(l)\right) = -RTV_{01}\Delta c_{\text{tot}} \tag{10}$$

where the summations include all mobile ionic species, namely the ions of the salt as well as the free counterions inside the gel, which may or may not be identical to the ions of the salt. Replacement of activities by concentrations can be justified even at rather high ionic strengths because the mean ion activity coefficients (for free ions) inside and outside the gel are probably within the same range.

The difference in total concentration of mobile ions, Δc_{tot}, is caused by the fact that the charges on the polymer, which are confined to the gel phase, require an equivalent number of counterions to stay within the gel to achieve electroneutrality. These counterions will always be there in excess of additional ions which may diffuse into the gel, but can do so only in equivalent portions of positively and negatively charged species. To calculate Δc_{tot}, one proceeds from the equilibrium condition for the salt, which in the case of a 1–1 electrolyte (for example NaCl) reads

$$c_{\text{Na}^+}(g)\cdot c_{\text{Cl}^-}(g) = c_{\text{Na}^+}(l)\cdot c_{\text{Cl}^-}(l) \tag{11}$$

Suppose now that the polyelectrolyte is anionic, like a salt of polyacrylic acid. For convenience we may write

$$c_{\text{Na}^+}(l) = c_{\text{Cl}^-}(l) = c_s \tag{12}$$

Na^+ and Cl^- ions have diffused into the gel such as to build up the chloride concentration

$$c_{\text{Cl}^-}(g) = c' \tag{13}$$

The sodium ion concentration inside the gel is then

$$c_{\text{Na}^+}(g) = c' + \rho\phi_2/M_2 \tag{14}$$

because an amount equal to the chloride ions has moved into the gel (c'), and an additional amount corresponding to $\rho\phi_2/M_2$ is there to neutralize the $\rho\phi_2/M_2$ charges on the polymer network. Here M_2 is the molar mass of polymer per free monovalent counterion, and ρ is the density of the dry polymer.

Putting Eqs (12)–(14) into the equilibrium Eq. (11) yields

$$(c' + \rho\phi_2/M_2)c' = c_s^2 \tag{15}$$

This equation has to be solved with respect to c'. Then Δc_{tot} is determined and the expression obtained from Eq. (10) is

$$\Delta\mu_{1,\text{Ion}} = 2RTV_{01}\left[c_S - \sqrt{\left\{c_S^2 + \left(\frac{\rho\phi_2}{2M_2}\right)^2\right\}}\right] \tag{16}$$

In the case of a 1–2 salt like Na_2SO_4, the corresponding equations are

$$c^2_{Na^+}(g)\cdot c_{SO_4^{2-}}(g) = c^2_{Na^+}(l)\cdot c_{SO_4^{2-}}(l) \qquad (11a)$$

$$\left.\begin{aligned} c_{Na^+}(l) &= 2c_s \\ c_{SO_4^{2-}}(l) &= c_s \end{aligned}\right\} \qquad (12a)$$

$$c_{SO_4^{2-}}(g) = c'/2 \qquad (13a)$$

$$c_{Na^+}(g) = c' + \rho\phi_2/M_2 \qquad (14a)$$

$$(c' + \rho\phi_2/M_2)^2\cdot c'/2 = 4c_s^3 \qquad (15a)$$

In the case of a 2–1 salt like $CaCl_2$, one obtains

$$c_{Ca^{2+}}(g)\cdot c^2_{Cl^-}(g) = c_{Ca^{2+}}(l)\cdot c^2_{Cl^-}(l) \qquad (11b)$$

$$\left.\begin{aligned} c_{Ca^{2+}}(l) &= c_s \\ c_{Cl^-}(l) &= 2c_s \end{aligned}\right\} \qquad (12b)$$

$$c_{Cl^-}(g) = 2c'' \qquad (13b)$$

$$c_{Ca^{2+}}(g) = c'' + \rho\phi_2/2M_2 \qquad (14b)$$

$$(c'' + \rho\phi_2/2M_2)4c''^2 = 4c_s^3 \qquad (15b)$$

c'' being the calcium ion concentration in excess of that needed to neutralize the polyacrylic acid.

Equations (15a) and (15b) are cubic equations, hence an analytical solution such as Eq. (16) cannot be given. However, the concentration of cations inside the gel phase, c' or c'', is most conveniently computed by numerical iteration in order to calculate $\Delta\mu_{1,\text{Ion}}$.

In all cases, it was assumed that the amount of swelling agent is sufficiently large such that the salt concentration in the liquid phase does not change appreciably during swelling (infinite bath). Also a complete ion exchange should have occurred when the counterion provided by the salt is different from that originally present in the polyelectrolyte network.

A comment on the physical significance of the parameter M_2, the molar mass of polymer per free monovalent counterion, is expedient. If there were complete dissociation of the polyelectrolyte, M_2 would be equal to the molar mass of the monomeric repeat unit in the case of a charged homopolymer. Counterion condensation, however, can reduce the number of free counterions, so that M_2 is then markedly larger. We recall at this point the remarks on the extent of counterion condensation in section 2.2 stating that a quantitative consideration seems to be hardly possible.

If one wants to compare the consequences of counterion condensation for different types of gels, the parameter M_2 is not very suitable. For this purpose it could be more convenient to introduce a relative effective charge density, ψ, which expresses the fraction of free counterions with respect to the nominal charges present.

$$\psi = M_M/M_2 \tag{17}$$

where M_M is the molar mass of the monomeric unit.

An additional contribution to the chemical potential which may arise from electrostatic repulsive forces between charges fixed on the polymer chain and which was first mentioned by Ilavsky (Hasa, Ilavsky and Dusek, 1975), is not included in our treatment. It is our opinion that such forces influence the shape of a polyion only when the solution is very dilute and no additional electrolyte is present. This could be confirmed by measurements of the Kerr effect and by viscosity measurements (Oppermann, 1988; Vink, 1970; Cohen, Priel and Rabin, 1988). Gels are actually semi-dilute systems, so that an electrostatic contribution should be of minor importance compared to the strong Donnan-type mixing contribution considered here.

We now turn to the modification of the elastic term of the chemical potential, $\Delta\mu_{1,El}$. As polyelectrolyte gels can reach very high degrees of swelling, the network chains are highly extended. In this case the Gaussian statistics fail to describe the statistical properties of the chains properly. Instead, non-Gaussian statistics have to be introduced. We have used the well-known treatment based on the inverse Langevin function, $\mathcal{L}^{-1}$, to derive a formula for the change of the chemical potential due to swelling.

$$\Delta\mu_{1,El} = \tfrac{1}{3} A v^* RT V_{01} q_c^{-1} q_r^{-2/3} n^{1/2} \mathcal{L}^{-1}(q_r^{1/3} n^{-1/2}) \tag{18}$$

The derivation of Eq. (18) as well as the derivation of the non-Gaussian elastic properties (cf. section 2.4) rests on the assumption of affine displacement of crosslinks. Therefore, the incorporation of the structure factor A is somewhat inconsistent. The justification is due to the view that highly swollen gels are probably showing phantom-network behaviour (Flory, 1977).

Furthermore, we have assumed that $\langle r^2 \rangle / \langle r^2 \rangle_0$ is unity under the conditions of network formation. Then $\langle r^2 \rangle / \langle r^2 \rangle_0$ in the dry state can be expressed via the degree of swelling during crosslinking, q_c (Dusek and Prins, 1969). The relative degree of swelling, q_r, refers to this reference state, whereas the total degree of swelling, q_0, is given by the relation

$$q_0 = q_c \cdot q_r = 1/\phi_2 \tag{19}$$

As most of the reactions leading to polyelectrolyte networks are carried out in solution rather than in bulk, this is a convenient way of taking care of q_c.

The quantity n, the number of statistical segments per network chain, includes the non-Gaussian properties of the network; n depends on the stiffness of the polymer backbone and on the molar mass of the network chains. When n approaches infinity, Eq. (18) reduces to the Gaussian description (Eq. (5)).

We now have introduced four structural parameters that have to be known if the swelling behaviour of a polyelectrolyte network is to be described on the basis

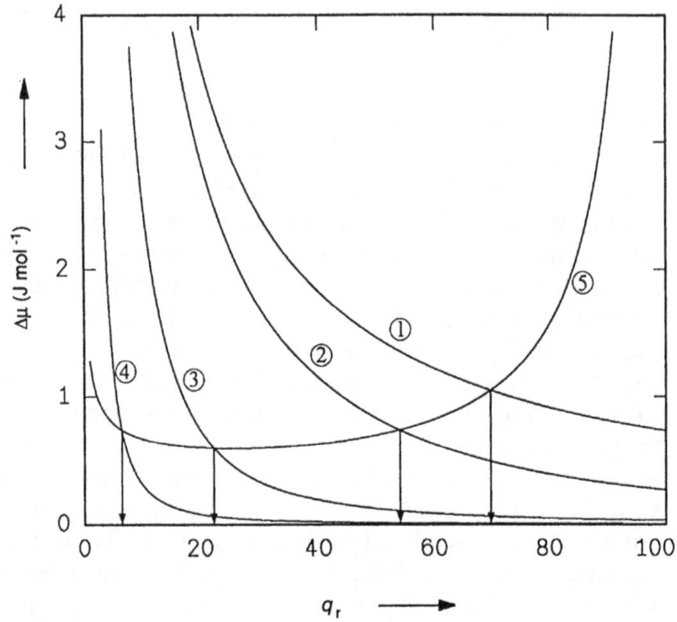

Figure 1. Determination of the equilibrium degree of swelling. (1–4) Ion concentration dependent term (negative), calculated with $\chi = 0.48$, $M_2 = 600\,\text{g mol}^{-1}$, $c_{NaCl} = 0$ (1), 0.01 (2), 0.1 (3) and 1 mol.l^{-1} (4); (5) elastic contribution, calculated with $\nu^*_{eff} = 200\,\text{mol m}^{-3}$, $n = 6$. Adopted with permission from Oppermann (1992)

of the foregoing treatment, namely ν^*, n, M_2 and χ. Figure 1 gives an example of the changes of $-(\Delta\mu_{1,M} + \Delta\mu_{1,Ion})$ on the one hand and $\Delta\mu_{1,El}$ on the other hand as a function of the degree of swelling for different concentrations of 1–1 salt in the liquid phase. The intersections of the two corresponding curves indicate the equilibrium degrees of swelling.

The number of four parameters which are treated independently of each other, is obviously too large to allow a distinct allocation of values to these parameters on the basis of experimental data obtained by one method of investigation. One has to carry out at least two different independent sets of experiments to obtain information about the network parameters. We realized this by

1. Measuring the shear modulus in dependence of the degree of swelling;
2. Measuring the equilibrium degree of swelling of the same gels in salt solutions of different concentrations.

2.4 Elastic properties

The theoretical approach to the shear modulus of a deformed, swollen network has also to be based on non-Gaussian chain statistics. The situation in this case is more complicated than in the case of calculating the elastic contribution to the change of the chemical potential during swelling, because the mechanical stress arises from a very small uniaxial deformation, which is superimposed on a very large deformation of the network chains due to the swelling process.

For simplification the derivation follows a three-chain model, where the actual set of randomly oriented chains is replaced by three independent sets of chains of equal length, which are oriented parallel to the three main directions of a rectangular coordinate system (Treloar, 1975). One set of chains is located in direction of the uniaxial deformation (z-direction) and the other two sets are oriented perpendicular to the deformation direction, parallel to the x-axis and the y-axis, respectively. A uniaxial deformation by λ with

$$\lambda = L/L_0 \tag{19}$$

changes the end-to-end distance of the chains in the z-direction by the factor λ and in the x- and y-direction by a factor of $\lambda^{-1/2}$; L is the length of the deformed sample and L_0 is the length of the undeformed, swollen sample.

In the first step of the derivation the change of entropy on swelling from the state of crosslinking to a certain degree of swelling is calculated. In this process, the end-to-end distance of each chain changes with a factor of $q_r^{1/3}$. This yields

$$\Delta S_{\mathrm{I}} = -knN\left[\left((r/nl)\mathscr{L}^{-1}(r/nl) + \ln\frac{\mathscr{L}^{-1}(r/nl)}{\sinh \mathscr{L}^{-1}(r/nl)}\right) - \left((r/nl)q^{1/3}\mathscr{L}^{-1}(q_r^{1/3}r/nl) + \ln\frac{\mathscr{L}^{-1}(q_r^{1/3}r/nl)}{\sinh \mathscr{L}^{-1}(q_r^{1/3}r/nl)}\right)\right] \tag{20}$$

where N is the number of network chains and k the Boltzmann constant. The relative end-to-end distance of the network chains is r/nl, where r is the actual end-to-end distance, n the number of statistical segments and l the length of a statistical segment.

In the second step the change of entropy for each set of chains when going from the reference state to the swollen, deformed state is calculated.

For the set of chains parallel to the z-direction, one obtains

$$\Delta S_{\mathrm{II(z)}} = -\tfrac{1}{3}knN\left[\left((r/nl)\mathscr{L}^{-1}(r/nl) + \ln\frac{\mathscr{L}^{-1}(r/nl)}{\sinh \mathscr{L}^{-1}(r/nl)}\right) - \left((r/nl)\lambda q_r^{1/3}\mathscr{L}^{-1}(\lambda q_r^{1/3}r/nl) + \ln\frac{\mathscr{L}^{-1}(\lambda q_r^{1/3}r/nl)}{\sinh \mathscr{L}^{-1}(\lambda q_r^{1/3}r/nl)}\right)\right] \tag{21}$$

and for the set of chains parallel to the x- or y-direction

$$\Delta S_{\mathrm{II(x,y)}} = -\tfrac{1}{3}knN\left[\left((r/nl)\mathscr{L}^{-1}(r/nl) + \ln\frac{\mathscr{L}^{-1}(r/nl)}{\sinh \mathscr{L}^{-1}(r/nl)}\right) - \left((r/nl)\lambda^{-1/2}q_r^{1/3}\mathscr{L}^{-1}(\lambda^{-1/2}q_r^{1/3}r/nl) + \ln\frac{\mathscr{L}^{-1}(\lambda^{-1/2}q_r^{1/3}r/nl)}{\sinh \mathscr{L}^{-1}(\lambda^{-1/2}q_r^{1/3}r/nl)}\right)\right] \tag{22}$$

The difference between ΔS_{II} and ΔS_{I}, where $\Delta S_{\text{II}} = \Delta S_{\text{II}(z)} + 2\Delta S_{\text{II}(x,y)}$, corresponds to the change of entropy when a swollen network is uniaxially deformed. Multiplication with T leads to the work of deformation, W, and differentiation with respect to λ gives the true mechanical stress σ.

$$\sigma = \tfrac{1}{3}A\nu^{*}RT\eta q_{c}^{-1}q_{r}^{-2/3}n^{1/2}[\lambda\mathscr{L}^{-1}(\lambda q_{r}^{1/3}n^{-1/2}) - \lambda^{-1/2}\mathscr{L}^{-1}(\lambda^{-1/2}q_{r}^{1/3}n^{-1/2})] \tag{23}$$

The relative end-to-end distance of the network chains, (r/nl) was replaced by $n^{-1/2}$. Other correction parameters are introduced in the same manner as in Eq. (18). The resultant formula is quite simple, because the complicated terms in Eqs (21) and (22) cancel after the differentiation.

The shear modulus is given by the initial slope of the plot σ versus $(\lambda^2 - \lambda^{-1})$. Calculations of the shear modulus in dependence of the degree of swelling were carried out with different numbers of statistical segments. The resultant plot shows the non-Gaussian influences on the elastic properties of a swollen, uniaxially deformed polymer network (Figure 2).

An infinite number of statistical segments per network chain yields a straight line with a slope of $-\frac{1}{3}$. This is the result for the Gaussian approach. The non-Gaussian derivation leads to curves which differ from this straight line, dependent on the number of statistical segments per network chain (Opperman, 1992). A variation of n moves the curve parallel to the Gaussian straight line, whereas

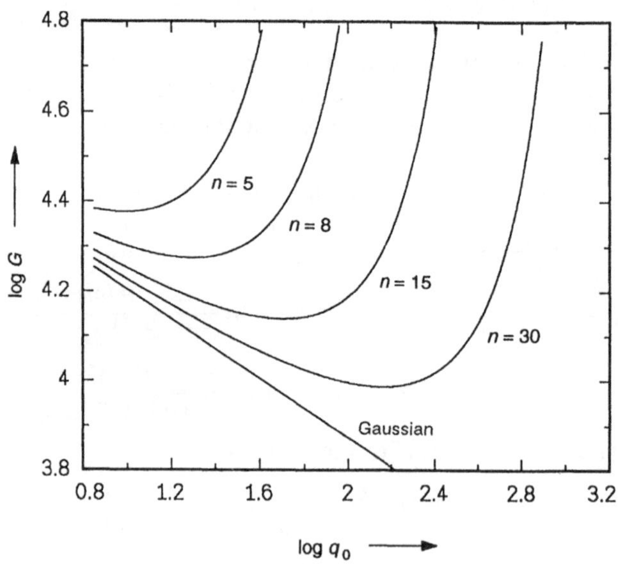

Figure 2. Dependence of the calculated shear modulus on the degree of swelling with networks having different numbers of statistical segments per network chain. Calculated with $\nu_{\text{eff}}^{*} = 40\ \text{mol m}^{-3}$. Adopted with permission from Oppermann (1992)

a variation of the network density ν^* causes a vertical shift of the curve. In this way a fit to experimental data is possible and the values of the parameters ν^* and n can be evaluated.

The next step is to use these two parameters for the calculations of the degree of swelling at equilibrium as described in section 2.3. The two remaining unknown quantities, M_2 and χ, are now accessible by a fit of theoretically calculated values of the degree of swelling to experimental data. It is important in this case that the swelling measurements are carried out with the same gels as are used for the elastic experiments.

2.5 Experimental results

We have collected data on elasticity and swelling for various polyelectrolyte gels. In this study, the results for gels made from partially hydrolysed poly-acrylamide (PAAm), poly-trimethylammoniumethyl-methacrylate-chloride (PTMAC) and a mixture of carboxymethyl-cellulose and hydroxyethyl-cellulose (CMC/HEC) are presented. For details concerning the preparation of these gels see Oppermann (1984) and Anbergen and Opperman (1990). The crosslink density of the networks was varied by different amounts of crosslinking agent.

A series of gels of constant crosslink density was prepared and the samples were brought to swelling equilibrium by swelling in aqueous salt solutions of different concentration and in pure water. In this way a large variation of degrees of swelling is available. The degree of swelling was determined gravimetrically. This is justified when the density of the gel varies only slightly with the degree of swelling. The shear modulus of the swollen networks was determined by static, uniaxial compression measurements of the immersed samples.

2.5.1 ELASTIC PROPERTIES

From the elasticity data found for each gel at different degrees of swelling, the parameters n and ν^*_{eff} were calculated according to section 2.4. To do so the shear modulus was plotted vs. the degree of swelling in a double logarithmic plot and theoretically calculated curves were fitted to the experimental data. This is shown in Figure 3 for PAAm gels of various crosslink density. We use the term ν^*_{eff} to denote the elastically effective density as determined from the experimental data, whereas ν^*_{ch} stands for the network density which is calculated on the basis of the amount of crosslinker and monomer used to prepare the gels. Generally, $\nu^*_{\mathrm{eff}} < \nu^*_{\mathrm{ch}}$.

Figure 3 shows, that the correlation between the theoretically calculated curves and the experimental data is quite good, even in the region of high degrees of swelling. This is generally found when other polyelectrolyte gels are regarded. Thus, the inverse Langevin approach seems to be a good tool to describe non-Gaussian elastic properties at high elongations of the network chains.

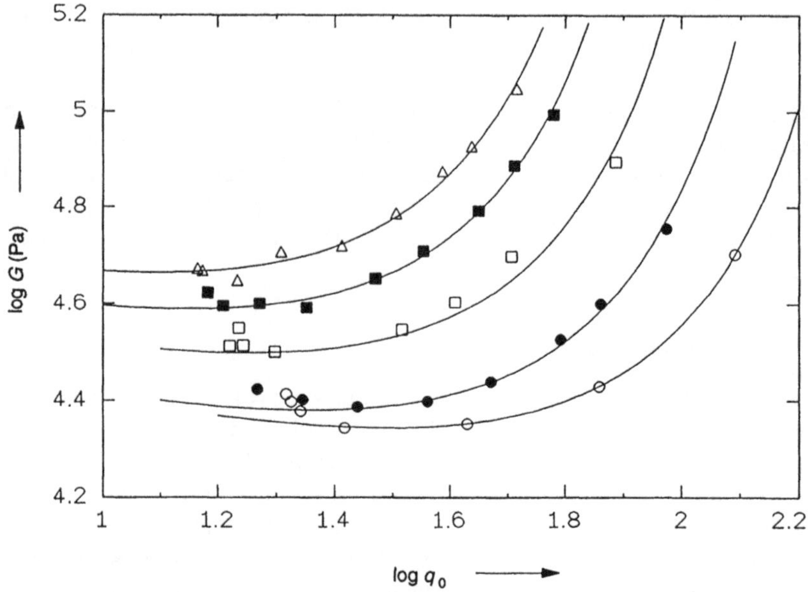

Figure 3. Dependence of the shear modulus on the degree of swelling. PAAm gels with different crosslinking density. (○) $\nu^*_{ch} = 260\,\text{mol m}^{-3}$; (●) $\nu^*_{ch} = 346\,\text{mol m}^{-3}$, (□) $\nu^*_{ch} = 433\,\text{mol m}^{-3}$; (■) $\nu^*_{ch} = 520\,\text{mol m}^{-3}$, (△) $\nu^*_{ch} = 606\,\text{mol m}^{-3}$. Fitting parameters see Table 1

At low degrees of swelling, deviations from the predicted course are arising. The shear modulus shows a steeper descent with increasing swelling than expected by the Gaussian theory (solid line), which should be valid in this region. These deviations are observed with all of the systems investigated, and were also observed by other workers (Buchholz, 1990; Bastide, Candau and Leibler, 1981), so that a general macroscopic influence can be assumed.

The reason for this behaviour could be that the effect of constraints restricting the chain motions is becoming more significant when the polymer concentration is increasing in the region of lower degrees of swelling. In this case a crossover from the phantom network to the affine network was postulated by Flory (1977). As a consequence, the structure factor A could increase from $A = 0.5$ to $A = 1$ for tetrafunctional crosslinks. This would explain the deviations at low degrees of swelling. It is worthwhile at this point to comment the 'independence' of the parameters n and ν^*_{eff}. For one series of gels prepared under identical conditions (like PAAm 1–5), ν^*_{eff} should be inversely proportional to n. This is in fact the case within experimental error. The number of monomers per statistical segment calculated on these grounds is 11 ± 1 for the hydrolized polyacrylamide networks. The results from the theoretical fits to the experimental elastic data for

Table 1. Structural network parameters of polyelectrolyte gels

Sample	Swelling agent	ν^*_{ch} (mol m^{-3})	ν^*_{eff} (mol m^{-3})	n	ν^*_{eff}/ν^*_{ch} (%)
CMC/HEC1	aq. NaCl	647	40.0	2.8	6.20
PAAm 1	aq. NaCl	260	137.0	10.9	52.80
PAAm 2	aq. NaCl	346	138.0	8.9	39.80
PAAm 3	aq. NaCl	433	166.0	7.6	38.30
PAAm 4	aq. NaCl	520	180.0	6.4	34.60
PAAm 5	aq. NaCl	606	204.0	5.8	33.70
PTMAC 1	aq. NaCl	173	11.0	13.5	6.35
PTMAC 2	aq. NaCl	260	58.0	6.4	22.30
PTMAC 3	aq. NaCl	346	45.0	6.5	13.00
PTMAC 4	aq. NaCl	433	31.0	7.2	7.20

n and ν^*_{eff} are listed in Table 1 for various polyelectrolyte gels of different crosslinking density.

2.5.1.1 Rigid polymers—flexible polymers

The influence of the stiffness of the polymer backbone on the non-Gaussian elastic properties is illustrated in Figure 4. The dependence of the shear modulus on the degree of swelling for a gel consisting of rather stiff polymer chains (CMC/HEC) is compared with a gel having flexible network chains (PAAm). A polymer with a more rigid chain backbone is expected to show non-Gaussian properties at lower extensions of the chains (i.e. degrees of swelling) than a flexible polymer because less statistical segments per network chain are expected. In Figure 4, it is demonstrated that this prediction is fulfilled. The upturn of the shear modulus of the polysaccharide gel occurs at lower degrees of swelling than that of the polyacrylate gel, despite the fact that the crosslink density is much lower, which is evident from the initial value of the shear modulus at low degrees of swelling. This high strength of gels based on rigid polymers at high degrees of swelling is a very interesting aspect and could be important for further technical applications.

2.5.1.2 Network inhomogeneities

In Figure 5 the shear modulus of swollen PTMAC gels containing different concentrations of crosslinking agent is plotted against the degree of swelling. The theoretically calculated curves are in good agreement with the experimental data. It is remarkable, however, that the shear modulus is not increasing monotonically with the concentration of the crosslinker. In a certain range of crosslinker concentration the shear modulus is decreasing with increasing crosslinker concentration. This effect has been reported earlier (Cohen *et al*, 1992) and is due to

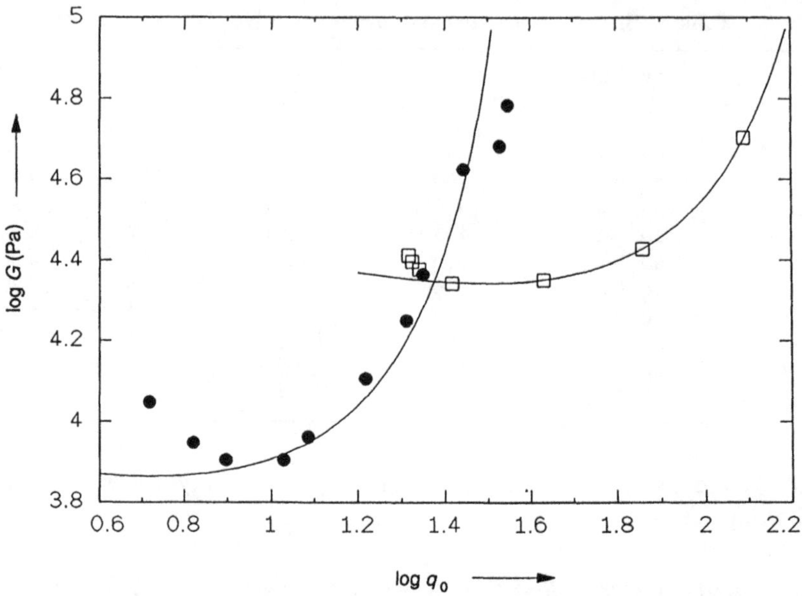

Figure 4. Dependence of the shear modulus on the degree of swelling. Comparison of a PAAm gel (□) with a CMC/HEC gel (●). Fitting parameters see Table 1

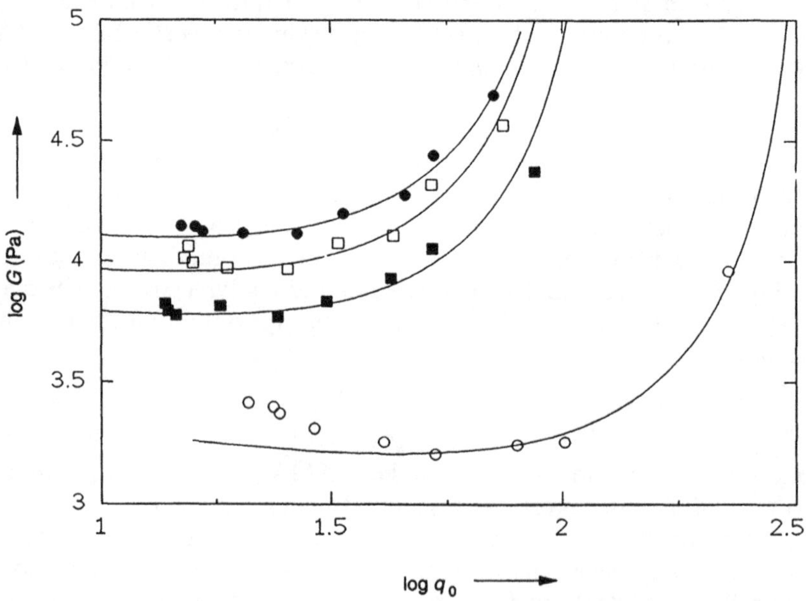

Figure 5. Dependence of the shear modulus on the degree of swelling (PTMAC gels). (○) $\nu^*_{ch} = 173$ mol m^{-3}, (●) $\nu^*_{ch} = 260$ mol m^{-3}, (□) $\nu^*_{ch} = 346$ mol m^{-3}, (■) $\nu^*_{ch} = 433$ mol m^{-3}. Fitting parameters see Table 1

a changing efficiency of the crosslinking agent with concentration. It seems that with rising crosslinker concentration the formation of strongly crosslinked network inhomogeneities becomes predominant. These inhomogeneities do not contribute to the elastic properties and are in this way responsible for the drop of the shear modulus at high crosslinker concentrations.

This phenomenon is further explored in Figure 6, where the efficiency of the crosslinking reaction, ν^*_{eff}/ν^*_{ch}, is shown as a function of ν^*_{ch}. Here ν^*_{ch} is the theoretical density of network chains, which can be evaluated by stoichiometric considerations and which represents the crosslinking agent concentration. Deviations from a horizontal line are indicating a change of the efficiency of the crosslinker. As expected from the modulus results, the course of the plot for PTMAC gels is showing a strong dependence on the crosslinker concentration. The efficiency of the crosslinker passes through a maximum at an intermediate crosslinker concentration and tends towards zero at higher concentrations, whereas the concentration dependence of the crosslinker efficiency for PAAm gels is less significant. This shows that the formation of network inhomogeneities can influence the elastic properties of a network in a decisive way, dependent on the system monomer/crosslinker.

2.5.2 SWELLING EXPERIMENTS

With the values of n and ν^*_{eff}, obtained from elasticity measurements, it is possible to make further evaluations of network parameters, which describe the depend-

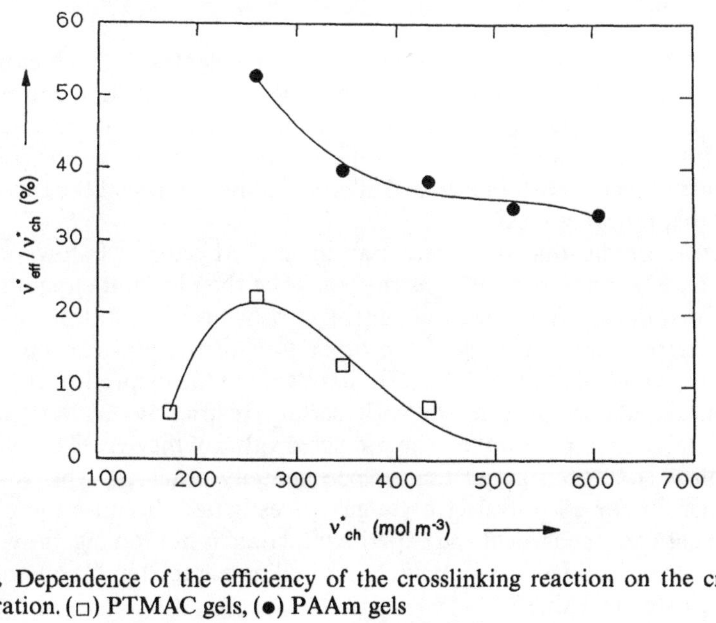

Figure 6. Dependence of the efficiency of the crosslinking reaction on the crosslinker concentration. (□) PTMAC gels, (•) PAAm gels

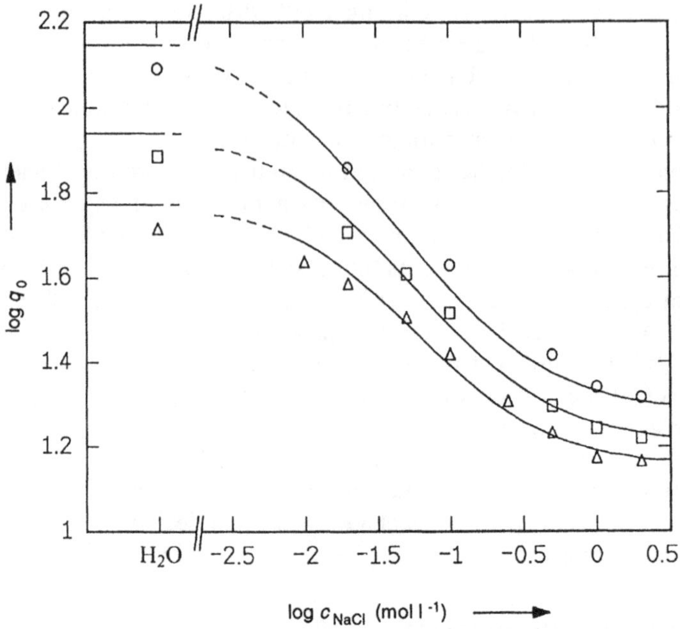

Figure 7. Dependence of the degree of swelling on the NaCl concentration (PAAm gels), (○) $\nu^*_{ch} = 260$ mol m^{-3}; (□) $\nu^*_{ch} = 433$ mol m^{-3}; (△) $\nu^*_{ch} = 606$ mol m^{-3}. Fitting parameters, see Table 2

ence of the equilibrium degree of swelling on salt concentration, as mentioned in section 2.3. Figure 7 shows results from swelling experiments in a double logarithmic plot of the degree of swelling against the salt concentration. The experimental data were obtained for three of the PAAm gels whose elastic properties were discussed on the basis of Figure 3. The solid lines represent the corresponding fitted theoretical curves.

The fitting of the remaining two parameters, M_2 and χ follows a simple procedure: the χ-parameter changes the course of the plot in the region of high salt concentrations, where the polyelectrolyte behaves more like an uncharged polymer. After fitting this region, the other part of the plot can be fitted by adjusting the value of M_2. It can be seen that the fit of the theoretical curve to the experimental data is quite good, with certain restrictions at very low salt concentration and in pure water. The predicted value of the degree of swelling in this region is somewhat larger than experimentally observed. This result was found generally for all polyelectrolyte gels investigated. Because the observed difference between theoretical and experimental data is not too big, the use of the theory should be justified for evaluating the swelling parameters. The results from the fits are listed in Table 2.

Table 2. Swelling parameters of polyelectrolyte gels

Sample	Swelling agent	ν^*_{eff}/ν^*_{ch} (%)	χ	M_2 (g mol^{-1})	ψ_{eff} (%)
CMC/HEC 1	aq. NaCl	6.2	0.49	3600	16.4
CMC/HEC 1	aq. Na_2SO_4	6.2	0.49	3600	16.4
CMC/HEC 1	aq. $CaCl_2$	6.2	(0.50)	7200	8.2
CMC/HEC 1	aq. $MgCl_2$	6.2	(0.50)	7200	8.2
PAAm 1	aq. NaCl	52.8	(0.46)	560	36.5
PAAm 2	aq. NaCl	39.8	0.475	560	36.5
PAAm 3	aq. NaCl	38.3	0.475	560	36.5
PAAm 4	aq. NaCl	34.6	0.475	560	36.5
PAAm 5	aq. NaCl	33.7	0.475	560	36.5
PTMAC 1	aq. NaCl	6.35	0.51	1640	12.6
PTMAC 2	aq. NaCl	22.3	0.51	1100	18.8
PTMAC 3	aq. NaCl	13.0	0.51	1200	17.3
PTMAC 4	aq. NaCl	7.2	(0.52)	1400	14.8

2.5.2.1 Rigid polymers—flexible polymers

Comparison of the swelling behaviour of gels made from relative stiff polymers with that of gels from flexible polymers is quite interesting. Figure 8 shows the dependence of the degree of swelling on the concentration of NaCl in a double logarithmic plot. Data from CMC/HEC gels are compared with data from PAAm gels. It can be shown that the stiffness of the polysaccharide chains is causing a much smaller variation of the degree of swelling with the salt concentration. This effect could be important for further applications as superabsorbent polymers, where a small influence of the salt concentration on the degree of swelling is desired.

2.5.2.2 Salt-type effects

The influence of the type of salt in the swelling of solution on the swelling properties of polyelectrolyte gels is shown in Figure 9. The degree of swelling is plotted against the ionic strength of the swelling solution for a CMC/HEC gel. The solid lines represent the fitted curves calculated from Eq. (9). It can be seen that the type of counterion in the swelling solution mainly determines the swelling properties of the gel. A change from NaCl to Na_2SO_4 has only a minor effect, whereas divalent counterions like Mg^{2+} or Ca^{2+} are leading to gels with much lower equilibrium degrees of swelling even at very low salt concentrations. This is predicted rather well by the concept of counterion condensation. With divalent counterions, the fraction of free counterions is only half as large as with monovalent counterions. Therefore, the M_2 value should be doubled when going from Na^+ to Ca^{2+} or Mg^{2+}. This is exactly what comes out of the simulations.

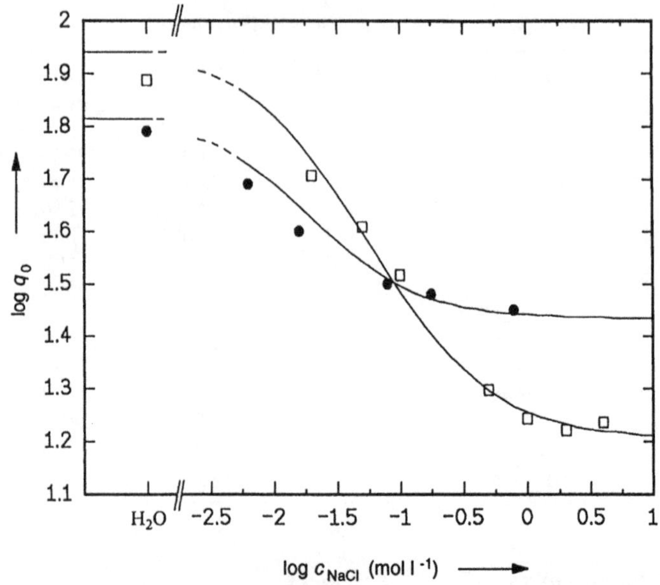

Figure 8. Dependence of the degree of swelling on the salt concentration. Comparison between a CMC/HEC gel (●) and a PAAm gel (□). Fitting parameters see Table 2

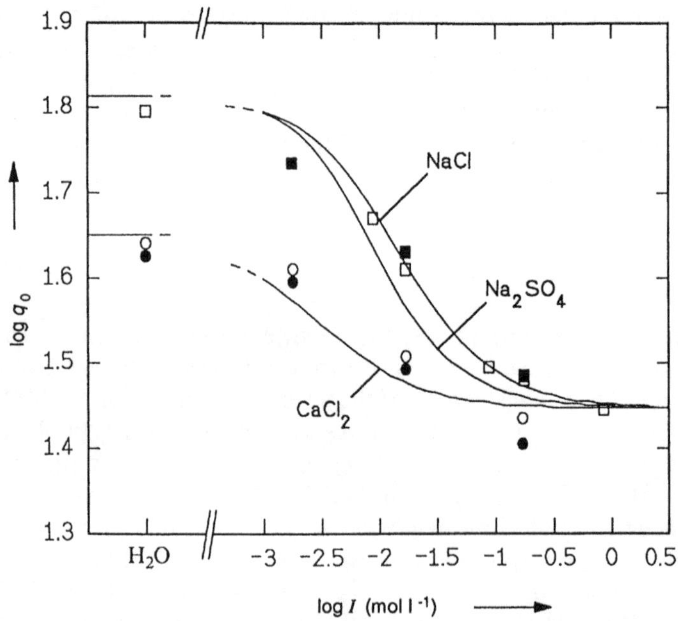

Figure 9. Dependence of the degree of swelling degree on the ionic strength (CMC/HEC gels). Different salt types in the swelling solution: (□) NaCl, (■) Na_2SO_4, (○) $MgCl_2$, (●) $CaCl_2$

The deviations from the theory at very high Ca^{2+} concentrations, where the observed degrees of swelling are lower than predicted, are not clear. An exact theoretical description of the swelling behaviour in this region is only possible if the χ-parameter is increased up to 0.5, which is questionable because the properties of the uncharged polymer are not expected to change in dependence of the type of the counterion.

There has been the perception in the literature that Ca^{2+} ions might act as crosslinkers by ionic binding to different chains. This would mean that ν^*_{ch} increases and that the degree of swelling decreases for that reason. However, this idea can be ruled out by elastic measurements. Identical networks, which are swollen in NaCl and $CaCl_2$ have the same moduli when they are brought to the same degree of swelling.

The parameter ψ in Table 2 reflects the effective fraction of free counterions. The quantity b used to express the linear charge density, is around 0.67 nm for CMC, 0.63 nm for PAAm (the degree of hydrolyzation is 40%), and 0.25 nm for PTMAC. If one uses the Manning theory to calculate the fraction of free counterions although the prerequisites are not fulfilled, one obtains 94% and 47% for CMC for monovalent and divalent counterions respectively, 89% for PAAm, and 35% for PTMAC. Obviously, the actual percentages are much lower. A systematic trend in these deviations is not apparent.

Regarding the results for the PTMAC gels it is remarkable that ψ is not even constant with the crosslinker concentration. It seems that ψ also depends on the occurrence of network inhomogeneities as expressed by the crosslinking efficiency ν^*_{eff}/ν^*_{ch}.

On the whole, the quantity M_2 can be used as a fit parameter. Its physical significance is clear, but it seems to be impossible so far to predict the actual value.

References

Alexandrowicz, Z., Katchalsky, A. (1963) *J. Pol. Sci. A1* 3231

Anbergen, U., Oppermann, W. (1990) *Polymer* **31**, 1854

Armstrong, R. W., Strauss, U. P. (1969) *Encyclopedia of Polymer Science & Techn.* **10**, 781

Bastide, J., Candau, S., Leibler, L. (1981) *Macromolecules* **14**, 719

Brannon-Peppas, L., Peppas, N. A. (1990) in *Absorbent Polymer Technology*, eds Brannon-Peppas, L., Harland, R. S., Elsevier, Amsterdam, p. 67

Buchholz, F. L. (1990) in *Absorbent Polymer Technology*, eds Brannon-Peppas, L., Harland, R. S., Elsevier, Amsterdam, p. 23

Cohen, J., Priel, Z., Rabin, Y. (1988) *J. Chem. Phys.* **88**, 7111

Cohen, Y., Ramon, O., Kopelman, J., Mizrahi, S. (1992) *J. Pol. Sci., Part B: Pol. Phys.* **30**, 1055

Dusek, K., Prins, W. (1969) *Adv. in Pol. Sci.* **6**, 1

Flory, P. J. (1953) *Principles of Polymer Chemistry*, Cornell University Press, Ithaca, NY

Flory, P. J. (1977) *J. Chem. Phys.* **66**, 5720

Fuoss, R. M., Katchalsky, A., Lifson, S. (1951) *Proc. Natl. Acad. Sci. USA*, **37**, 579

Harsh, D. C., Gehrke, S. H. (1990) in *Absorbent Polymer Technology*, eds Brannon-Peppas, L., Harland, R. S., Elsevier, Amsterdam, p. 103
Hasa, J., Ilavsky, M., Dusek, K. (1975) *J. Pol. Sci., Pol. Phys. Ed.* **13**, 253
Hooper, H. H., Baker, J. P., Blanch, H. W., Prausnitz, J. M. (1990) *Macromolecules* **23**, 1096
Manning, G. S. (1969) *J. Chem. Phys.* **51**, 924, 934
Manning, G. S. (1978) *Biophys. Chem.* **9**, 65
Oosawa, F., Imai, N., Kagawa, I. (1954) *J. Pol. Sci.* **13**, 93
Oosawa, F. (1971) *Polyelectrolytes*, Marcel Dekker, New York
Oppermann, W., Rose, S., Rehage, G. (1985) *Brit. Pol. J.* **17**, 175
Oppermann, W. (1988) *Makromol. Chem.* **189**, 927
Oppermann, W. (1992) in *Polyelectrolyte Gels*, eds Harland, Prud'Homme, ACS Symp. Ser. 480, Am. Chem. Soc., Washington DC
Oppermann, W. (1984) *Die Angew. Makromol. Chem.* **123/124**, 229
Prange, M. M., Hooper, H. H., Prausnitz, J. M. (1989) *AIChe J.*, **35**, 803
Ricka, J., Tanaka, T. (1984) *Macromolecules* **17**, 2916
Schmitz, K. S. (1993) *Macroions in Solution and Colloidal Suspension*, VCH, New York
Tanford, C. (1967) *Physical Chemistry of Macromolecules*, Wiley: New York
Treloar, L. R. G. (1975) *The Physics of Rubber Elasticity*, University Press, Oxford
Vink, H. (1970) *Makromol. Chem.* **131**, 133

3 NMR AND STATISTICAL STRUCTURES OF GELS

J. P. Cohen Addad
Université Joseph-Fourier, Grenoble, France

The Physical Properties of Polymeric Gels. Edited by J. P. Cohen Addad.

3.1 Introduction

Permanent polymeric gels can be highly strained above the glass transition temperature, and then they recover their initial shape when the external stress is interrupted. The description of gel properties is mainly focused on the understanding of the mechanisms of deformations of these giant molecules built from points of cohesion, distributed at random in space and connected to one another by flexible polymeric links, also called network chain segments. Whether or not they are time dependent, deformations of polymeric network structures are mainly governed by the presence of meshes or loops determined by several network chain segments. The characterization of loop sizes is the key to a better understanding of the mechanisms which govern the macroscopic responses of network structures to external stresses. Early descriptions of statistical structures of polymeric gels have been based on the use of a Cayley tree model which does not take the formation of loops into consideration (Flory, 1979; Stockmayer, 1943). However, this classical approach, based on a mean field theory, has been shown to apply to the description of properties of long linear molecules crosslinked at random (de Gennes, 1979); the process of formation of such polymeric systems is called a vulcanization. The random bond percolation approach predicts the formation of loops (de Gennes, 1976; Stauffer, 1976). The analogy of the gelation process with a percolation effect leads to the introduction of a variable of state of gelation which has proved to be a very useful tool of description of network structures; the assimilation of the gelation to percolation stresses the need to express any physical quantity, associated with a polymeric gel, as a function of this variable. This applies to the modulus of elasticity, the swelling ratio or the mean loop size; it will also be shown to apply to NMR. Furthermore, the assimilation permits the prediction of the critical behaviour of gels near the threshold of percolation.

3.1.1 PURPOSE OF NMR INVESTIGATIONS: STATISTICAL TEXTURE OF GELS

The purpose of NMR investigations into gel properties is to attempt to reveal the coarse-grained statistical structure of these complex polymeric systems; the approach is manifold.

1. The first step of the approach is a search for a variable which is used to define the state of gelation of a network structure, whether or not it is slightly deformed. This variable will depend on initial conditions which determine the polymeric system; parameters usually considered prior to the gelation process may be the chain length, the concentration of crosslinking agent, or the specific area and the mean size of aggregates when filled polymers are studied.
2. The statistical framework of description of fluctuations which occur within network chain segments is defined in the second step. In dry gels, network chain segments obey Gaussian statistics while the maximum swelling of network structures is governed by the excluded volume effect which may be partly screened.
3. The third step concerns the distribution of anchoring points represented by ends of network chain segments throughout the polymeric gel. This distribution depends on macroscopic strains applied to network structures.

The determination of network structures comes up against the necessity to recognize coupling junctions which represent a mechanical cohesion established between chains. It will be considered throughout the present study that the concentration of permanent connections between chains, resulting from random crosslinks, enters in the definition of the relevant variable that determines the state of a network structure. More precisely, the presence of slipping links will be ignored although they can result in a real coupling of chains. The exact picture of a network structure may depend on the experimental method used to investigate its properties. Both the nature and the strength of the constraint applied to the polymeric gel are crucial. For instance, this can be illustrated by comparing the effects of two zero-frequency constraints with each other: the swelling phenomenon of a gel, observed under the effect of osmotic pressure, and the elongation induced by an uniaxial stress; these are currently applied to network structures. Chain segments are disengaged from one another under the effect of swelling, whereas they may slip along one another during the deformation process caused by an uniaxial force. Thus, chain intertwinings probably disappear during the process of swelling except if they are trapped, while they can grapple chain segments during the process of stretching. Trapped loops slide along chain segments, a mechanism usually evoked to attempt to interpret the effect of elasticity observed under strong deformations.

3.1.2 NETWORK STRUCTURES AND TIME SCALES

The recovery of the initial shape of a gel is not instantaneous when the application of an external constraint is interrupted; this stresses the need also to take the time-scale of observation into consideration to define network structures. Experiments performed to measure the modulus of elasticity of a polymeric gel or its swelling involve a time average defined over a long period which enables all fluctuations to be accomplished. However, the average of any time-dependent physical quantity, whatever its nature, may be a partial average only; this effect is due to the broad spectrum of relaxation displayed by the mechanical behaviour of gels. In that case, the picture of the network structure may not coincide with that of the gel which results from the observation of all fluctuations. Consequently, the physical determination of a network structure can eventually be specific to the frequency range which is explored by applying a given external constraint to a gel.

3.1.3 NMR APPROACH

This NMR approach to physical properties of polymeric gels is based on a simple material picture. Nodes, distributed at random in space, are connected by polymer segments forming random paths. Retractive forces due to the elasticity of chains act on nodes; they are counterbalanced by all non-interpenetration forces existing between monomeric units. Nodes hinder fluctuations of ends of segments and conformational fluctuations which occur within neighbouring segments. A reduction of entropy results from the restriction of fluctuations; it is related to the property of elasticity which characterizes polymeric gels. Correspondingly, the magnetic relaxation process of nuclei linked to network chain segments is induced by conformational fluctuations originating in the effect of elasticity or the effect of swelling. Spin–spin interactions are tensorial functions of orientational coordinates of monomeric units; they are submitted to long-range conformational fluctuations which give rise to an irreversible dynamics of the nuclear magnetization specific to permanent or temporary gels.

This behaviour was first observed in polymer melts (Cohen Addad, 1974; Cohen Addad and Vogin, 1974). Then, it was used to investigate swelling properties of permanent gels (Cohen Added, Domard and Herz, 1978; Doskocilova, Schneider and Jakes, 1978). This specific mechanism of relaxation is easily perceived because of the great NMR sensitivity to proton magnetic interactions. It has been shown that it characterizes protons attached to crosslinked chains as well as protons located on small molecules diffusing through a polymeric gel (von Meerwall and Ferguson, 1981). In this last case the intimate texture of network structures is thus explored from NMR properties of molecular penetrants, including short polymer chains. The diffusional motion of these guest molecules yields a magnetic property of mobile protons naturally contrasted to properties of host nuclei linked to the network structure (Garrido *et al*, 1988; Cohen Addad and Girard, 1992).

The existence of a specific proton relaxation mechanism has proved to be a convenient and powerful approach to a broad variety of phenomena, mixing both elasticity and viscosity in polymers. It applies to the sharp characterization of the effect of crosslinking (Cohen Addad and Schmit, 1987). The kinetics of crystallization is easily observed down to a time-scale equal to about 1 s, at any temperature ranging from 100 to 500 K (Feio and Cohen Addad, 1988). The kinetics of gelation is conveniently perceived from the effect of topological constraints which arise from the growing network structure (Cohen Added and Schmit, 1987).

In addition to observations disclosed from protons, the magnetic relaxation specific to the existence of a network structure can be reflected from the irreversible dynamics of the magnetization of other nuclei. The magnetic resonance of deuterium nuclei observed in liquid crystals is a well-known tool for analysis of orientational molecular properties. It permits chemical groups to be differentiated from one another in short chains. This approach has also been extended to polymers, to study deformations of chain segments. The magnetic resonance observed in perdeuterated gels gives an insight into these systems as extensive as that revealed from the relaxation of protons (Dubault, Deloche and Herz, 1984; Gronski, Stadler and Jacobi, 1984; Cohen Addad, 1982). More precisely, it will be shown throughout this study that systems formed by two spins ($I = \frac{1}{2}$: ^1H–^1H or ^{13}C–^1H) or systems formed by three spins ($I = \frac{1}{2}$) located on a regular triangle give information about chain segment deformations similar to those obtained from systems of one spin $I = 1$ (^2H) (Cohen Addad and Dupeyre, 1983; Gronski, Stadler and Jacobi, 1984). Relaxation mechanisms of these systems are governed by a quadrupolar interaction or by dipole–dipole interactions which have tensorial expressions rigorously identical to one another. In addition to difficulties encountered in the synthesis of deuterated samples, the NMR sensitivity of deuterium is poor compared with that of protons.

A promising powerful method of investigation concerns the magnetic relaxation observed in gels partially labelled with specific nuclei. The presence of phosphorus (Dickinson *et al*, 1988a), or fluorine (von Meerwall and Ferguson, 1981), or deuterium (Sotta *et al*, 1987), or ^{13}C in natural abundance (Curran and Paelira, 1987; Dickinson *et al*, 1988b; Vanderhart, Perez and McKenna, 1988), located on well-defined parts of chain segments leads to an accurate description focused on sharp properties, including fluctuations of crosslinks for example. Similarly, deuterated small molecules or short chains are convenient probes which can be used to study the anisotropy induced by deformations of gels (Deloche and Samulski, 1981; Sotta, Deloche and Herz, 1988). It is of interest to mention that the coarse structure (over a scale of 1 μm) of polybutadiene rubber is nowadays observed using NMR imaging (Clough and Koenig, 1989).

It will be shown that NMR is sensitive to both the asymmetry and the dynamics of random motions of monomeric units; these fluctuations also give rise to the effect of elasticity. This interrelationship is specific to the NMR approach.

3.2 Principle of the NMR approach

An axial symmetry characterize naturally the dynamics of the nuclear magnetization when a strong steady magnetic field is applied to a spin system. This axial symmetry leads to two quite distinct irreversible behaviours of the transverse and the longitudinal components of the nuclear magnetization respectively (Abragam, 1961). The irreversible dynamics of the component parallel to the direction of the magnetic field is only sensitive to the quasi-resonant exchange which exists between the Zeeman energy of the spin system and the thermal energy which gives rise to molecular motions; the reference scale, used to characterize random molecular motions, is the Larmor frequency of nuclear spins. The dynamics of the transverse component may be sensitive to properties of quantum coherence of spins, only; the absence of quasi-resonant exchange of energy permits the detection of slow molecular processes. More precisely, it will be shown that the dominant mechanism of relaxation of this component may reflect asymmetry properties of the local diffusional motions of monomeric units, whereas the dynamics of these random motions may give a negligible contribution to the relaxation process. This property is specific to polymeric gels and high molecular weight molten polymers; it will be illustrated throughout this study.

It may be worth emphasizing that the dual NMR response parallels the mechanical behaviour of polymeric systems.

3.2.1 DUAL MECHANICAL BEHAVIOUR

Polymeric gels exibit a property of mechanical cohesion; consequently, highly strained network structures can recover their initial shape over a long time interval of relaxation when the application of an external constraint is interrupted. The time interval, elapsed during the recovery, characterizes a relaxation process caused by the effect of friction which affects the displacements of monomeric units. The dual mechanical behaviour of gels is well revealed through two time-dependent key functions (Ferry, 1983). One of these functions is the compliance, called $J(t)$. Contrary to the case of uncrosslinked chains, the compliance does not exhibit the known $t\mathrm{F}/\eta(0)$ limit behaviour of a flowing liquid when a permanent external force F, per unit area, is applied to a network structure; $\eta(0)$ is the zero shear rate viscosity: it is infinite in gels. The other key function is the modulus of stress relaxation, called $G(t)$. The stress release observed at a constant strain, after a sudden deformation of a gel, does not present any terminal spectrum of relaxation, unlike polymeric liquids. The frequency-dependent modulus of relaxation is written as $\eta^*(\omega) = \eta'(\omega) + i\eta''(\omega)$, it defines the linear response to a sinusoïdal rate of deformation:

$$\eta^*(\omega) = \frac{V}{kT}\int_0^\infty \mathrm{e}^{i\omega\tau}\langle\, \mathbf{f}(t)\mathbf{f}(t+\tau)\rangle \mathrm{d}\tau \tag{1}$$

where

$$G(\tau) = \langle\, \mathbf{f}(t)\mathbf{f}(t+\tau)\rangle \tag{2}$$

$G(\tau)$ is the correlation function associated with the scattering of molecular momenta; $\mathbf{f}(t)$ is a random force, per unit area, acting in a molecular scale, in the volume V of the sample. The storage modulus, $\eta''(\omega)$, is equal to the modulus of elasticity E when ω is set equal to zero. Neither $\eta''(\omega)$ nor the loss modulus $\eta'(\omega)$ have been easily calculated until now. Fluctuations cannot be described from a single mode of relaxation. The complex structure of the relaxation spectrum involved in the definition of $J(t)$ or $G(t)$ stresses the need to consider carefully the time-scale of observation. For many years the characterization of polymer gels has been mainly referred to their mechanical behaviour. The intimate texture of these systems has been investigated from their macroscopic responses to external stresses or strains, whether these were static or time dependent (oscillatory strains).

3.2.2 NMR: DUAL BEHAVIOUR

For several years, light scattering, NMR, fluorescence correlation, and neutron scattering expriments have been used to obtain a through insight into the statistical structure of gels and to correlate it to its mechanical behaviour. It will be shown, throughout this study, that the irreversible dynamics of the nuclear magnetization, observed in polymeric gels, exhibits a dual behaviour analogous to the viscoelastic property. The mechanical limit $\eta''(\omega) = E$, observed when ω goes to zero reflects an elasticity property, i.e. a solid-like behaviour. It implies a parallel NMR property described in the following way. Spin–spin interactions are tensorial functions of space variables of nuclei which are randomly time dependent; the spin-hamiltonian associated with these interactions will be hereafter called $\mathscr{H}_T(t)$. It will be shown that the observation of a modulus of elasticity, E, is closely related to a non-zero time average of $\mathscr{H}_t(t)$. Let $\langle \mathscr{H}_T(t) \rangle$ denote this time average; it is permanently different from zero in gels because of the presence of nodes. Therefore, $\mathscr{H}_T(t)$ consists of two parts: $\langle \mathscr{H}_T(t) \rangle$ and $\mathscr{H}_T(t) - \langle \mathscr{H}_T(t)) \rangle$. The $\mathscr{H}_T(t) - \langle \mathscr{H}_T(t) \rangle$ part governs a magnetic relaxation behaviour characterized by a spectral density of energy:

$$\mathscr{J}_v(\omega) = \mathscr{R} \int_0^{\infty} \langle (\mathscr{H}_T(t+\tau) - \langle \mathscr{H}_T \rangle)(\mathscr{H}_T(t) - \langle \mathscr{H}_T \rangle) \rangle e^{i\omega_L \tau} d\tau \tag{3}$$

analogous to that observed in ordinary liquids; ω_L is the Larmor frequency. The behaviour of $\mathscr{J}_v(\omega_L)$ is comparable to that of the viscosity $\eta'(\omega)$ with respect to the frequency ω (Ferry, 1983). The part $\langle \mathscr{H}_T \rangle$ induces a relaxation process which does not depend on the dynamics of motions provided these occur within a time-scale longer than about 10^{-3} s; the magnetic relaxation is sensitive to deviations from isotropic motions only. This solid-like behaviour is easily disclosed from specific NMR observations. It is closely related to the existence of a network structure whether it is permanent or temporary, as in the case of a high molecular weight polymer melt.

3.2.3 ELEMENTARY CHAINS: RESTRICTED FLUCTUATIONS

The quantitative interpretation of NMR properties requires the building of a model which attempts to describe the statistical structure of polymeric gels. Properties of a polymeric gel are not entirely disconnected from those of a melt which arise from fluctuations associated with an ensemble $\mathscr{E}_R$ of chain configurations. In a melt, a given chain explores all configurations of the ensemble within a time interval T_R, also called terminal relaxation time. Starting from monomers forming an ordinary liquid, the macromolecular structure induces a restriction of fluctuations which originates a non-negligible elasticity component, observed at frequencies different from zero. Fluctuations which occur within long polymer chains, in a melt, are known to spread over a broad time-scale ranging from about 10^{-11} s to 1 s or more. Starting from an initial time t_0, it is now well established that there is a break which occurs during the time evolution of correlations of configurational fluctuations in one long chain. The time dependence of the loss of memory represented by $G(t)$ shows several characteristic steps. The first step concerns short segments; it is associated with a dynamic screening effect which is reflected by the independence of fluctuations with respect to the chain molecular weight. The dynamic screening effect lasts for a time intervaΓ_S. Then, the time dependence of $G(t)$ exhibits a plateau behaviour which lasts for a time interval Γ_T; it reflects the break of the time evolution of correlations of chain fluctuations. To give a simple picture, the first step is associated with the existence of a temporary network structure which arises from the presence of non-covalent links, also called entanglements. This temporary gel is formed by characteristic chain segments comprising N_S skeletal bonds which determine a correlation length, called ξ_s.

Long-range dynamic properties of chains are screened within these domains, characterized by the mean size ξ_s. During the time interval $\Gamma_S + \Gamma_T$ the ensemble of configurational fluctuations $\mathscr{E}_T$ is much more restricted than $\mathscr{E}_R$: the polymeric system exhibits a property of temporary elasticity. A relationship is established between the break of time fluctuations and a spatial break which occurs along one chain. The third step of configurational fluctuations which occur in one chain corresponds to long-range properties described by the terminal relaxation; this is pictured from the reptational motion of chains. In a pure polymer, dynamic screening domains are defined from at least 200 skeletal bonds while the persistence length is reduced to a few monomeric units.

It has been already shown that NMR is sensitive to local motions and to the break of time correlations of fluctuations which characterize the dynamical behaviour of a melt. There is a close relationship which exists between the nuclear magnetic relaxation of protons attached to polymer chains and the temporary elasticity of a melt.

In concentrated solutions, static domains of correlations are formed; they obey Gaussian statistics while in static domains of correlations formed in semi-dilute solutions, the excluded volume effect is not screened. The collective behaviour of statistical domains along a given chain is Gaussian. It is currently considered that the average size of domains associated with dynamic screening effects in a semi-dilute solution is not very different from the size of static domains (Muthukumar and Edwards, 1982). In this case the break of correlations in dynamical fluctuations as well as static domains of correlations are perceived from NMR.

In a polymeric gel, chains are linked to one another; they are subjected to different kinds of constraints which hinder configurational fluctuations. The ensemble of chain configurations in a gel is called $\mathscr{E}_G$; it is still much more restricted than $\mathscr{E}_T$ because of the presence of permanent constraints: $\mathscr{E}_G C \mathscr{E}_T C \mathscr{E}_R$. It is considered that nodes created between chain segments induce a break of time correlations of configurational fluctuations. Therefore, they determine so-called elementary network chain segments which are observed from NMR. In addition to covalent crosslinks, elementary network chains observed from NMR can be defined by slipping links or entanglements as long as these can be still defined from a dynamical screening effect which occurs in a gel. The fundamental concept of entanglement must be probably revised in the case of a polymeric gel submitted to a sudden external constraint and domains associated with dynamic screening effects in a melt cannot be considered as trapped by a crosslinking reaction. However, static domains of correlations in solutions can be trapped by the formation of crosslinks during a gelation process; these domains can contribute to the determination of the gel structure. As a main consequence of this effect, static screening domains which are not detected, from NMR in solutions, can be revealed from this experimental approach, after a crosslinking reaction which connects permanently these domains to one another. They give rise to a statistical structure of gels which can be controlled from the size of domains determined prior to the crosslinking reaction, i.e. from the polymer concentration. Such an experimental procedure yields a wide variety of polymeric gel structures.

It is considered that the main consequence of the gelation process is to give rise to the partition of one gel into elementary chains determined by breaks of time correlations which are not necessarily represented only by crosslinks.

3.2.4 LOCAL DIFFUSIONAL ASYMMETRY

In accordance with all NMR results observed first on high polymer melts and then on polymeric gels, it is considered that skeletal bonds undergo non-isotropic motions even in the absence of any deformation: the orientation diffusion of monomeric units which is observed within any time-scale is not fully isotropic because of the presence of a network structure. This is the basic property which underlies the interpretation of NMR investigations.

1. It is considered that the non-isotropic rotational diffusion of skeletal bonds is related to the existence of elementary network chains. It is supposed that

elementary segments are determined from nodes which have well-defined positions in space whatever their exact physical nature. Let $\mathbf{R}_j$ denote the position of node j. elementary network chains are defined from a characteristic number of skeletal bonds N_{ij} between two consecutive nodes i and j. There are ν_e elementary chains in a given gel.

2. Then, it is supposed that the state of an elementary chain segment joining any two consecutive coupling junctions $\mathbf{R}_j$ and $\mathbf{R}_i$ can be characterized by its end-to-end vector $\mathbf{r}_{ij} = \mathbf{R}_j - \mathbf{R}_i$, associated with N_{ij} bonds. End vectors are not equal to zero, even on average, in a real gel. The difference which exists between a polymer melt and a permanent gel is reflected by the relaxation of $\mathbf{r}_{ij}$ vectors; in a gel, components of most of them never reach the value zero, whereas in a melt all components of elementary chain vectors go to zero provided the time-scale of observation is long enough compared with the reptation time of one chain. The permanent state of small elongation which characterizes elementary chains in a gel arises from all restrictions affecting the ensemble $\mathscr{E}_R$ of configurations which would be available to polymer chains in a melt.

Consequently, considering the time average of a given end-to-end vector $\mathbf{r}_{ij} = \mathbf{R}_j - \mathbf{R}_i$ which consists of $N_{i,j}$ bonds $\mathbf{a}_n^{ij}$ ($n = 1, 2, \ldots, N_{i,j}$), the time average of orientations of these skeletal bonds must lead to

$$\sum_n \langle \mathbf{a}_n^{ij} \rangle = \langle \mathbf{r}_{ij} \rangle \tag{4}$$

with

$$\langle \mathbf{r}_{ij} \rangle \neq 0 \tag{5}$$

hence, for any $\mathbf{a}_n^{ij}$ bond

$$\langle \mathbf{a}_n^{ij} \rangle \neq 0 \tag{6}$$

A partial orientational ordering is induced, on average, along the direction of the non-zero end-to-end vector of the chain segment. The degree of anisotropy of the diffusional rotation depends upon the elongation which affects the chain segment. The time average corresponds to all rotational isomerization processes which occur along a chain segment and which are compatible with $\langle \mathbf{r}_{ij} \rangle$ (Figure 1). The spatial distribution of $\langle \mathbf{r}_{ij} \rangle$ vectors is isotropic, in the absence of any gel deformation.

The $\langle \mathbf{a}_n^{ij} \rangle$ time average can be associated with a distribution function of stochastic orientations of the skeletal bonds which form each chain segment. This distribution is necessarily anisotropic, in accordance with Eq. (5); it depends on the length and the orientation of $\mathbf{r}_{ij}$. It is convenient to characterize a given elementary chain by its end-to-end vector or, equivalently, by a force $\mathbf{f}_{ij}$ which acts on the ends of the segment; the $\mathbf{f}_{ij}$ force and the $\mathbf{r}_{ij}$ vector are related to each other through the state equation of the polymer chain. The use of $\mathbf{f}_{ij}$ permits the $\mathbf{r}_{ij}$

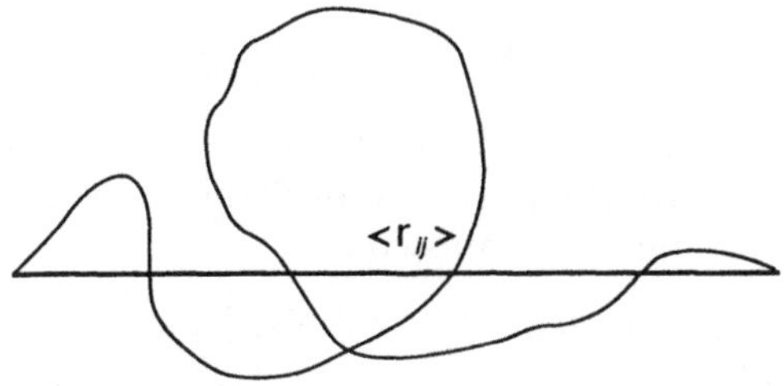

Figure 1. Schematic representation of one chain segment submitted to a weak mean elongation $\langle \mathbf{r}_{ij} \rangle$

end vector fluctuate in time. Taking all orientational correlations between monomeric units into consideration, whether these are located on different segments or not, the partition function of the ensemble of elementary chains is written as

$$\mathscr{Z} = \int \exp\left\{\sum_{ij}^{\nu_e} \mathbf{r}_{ij}\cdot\mathbf{f}_{ij}/kT\right\} \mathscr{Q}(\mathbf{a}_1^{ij}, \ldots, \mathbf{a}_{N_{ij}}^{ij})\mathrm{d}\mathbf{a}_1^{ij}\cdots\mathrm{d}\mathbf{a}_{N_{ij}}^{ij} \tag{7}$$

where $\mathscr{Q}\,(\mathbf{a}_1^{ij}, \ldots, \mathbf{a}_{N_{ij}}^{ij})$ is the probability distribution function which describes all rotational isomerization processes of monomeric units. The partition function associated with one elementary chain is written as

$$\mathscr{Z}_{ij} = \int \exp\left\{\sum_{i'j'\neq ij}^{\nu_e} \exp(\mathbf{r}_{i'j'}\cdot\mathbf{f}_{i'j'}/kT\right\} \mathscr{Q}(\mathbf{a}_1^{i'j'}, \ldots, \mathbf{a}_{N_{i'j'}}^{i'j'})\mathrm{d}\mathbf{a}_1^{i'j'}\cdots\mathrm{d}\mathbf{a}_{N_{i'j'}}^{i'j'} \tag{8}$$

mutual orientational interactions of monomeric units located on different segments are taken into consideration in this integral. The partition function is also written as

$$\mathscr{Z}_{ij} = \int \exp(\mathbf{r}_{ij}\cdot\mathbf{f}_{ij}/kT)\mathscr{W}(\mathbf{a}_1^{ij}, \ldots, \mathbf{a}_{N_{ij}}^{ij})\mathrm{d}\mathbf{a}_1^{ij}\cdots\mathrm{d}\mathbf{a}_{N_{ij}}^{ij} \tag{9}$$

$\mathscr{W}(\mathbf{a}_1^{ij}, \ldots, \mathbf{a}_{N_{ij}}^{ij})$ is the probability distribution function of orientations of skeletal bonds within a given elementary network chain. For weak elongations of elementary segments, corresponding to a gel without any deformation, the partition function reads

$$\mathscr{Z}_{ij} \approx 1 + \tfrac{1}{2}(\mathbf{f}_{ij}/kT)^2 \sum_{p,n} \langle z_p^{ij} z_n^{ij} \rangle_0 + \cdots \tag{10}$$

the $\mathbf{f}_{ij}$ force is parallel to the z-axis of the reference frame. The state equation of the chain segment is given by

$$\langle \mathbf{r}_{ij} \rangle = -kT\frac{\partial \mathscr{Z}_{ij}}{\partial \mathbf{f}_{ij}}/\mathscr{Z}_{ij} \tag{11}$$

or

$$\langle \mathbf{r}_{ij} \rangle \approx -\langle \mathbf{r}_{ij}^2 \rangle_0 \mathbf{f}_{ij}/3kT \tag{12}$$

$\langle \mathbf{r}_{ij}^2 \rangle_0$ is the mean square end-to-end distance of the elementary chain network in the absence of any gel deformation. The probability distribution function of orientations of skeletal bonds is expressed as

$$\mathscr{P}(\mathbf{a}_1^{ij},\ldots,a_n^{ij},\langle \mathbf{r}_{ij} \rangle) = \exp(3\mathbf{r}_{ij}\cdot\langle \mathbf{r}_{ij} \rangle/\langle \mathbf{r}_{ij}^2 \rangle_0\, \mathscr{W}(\mathbf{a}_1^{ij},\ldots,\mathbf{a}_n^{ij})/\mathscr{Z}_{ij} \tag{13}$$

There is a local diffusional asymmetry which affects all skeletal bonds which pertain to network chain segments. For example, for freely jointed chains, the probability distribution function is a simple exponential function

$$\mathscr{P}_0(\mathbf{a}^{ij},\langle \mathbf{r}_{ij} \rangle) = \exp(3\mathbf{a}^{ij}\cdot\langle \mathbf{r}_{ij} \rangle/\langle \mathbf{r}_{ij}^2 \rangle_0)/4\pi \tag{14}$$

or

$$\mathscr{P}_0(\mathbf{a}^{ij},\langle \mathbf{r}_{ij} \rangle) = 4\pi \sum_{l,m} \left(\frac{\pi}{2w}\right)^{1/2} I_{l+1/2}(w)\, Y_l^m(\mathbf{a}^{ij})\, Y_l^{m*}(\langle r_{ij} \rangle) \tag{15}$$

with

$$w = \frac{3\mathbf{a}^{ij}\langle \mathbf{r}_{ij} \rangle}{\langle \mathbf{r}_{ij}^2 \rangle_0};$$

Y_l^m spherical harmonics $(l = 0, 1, 2, \ldots; m = -l, \cdots + l)$ are associated with $I_{l+1/2}(w)$ Bessel function of the second kind.

3.2.5 RESIDUAL SPIN–SPIN INTERACTIONS

Spin–spin interactions established between nuclei attached to chain segments are tensorial functions which depend on orientational properties of skeletal bonds. Any $\mathbf{a}_n^{ij}$ skeletal bond is slightly polarized, along the non-zero average of the $\mathbf{r}_{ij}$ vector which joins two consecutive coupling junctions i and j.

Typical orientational quantities are

$$(\mathbf{z}_n^{ij}/a_n^{ij})^2 - \tfrac{1}{3}) \quad \text{or} \quad ((x_n^{ij})^2 - (y_n^{ij})^2) \quad \text{and} \quad (x_n^{ij} z_n^{ij}), (y_n^{ij} z_n^{ij})$$

The angular average of $(\mathbf{z}_n^{ij}/\mathbf{a}_n^{ij})^2 - \frac{1}{3})$ is expressed by taking (13) into consideration

$$\langle (\mathbf{z}_n^{ij}/\mathbf{a}_n^{ij})^2 - \tfrac{1}{3}) \rangle_{ij}^0 = \frac{1}{2}\left(\frac{f_{ij}}{kT}\right)^2 \left(\sum_{p,q} \langle z_p^{ij} z_q^{ij} (z_n^{ij}/a_n^{ij})^2 \rangle_0 - \langle \mathbf{r}_{ij}^2 \rangle_0/9\right) \tag{16}$$

Such a description holds whatever the complexity of $\mathscr{W}$; it applies whether or not the effect of excluded volume is screened. The statistical framework of description is revealed from the state equation

$$\frac{f_{ij}}{kT} = \frac{3\langle \mathbf{r}_{ij} \rangle}{\langle \mathbf{r}_{ij}^2 \rangle_0}$$

where

$$\langle \mathbf{r}_{ij}^2 \rangle_0 = (N_{ij})^{2v} \lambda_c$$

where λ_c is the Kuhn step length and v is a critical exponent equal to 0.5 or 0.588 (excluded volume effect). Mean values of other orientational quantities are set equal to zero because of the axial symmetry of fluctuations around the $\mathbf{r}_{ij}$ end-to-end vector.

The presence of an external constraint gives rise to a modified time average of orientations of the $\mathbf{a}_n^{ij}$ skeletal bond. The time average of the $(\mathbf{z}_n^{ij}/\mathbf{a}_n^{ij})^2 - \frac{1}{3})$ angular quantity is also modified. Consequently, the observation of the residual interaction of spins is expected to reveal the law of deformation of the $\langle \mathbf{r}_{ij} \rangle$ mean chain segment connecting two consecutive coupling junctions in the network.

The main question which is raised about the NMR approach concerns the exact identification of (i, j) chain segments which are involved in the determination of residual spin–spin interactions. The description of a network structure requires the use of a distribution function for anchoring points. It is assumed that any change of the residual interaction detected from NMR is closely related to deformations of network chain segments or units of percolation.

3.3 NMR structural and statistical parameters

In this section, it is shown how the polarization of orientations of skeletal bonds along the direction of elongation of a chain segment induces a residual energy of tensorial interactions established between nuclear spins located on this segment. More precisely, the attention is focused on the close relationship which exists between the state of elongation of a chain segment and the residual interaction of nuclear spins. Correspondingly, any chain segment may be considered as a sharp probe which applies conveniently to the exploration of the effect of topological constraints exerted in a polymeric gel. The purpose of this section is to show that a parameter can be conveniently determined from experimental transverse relaxation curves to characterize statistical structures of polymeric gels. The integral definition of this parameter avoids the necessity to give a detailed mathematical description of relaxation curves.

3.3.1 NUCLEAR MAGNETIC INTERACTIONS

In addition to field inhomogeneities and diamagnetic heterogeneities induced by the synthesis of most gels, three main interactions govern the magnetic relaxation of nuclear spins in polymers. The dipole–dipole interaction is expressed in the atomic scale by a formula analogous to that used in the macroscopic description of magnetic properties (Abragam, 1961). The energy of interaction of two dipoles

μ_1 and μ_2 is represented by

$$\hbar\mathscr{H}_{12} = \left(\mu_1 \cdot \mu_2 - 3\frac{(\mu_1 \cdot \mathbf{r}_{12})(\mu_2 \cdot \mathbf{r}_{12})}{r_{12}^2} \right) \Big/ r_{12}^3 \tag{17}$$

where $\mathbf{r}_{12}$ is the vector joining nuclei 1 and 2, $\mu_1 = \gamma\hbar\mathbf{I}_1$ and $\mu_2 = \gamma\hbar\mathbf{I}_2$ are represented by spin operators $\mathbf{I}_1$ and $\mathbf{I}_2$ respectively where γ is the gyromagnetic ratio. Quadrupolar interactions concern nuclei characterized by a spin I larger than one

$$\hbar\mathscr{H}_{\mathrm{Q}} = \frac{e^2 qQ}{4I(2I-1)}(3I_z^2 - I)(I+1))(3\cos^2\theta - 1) \tag{18}$$

for an axial symmetry of the electric field gradient acting on the nucleus; Q and q parameters represent electrical properties of the nucleus and of the electronic medium surrounding the nucleus respectively, θ is the angle between the magnetic field direction and the axis **a** of the electric field gradient and e is the electron charge. The chemical shift is induced by the deformation of electronic orbitals in the presence of a strong steady magnetic field

$$\hbar\mathscr{H}_{\mathrm{cs}} = \varpi_0 I_z \left\{ \sigma + \delta\frac{3\cos^2\theta - 1}{2} \right\} \tag{19}$$

where σ is the classical chemical shift observed in liquids, δ is a constant of anisotropy characterizing orbital motions of electrons within an axial symmetry and θ is the angle between the magnetic field and the principal axis of the molecular system.

The transverse magnetic relaxation of all nuclear spin systems embedded in a polymeric medium have been described within the adiabatic approximation, until now. In other words, the part of $\mathscr{H}_{\mathrm{D}}$ which does not commute with the Zeeman energy is given up. The adiabatic part $\mathscr{H}_{\mathrm{T}}$ can be always defined as a product of two terms.

$$\mathscr{H}_{\mathrm{T}} = \{3\cos^2[\theta(a)] - 1\}\mathscr{A}(\text{spins}) \tag{20}$$

where $\mathscr{A}$ (spins) is a linear or a bilinear function of spin component operators. The first term of the product depends on angular coordinates of a vector **a** characterizing the orientation of the molecular probe. It could be measured in the polymer system in the glassy state. Equation (20) applies to dipole–dipole interactions established within C–H, $-CH_2-$ or $-CH_3$ groups where the distances between nuclei are kept constant; these groups are very convenient probes sensitive to orientational properties of chain segments only.

3.3.2 PARTIAL MOTIONAL AVERAGING

According to Eq. (16), NMR investigations into the statistical structure of polymeric networks are based on the simple following effect. Whenever a force is applied to the ends of a chain segment, the rotational diffusion of skeletal bonds which belong to this segment undergoes a hindrance due to the presence of the force. This non-isotropic motion of monomeric units induces a non-zero average of tensorial interactions which exist between nuclei located on this segment. A chain segment defined by two consecutive i and j coupling junctions is considered; it consists of N_{ij} skeletal bonds. It is subjected to a force $\mathbf{f}_{ij}$. The residual interaction associated with this segment is called $\mathscr{H}_{\mathrm{R}}$ hereafter. The $\mathbf{f}_{ij}$ force controls configurational fluctuations which, in turn, determine the residual energy.

The residual interaction between nuclear spins located on a real chain is easily calculated in the case of a small elongation $\mathbf{a}_n^{ij}\cdot\mathbf{f}_{ij} < kT$). Using Eq. (16), the general expression of the residual dipole–dipole interactions of nuclear spins which commute with the Zeeman energy is

$$\mathscr{H}_{\mathrm{R}}(\langle \mathbf{r}_{ij}\rangle, L_{ij}) = 0.3(3\cos^2\theta_r - 1)(a\langle \mathbf{r}_{ij}\rangle/\zeta(L_{ij})^2)^2\,\Lambda \sum_{qq'} A_{qq'}B_{qq'} \tag{21}$$

where θ_r is the angle between the $\langle \mathbf{r}_{ij}\rangle$ end-to-end vector and the steady magnetic field direction and $(\zeta(L_{ij}))^2$ is the mean–square end-to-end vector of the segment of L_{ij} contour length. The sum is extended to all pairs of protons located along one chain segment. The angular factor $B_{qq'}$ results from the transformation of spherical harmonics associated with orientations of vectors joining any two nuclei q and q' and $A_{qq'}$ is a spin operator of the dipole–dipole interaction of nuclei q and q'. The average value is calculated over all configurational fluctuations compatible with the $\mathbf{f}_{ij}$ force.

More elongation induces more ordering and a stronger residual interaction. More contour length induces more disorder and a smaller residual interaction. the Λ factor is a parameter of second-order stiffness which describes the correlation of orientations of three skeletal bonds, along one chain segment,

$$\Lambda = \frac{15}{4}\sum_{nm}\langle\{3(z_p^{ij})^2 - a^2\}z_n^{ij}z_m^{ij}\rangle_0/a^4 \tag{22}$$

the average is calculated in the absence of any force; Λ is cancelled whenever the three monomeric units p, m and n are too far apart from one another along the polymer chain. This second-order stiffness is revealed from the effect of elongation detected from NMR. It parallels the first-order stiffness revealed from the state equation and represented by $\lambda_c = aC_\infty$ (C_∞ is the characteristic ratio of the polymer chain). For an ideal chain $\Lambda_0 = 1$. The second-order term Λ depends on the properties of the medium surrounding the chain segment. In a dilute solution, it reflects mainly intrinsic properties of the polymer chain induced by rotational isomerizations. Considering one single polymer chain, the second-order stiffness Λ can be calculated in a simple way, in the case where $L_{ij}\lambda_c \ll L_{ij}^2$. The correlation of orientations of skeletal bonds, in the absence of an external force, is described from a rotational diffusion equation (Yamakawa, 1971). Then Λ is expressed as $\Lambda = \beta\lambda_c^2$, with β a numerical factor nearly equal to 1.

3.3.3 RENORMALIZATION OF SPIN–SPIN INTERACTIONS

The main interest of Eq. (21) is to show that the residual interaction established between nuclear spins is reduced by non-isotropic motions of monomeric units down to a value $\mathscr{H}_T(a\langle \mathbf{r}_{ij}\rangle/\zeta(L_{ij})^2)^2$. The renormalization amounts to considering that the nuclear magnetic interaction which exists in the glassy state must be multiplied by the mean value of the $(a\langle \mathbf{r}_{ij}\rangle/\zeta(L_{ij})^2)^2$ ratio when the magnetic relaxation is observed in a gel. For a Gaussian polymeric system, the renormalization term is proportional to the L_{ij}^{-1} contour length. The order of magnitude of the renormalized value of the residual spin–spin interaction is about 10^3 rad s^{-1}, for $N_{ij} = 10^2$, in agreement with most experimental results. The effect of reduction of the strength of interactions established between nuclear spins is specific to the existence of a network structure. It is the key interaction which originates the irreversible dynamics of the transverse magnetization of nuclei linked to chain segments or to free molecules diffusing through the gel system. Thus, the nuclear relaxation of the transverse magnetization component behaves like in a conventional solid. Tensorial interactions are fully averaged to zero in the absence of crosslinks, assuming that the diffusion process of chains is isotropic: $\langle \mathscr{H}_T \rangle = 0$.

Equation (21) can be also interpreted as reflecting a transfer of NMR properties from the space scale of one monomeric unit to the scale which characterizes the state of average elongation of all network chain segments throughoutthe polymeric gel. Monomeric units become indistinguishable from one another along a given elementary chain segment. Equation (21) shows that NMR is actually very sensitive to the effect of partial orientational ordering of skeletal bonds, although the elongation of chain segments may be small. This illustrates the profound reason why NMR is sensitive to the property of temporary elasticity in a melt or to the effect of swelling or of stretching of a gel.

3.3.4 TRANSVERSE MAGNETIZATION: TIME SCALING

Comparisons of observations made from ^{1}H nuclei with those made from D or ^{13}C ones have clearly shown that magnetic interactions between nuclei located on different segments have negligible effects on the transferse relaxation. Consequently the determination of the relaxation function relies on the assumption that a linear magnetization $m_x(t, \mathbf{r}_{ij}, L_{ij})$ can be defined from all nuclear spins located along any segment determined by two consecutive network junctions and characterized by a contour length L_{ij} and a chain vector $\mathbf{r}$ (Cohen Addad and Dupeyre, 1983).

The $\mathscr{H}_{\mathrm{R}}(\mathbf{r}_{ij}, L_{ij})$ interaction induces a pseudo-solid behaviour of the magnetization

$$m_x(t, \mathbf{r}_{ij}, L_{ij}) = \mathscr{T}_r\{\exp[i\mathscr{H}_{\mathrm{R}}(r_{ij}, L_{ij})]t)\mathscr{M}_x \exp[-i\mathscr{H}_{\mathrm{R}}(\mathbf{r}_{ij}, L_{ij})t]\mathscr{M}_x\}/\mathscr{T}_r\{\mathscr{M}_x^2\} \tag{23}$$

where $\mathscr{M}_x$ is the transverse spin operator associated with all spins located along one chain segment. Two simple assumptions about fluctuations of network junctions underlie the calculation of the above equation. Either the rate of fluctuations is higher than the magnetic relaxation rate and only mean positions of junctions are observed, or the rate of fluctuations is lower and NMR is sensitive to the actual positions of junctions. Expression (3) is usually expanded in the basis of eigenfunctions of the Zeeman energy

$$m_x(t, \mathbf{r}_{ij}, L_{ij}) = \sum_{nn'} |h_{nn'}^2| \cos[0.3\Lambda\omega_{nn'}t(3\cos^2\theta_r - 1)a^2\langle r_{ij}\rangle^2/\zeta^4(L_{ij})] \tag{24}$$

with

$$\omega_{nn'} = \sum_{qq'} B_{qq'}(\langle n'|A_{qq'}|n'\rangle - \langle n|A_{qq'}|n\rangle) \tag{25}$$

where $\omega_{nn'}$ is a broadening frequency induced by the dipole–dipole interaction and $|h_{nn'}^2|$ is a matrix element of $\mathscr{M}_x^2$ which plays the role of a weighting factor. It is clearly seen from Eq. (24) that the ratio plays the role of a time-scaling factor; the statistical structure of polymeric gels is predicted to govern the transverse magnetic relaxation of nuclei attached to network chain segments. For a Gaussian polymeric system the N_{ij} number of skeletal bonds in one segment is a scaling factor acting upon the time-scale of the magnetic relaxation.

3.3.5 GEL DISTRIBUTION FUNCTION

A distribution function associated with the statistical structure of a polymeric gel must be introduced to express the irreversible behaviour of the transverse magnetization observed over the whole polymer gel and reads

$$M_x^{\mathrm{R}}(t) = \int \mathrm{d}L_{ij} \int \mathrm{d}\mathbf{r}_{ij} m_x(t, \mathbf{r}_{ij}, L_{ij})\mathscr{L}(L_{ij}/L_c)\mathscr{G}(\mathbf{r}_{ij}/\zeta(L_{ij})) \tag{26}$$

where $\mathscr{G}(\mathbf{r}_{ij}/\zeta(L_{ij}))$ is the probability distribution function of end-to-end vectors associated with chain segments of L_{ij} contour length, the $\mathscr{L}(L_{ij}/L_c)$ probability function accounts for a possible distribution of contour lengths of active segments and L_c is the width of the distribution.

3.3.6 MOMENTS OF THE GEL DISTRIBUTION FUNCTION

A very convenient treatment of transverse magnetic relaxation curves can be developed by considering two integrals which can be easily calculated:

$$\Phi_1 = \int_0^\infty t^{-1/2} M_x^{\mathrm{R}}(t)\mathrm{d}t \tag{27}$$

and

$$\Phi_3 = \int_0^\infty t^{-1/2}(\mathrm{d}M_x^{\mathrm{R}}(t)/\mathrm{d}t)\mathrm{d}t \tag{28}$$

Then it is easily shown that

$$\Phi_1 = 3.6\Lambda^{-1/2}\langle\langle(\Delta\omega)^{-1/2}\rangle\rangle\langle\zeta(L_{ij})/a\rangle m_1/m_2 \tag{29}$$

and

$$\Phi_3 = 0.4\Lambda^{1/2}\langle\langle(\Delta\omega)^{1/2}\rangle\rangle\langle a/\zeta(L_{ij})\rangle m_3/m_2 \tag{30}$$

where a is the mean skeletal bond length and with the following definitions of m_β ($\beta = 1, 2$ and 3) and $\langle\langle(\Delta\omega)^\beta\rangle\rangle(\beta = \frac{1}{2}$ or $-\frac{1}{2})$.

$$m_\beta = \int_0^\infty (r_{ij})^\beta \mathscr{G}(\mathbf{r}_{ij}/\zeta(L_{ij}))\mathrm{d}r_{ij} \tag{31}$$

and

$$\langle\langle\Delta\omega^\beta\rangle\rangle = \sum_{nn'} |h_{nn'}^2|(\omega_{nn'})^\beta \tag{32}$$

$\langle\langle\Delta\omega^\beta\rangle\rangle$ ($\beta = 1, 2$) average values can be considered as being specific to a given polymeric system. The definition of the $\langle\zeta(L_{ij})\rangle$ average is straightforward considering the function $\mathscr{L}(L_{ij}/L_c)$. Let γ_c denote the mean number of crosslinks per unit mass of polymer, then the total number of segments per unit mass is given by

$$\int \mathrm{d}L_{ij}\,\mathscr{L}(L_{ij}/L_c) = 2\gamma_c \tag{33}$$

while the average length of segments is expressed as

$$\int \mathrm{d}L_{ij}\,\mathscr{L}(L_{ij}/L_c)L_{ij} = L_c \tag{34}$$

with $L_c = a/2M_m\gamma_c$; M_m is the mean molecular weight of one skeletal bond. Equations (33) and (34) applies to the case where the average chain segment detected from NMR is defined by the crosslink density only.

The ϕ_3/ϕ_1 ratio defines the NMR structural parameter, it is called χ_c and is expressed as

$$\chi_c = 0.1\Lambda\Delta_g\xi_n^{-2}a^2 \tag{35}$$

with

$$\xi_n^2 = \langle\zeta(L_{ij})\rangle/\langle\zeta(L_{ij})^{-1}\rangle \tag{36}$$

and

$$\Delta_g = \langle\langle\Delta\omega^{1/2}\rangle\rangle/\langle\langle\Delta\omega^{-1/2}\rangle\rangle$$

The Δ_g ratio represents an intrinsic spin–spin interaction which results from dipole–dipole interactions established along one given polymer; it would be observed in a glassy state. The parameter, χ_c, plays the role of a relaxation rate. The $\langle\zeta(L_{ij})^{-1}\rangle\langle\zeta(L_{ij})\rangle^{-1}a^2$ ratio is a time-scaling factor governed by the network structure. The main advantage of the χ_c parameter is that its easy use does not require the establishment of an exact mathematical expression of the observed transverse relaxation function. A statistical NMR parameter is defined by the $\phi_3\phi_1$ product, it is called π_c and is expressed as

$$\pi_c = 1.44(\langle\langle\Delta\omega^{1/2}\rangle\rangle/\langle\langle\Delta\omega^{-1/2}\rangle\rangle)\langle\zeta(L_{ij})^{-1}\rangle/\langle\zeta(L_{ij})\rangle^{-1}m_3m_1/m_2^2 \qquad (37)$$

This NMR parameter reflects mainly the statistical framework of description of polymeric gels, taking the excluded volume effect into consideration, if necessary.

Properties of the gel distribution function are revealed from its moments in χ_c and π_c.

3.3.7 PSEUDO-SOLID BEHAVIOUR

The presence of residual spin–spin interactions is easily evidenced by forming pseudo-solid spin echoes or by observing the effect of a sample rotation on the relaxation curve. The general expression of the relaxation function is

$$M_x(t) = M_x^{\mathrm{R}}(t)\cdot\phi_{\mathrm{R}}(t)\,. \qquad (38)$$

where $M_x^{\mathrm{R}}(t)$ is given by Eq. (26), while $\phi_{\mathrm{R}}(t)$ results from fast rotational isomerization of skeletal bonds which give rise to residual dipole–dipole interactions. A solid-like echo can be formed at a time τ, after the beginning of the transverse relaxation process; let $E_x^{\mathrm{R}}(\tau, t)$ denote the function which represents this echo. Then the spin echo which is observed is described by the product

$$E_x(\tau; t) = E_x^{\mathrm{R}}(\tau, t)\cdot\phi_{\mathrm{R}}(t)$$

The pulse sequence which is applied to the spin system, to observe the pseudo-solid echo at a time τ, is determined by

$$90°/y - \tau/2 - 180°/y - \tau/2 - 90°/x - (t-\tau)/2 - 180°/x - (t-\tau)/2$$

The formation of one pseudo-solid spin echo results from the following simple property. Let $\mathcal{R}(n\pi/2; x)$ denote the rotation matrix which applies to spin operators in the frame rotating at the Larmor frequency; the angle of rotation is $n\pi/2$ ($n = 1, 2, 3 \ldots$) while the axis of rotation is called x. Then, the residual dipole–dipole interaction represented by the $\mathscr{H}_{\mathrm{R}}$ $(\langle\mathbf{r}_{ij}\rangle, L_{ij})$ hamiltonian obeys the following equation:

$$[\mathcal{M}_x, \mathcal{R}(n\pi/2; x)\mathscr{H}_{\mathrm{R}}(\langle\mathbf{r}_{ij}\rangle, L_{ij})\mathcal{R}^{-1}(n\pi/2; x)] = (-1)[\mathcal{M}_x, \mathscr{H}_{\mathrm{R}}(\langle\mathbf{r}_{ij}\rangle, L_{ij})]$$

According to this equation, the time derivative of $M_x^{\mathrm{R}}(t)$ at $t = \tau_-$ is equal to the opposite of the time derivative of $M_x^{\mathrm{R}}(t)$ at $t = \tau_+$ when a 90°/x rotation is applied at a time τ after the beginning of the transverse relaxation process. This corresponds to the formation of one echo.

Furthermore, the time derivative of $M_x(t)$, at $t = \tau$, is written

$$\frac{dM_x(t=\tau)}{dt} = \frac{dM_x^R(t=\tau)}{dt}\phi_R(\tau) + M_x^R(\tau)\frac{d\phi_R(t=\tau)}{dt}$$

and the time derivative of $E_x(t)$, at $t = \tau$, is expressed as

$$\frac{\partial E_x(t;t=\tau)}{\partial t} = -\frac{dM_x^R(t=\tau)}{dt}\phi_R(\tau) + M_x^R(\tau)\frac{d\phi_R(t=\tau)}{dt}$$

let p_s and p'_s denote these two time derivatives respectively. The addition and the substraction of the above two equations lead to the determination of the times derivatives of $M_x^R(t)$ on the one hand and $\phi_R(t)$ on the other hand, at $t = \tau$

$$\frac{dM_x^R(t=\tau)}{dt}/M_x^R(t) = (p_s - p'_s)/2M_x(t)$$

and

$$\frac{d\phi_R(t=\tau)}{dt}/\phi_R(t) = (p_s + p'_s)/2M_x(t)$$

This analysis allows the differentiation of the $M_x^R(t)$ function from the $\phi_R(t)$ function when it is necessary to determine the structural parameter χ_c with great accuracy.

The observation of pseudo-solid spin echoes has been proposed to determine clearly the nature of the relaxation process of protons which are attached to chain segments. In no case can any dynamic characterization of random motion be given when observed echoes are well shaped; the characterization concerns topological constraints exerted on chains only (Cohen Addad, 1993a).

3.3.8 CONCLUSION

It is clearly shown that two simple NMR parameters can be derived in a standard way from relaxation curves. They have a quite general meaning because no specific hypothesis underlies their definition. One of these parameters gives an estimate of the mesh size of network structures, while the other reflects the statistical framework of description of these polymeric systems.

3.4 Randomly crosslinked chains

In this section, gels formed by long chains which are crosslinked at random are considered. The approach to the analysis of characteristic properties of these network structures is illustrated from polybutadiene chains which are linked by sulphur bridges. The first step of the approach is the search for a variable for the state of gelation. The analogy between the gelation process and the percolation effect underlies the following description. It has been already shown that the sol–gel transition is correctly described within the Cayley tree approximation (de Gennes, 1979). From this framework of classical percolation, it is asserted that the gel fraction G, the modulus of elasticity E and the swelling ratio Q_m are thermodynamic quantities which must be expressed as functions of the variable of state of gelation ε. Furthermore, the NMR parameter, χ_c, is also expected to be a function of ε.

3.4.1 VARIABLE OF GELATION

It has been shown that the variable ε of state of gelation of long chains, crosslinked at random, can be defined by using a mean field approach to the gelation process (Cohen Addad and Soyez, 1992). The functionality f of long chains is equal to the number of monomeric units which are supposed to react with the crosslinking agent; for simplicity, f is set equal to N, the number of monomeric units per chain. Let γ_c denote the concentration of crosslinks per gram of polymer and γ_c^s the sulphur concentration. It is assumed that the mean number of sulphur atoms which participate in the formation of one bridge is x_s; then

$$\gamma_c = \gamma_c^s \mathcal{N} / x_s M_s$$

where M_s is the molar weight of one sulphur atom and $\mathcal{N}$ is the Avogadro number. Any sulphur bridge is considered as a bifunctional unit. The principle of the description is to calculate the probability, p, that one link is formed between two polymer molecules. Then the variable ε is defined as

$$\varepsilon = (p - p_c)/p_c \tag{39}$$

with $p_c = N^{-1}$ and

$$p = \phi_p \phi_s \tag{40}$$

where ϕ_p is the fraction of polymer functions which participate in the network structure and ϕ_s is the corresponding fraction of sulphur functions; ϕ_p and ϕ_s are related to each other by the equation

$$\phi_p = 2\phi_s \gamma_c^s M_m / x_s M_s$$

where M_m is the molar weight of one monomeric unit; p is easily defined by supposing that the crosslinking reaction is completed

$$p = 2 \frac{\gamma_c^s M_m}{M_s x_s}$$

and

$$\varepsilon = 2\frac{\gamma_c^s M_w}{M_s x_s} - 1 \tag{41}$$

where M_w is the weight average molecular weight of polybutadiene. It has already been shown that G, E and Q_m depend on ε; M_w was varied from 77 000 to 354 000 g.mol.$^{-1}$, while γ_c^s was varied from 0.002 to 0.01 g g^{-1} (Cohen Addad and Soyez, 1992). All samples were prepared with stearic acid, Santoflex 13, N-cyclohexyl-2-benzothiazylsulphenamide (CBS), and zinc oxide (ZnO) in addition to sulphur.

Considering the gel fraction, the modulus of elasticity, the swelling ratio or the structural NMR parameter, the threshold of gelation, obtained for $\varepsilon = 0$, was found to occur when the reduced variable $\gamma_c^s M_w$ is about equal to 500 g.mol.$^{-1}$. The square of the modulus of elasticity exhibits a linear dependence on $\gamma_c^s M_w$; this empirical law holds for values of $\gamma_c^s M_w$ lower than about 6000 g.mol.$^{-1}$. Equation (41) does not hold in the high sulphur concentration range ($\gamma_c^s M_w > 10\,000$ g.mol.$^{-1}$) where physical properties of network structures are nearly independent of the chain molecular weight. Typical proton relaxation curves observed from vulcanized chains, with different sulphur concentrations, are reported in Figure 2 (Soyez, 1992). NMR measurements were performed at

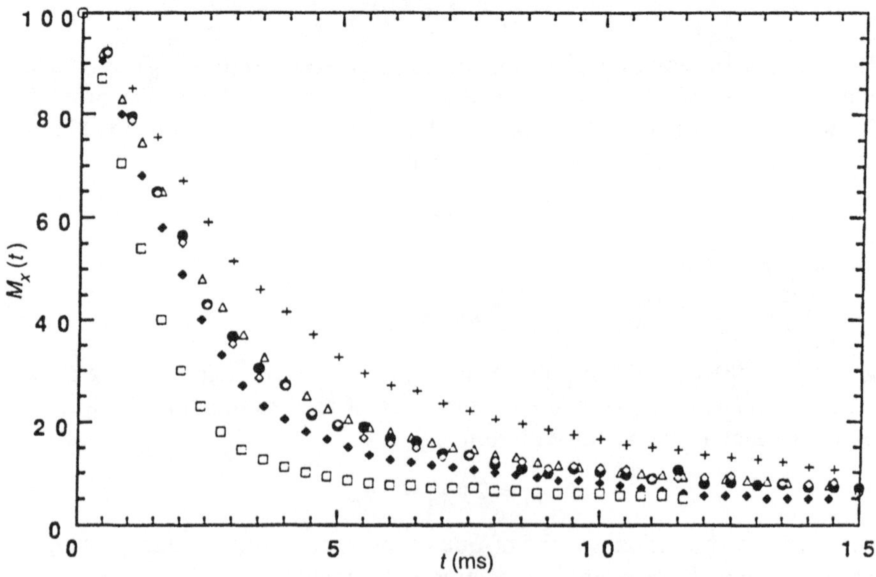

Figure 2. Typical proton transverse relaxation curves recorded from randomly crosslinked polybutadiene chains. Sulphur concentration γ_c^s: 0.002 g/g (+); 0.0035 g g^{-1} (△); 0.004 g g^{-1} (▲); 0.0045 g g^{-1} (☆); 0.005 g g^{-1} (★); 0.01 g g^{-1} (□)

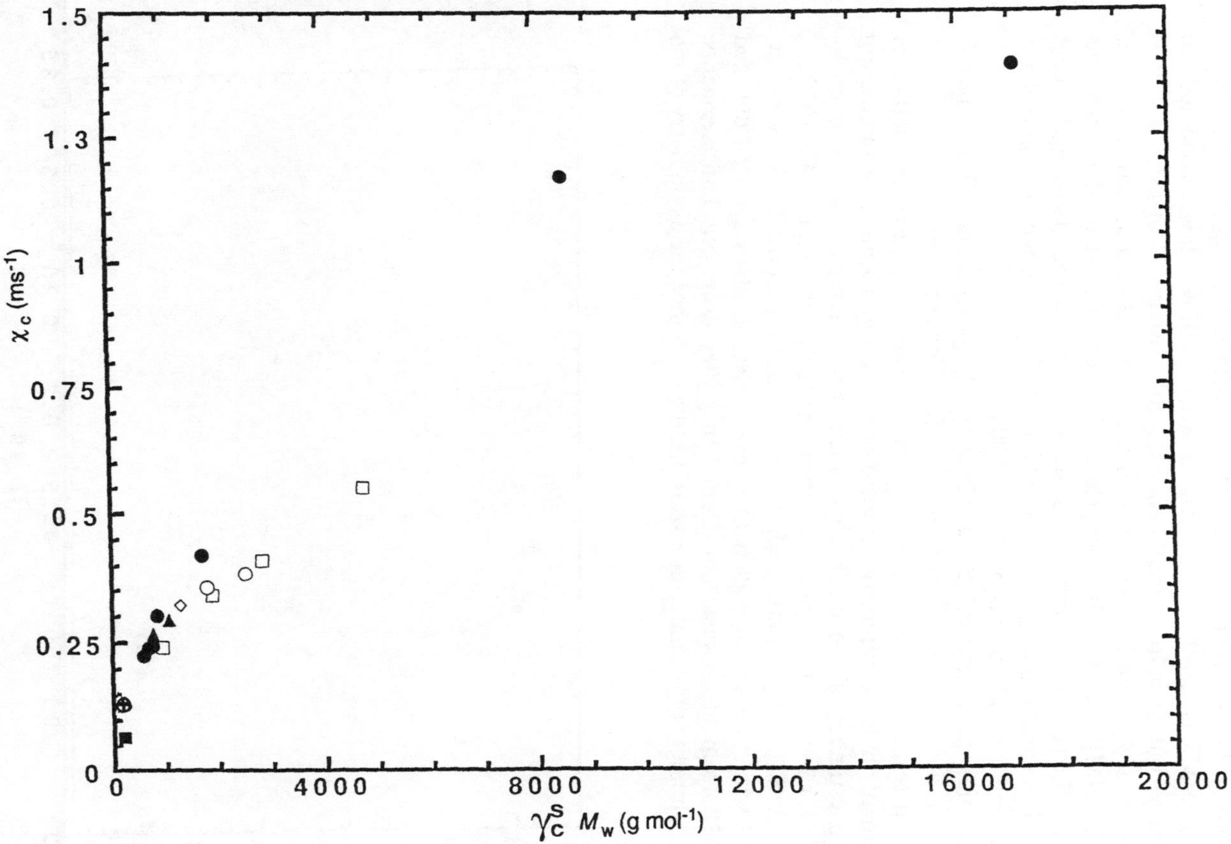

Figure 3. Variations of the structural NMR parameter as a function of the reduced variable $\gamma_c^s M_w$, at 150 °C: $M_w = 0.77 \times 10^5$ g mol^{-1} (▲); 1.3×10^5 g mol^{-1} (☆); 1.7×10^5 g mol^{-1} (●), 1.8×10^5 g mol^{-1} (○); 1.9×10^5 g mol^{-1} (□). Pure melt ($\gamma_c^s = 0$); $M_w = 0.77 \times 10^5$ g mol^{-1} (■); 1.8×10^5 g mol^{-1} (⊕). (From Cohen Addad and Soyez, 1992. Reprinted with permission from *Macromolecules* **25**, 6855, copyright 1992, American Chemical Society)

150 °C to induce fast segmental motions; these give rise to a motional average which is well contrasted to the anchoring effect affecting crosslinks. The dependence of the NMR parameter χ_c on $\gamma_c^s M_w$ is illustrated in Figure 3. This curve shows that NMR is sensitive to a characteristic length which is determined not only from γ_c^s but also from the chain molecular weight. Consequently, it is considered that χ_c is sensitive to the mesh size of the network structure; in other words, the NMR characteristic length, ξ_n, is proportional to the mesh size. Equation (33) in section 3.3, applies only to the case where the chain length is kept constant. The variation of χ_c, as a function of γ_c^s only, is shown in Figure 4; the chain molecular weight was equal to 170 000 g.mol.$^{-1}$.

Molten high polymers exhibit a property of dynamic screening which is reflected by the effect of temporary elasticity. Correspondingly, the nuclear spin-system response behaves as in a permanent gel, provided the lifetime of entanglements is longer than the inverse of the residual spin–spin interaction $\langle \mathscr{H} \rangle$. An estimate of the mean concentration of entanglements is given by $5 \times 10^{-4} \mathscr{N} \mathrm{g}^{-1}$ while the mean concentration of sulphur bridges is $4 \times 10^{-5} \mathscr{N}$, if γ_c^s is equal to 10^{-2} g g^{-1} and $x_s = 8$. The concentration of crosslinks is very low, compared with that of entanglements; nevertheless, bridges are well detected from NMR from topological constraints that they exert on chain segments. Crosslinks hinder rotational isomerizations; they enhance the asymmetry of the

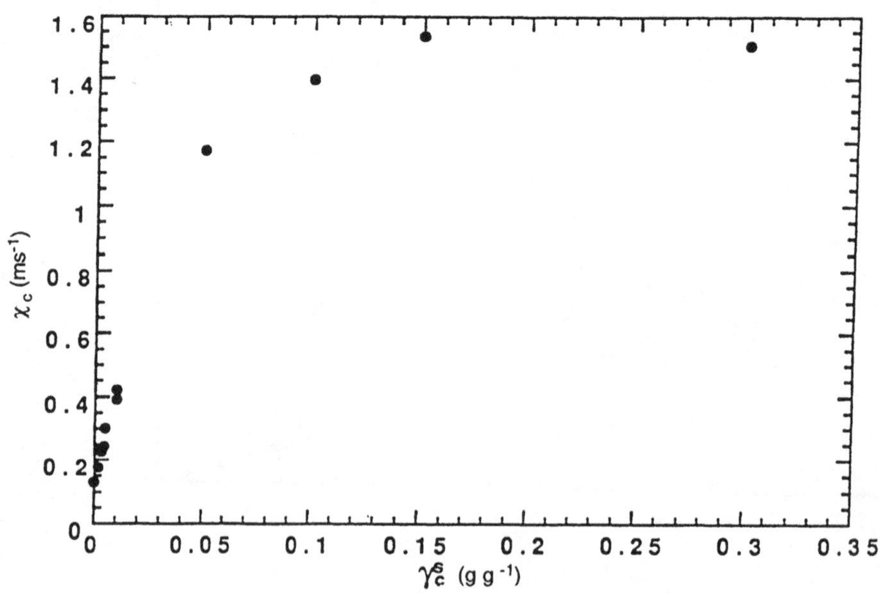

Figure 4. Variations of the structural NMR parameter as a function of the sulphur concentration, for a given chain molecular weight ($M_w = 170000$ g mol^{-1})

random rotation of skeletal bonds. This hindrance effect concerns about 250 monomeric units per crosslink. For short chains, the limiting value of χ_c, near the threshold of gelation, is due to the dynamics of chain migration; for long chains, the limiting value of χ_c is determined from the temporary network structure created by entanglements.

3.4.2 EFFECT OF SWELLING

Considering vulcanized chains, NMR investigations into properties of fully swollen networks show that the structural parameter χ_c is related to the swelling ratio, in a simple way. Let χ_c^Q denote the measured NMR parameter in a state of swelling. The state of maximum swelling is described according to the following schematic picture. The swollen polymeric system is supposed to consist of domains which obey a packing condition. In a given state of gelation, the mesh size of most domains is determined by a contour length L_Q which is a function of ε; L_Q may contain several consecutive strands defined by crosslinks. The mean concentration of monomeric units in one domain is given by

$$Q_M^{-1} \alpha (L_Q/a)^{1-3\alpha} \tag{42}$$

Strands which are possibly longer than L_Q are divided into blobs of size $(L_Q)^{3\alpha}$ while strands which are shorter than L_Q are in solution in domains and contribute to the partial screening of the excluded volume effect. The structural parameter χ_c^Q is proportional to $(L_Q/a)^{-2\alpha}$; then χ_c^Q must obey the equation

$$\chi_c^Q \alpha Q_M^{-2\alpha/3\alpha-1} \tag{43}$$

The exponent is equal to -2 for $\alpha = 0.5$. Figure 5 illustrates the dependence of χ_c^Q given by equation (43), this figure shows that the characteristic length detected from NMR is also involved in the swelling effect (the swelling agent was cyclohexane). This result reinforces the interpretation of NMR properties which are considered as induced by a collective segmental behaviour.

It has been shown experimentally that $\alpha = 3/5$ and $\chi_c^Q \alpha Q_M^{-1.5}$, for networks formed from end-linked polydimethylsiloxane strands (Soyez, 1992). For these systems, after completion of the chemical reaction, the random aspect lies in the formation of crosslinks which are nearest neighbours in space, but which are not necessarily nearest neighbours along one polymeric path. This relationship between χ_c^Q and Q_m also applies with $\alpha = \frac{1}{2}$ to crosslinked systems which were prepared from the reaction of ethylene vinyl acetate and ethylene methylacrylate polymer blends in the presence of dibutyltin oxide. The transesterification occurred at high temperature, leading to the formation of methyl acetate; the extent of the reaction was controlled by measuring the amount of methyl acetate produced during the crosslinking process. The mean lengths of ethylene sequences between two comonomeric units (EVA or EMA) were equal to 7.9 and 8.1 monomeric units respectively. The weight average molecular weights of the

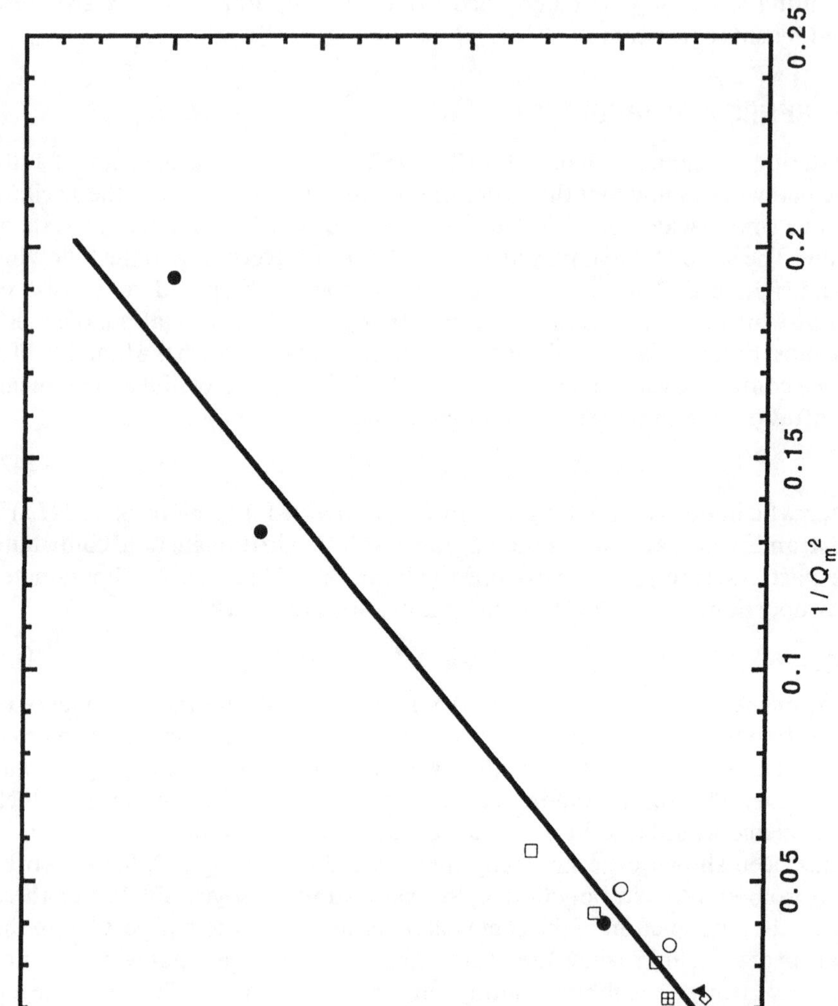
0
0.4
0.8
1.2
1.6
2
χ_σ^c (ms^-1)
0
0.05
0.1
0.15
0.2
0.25
1/Q_m^2

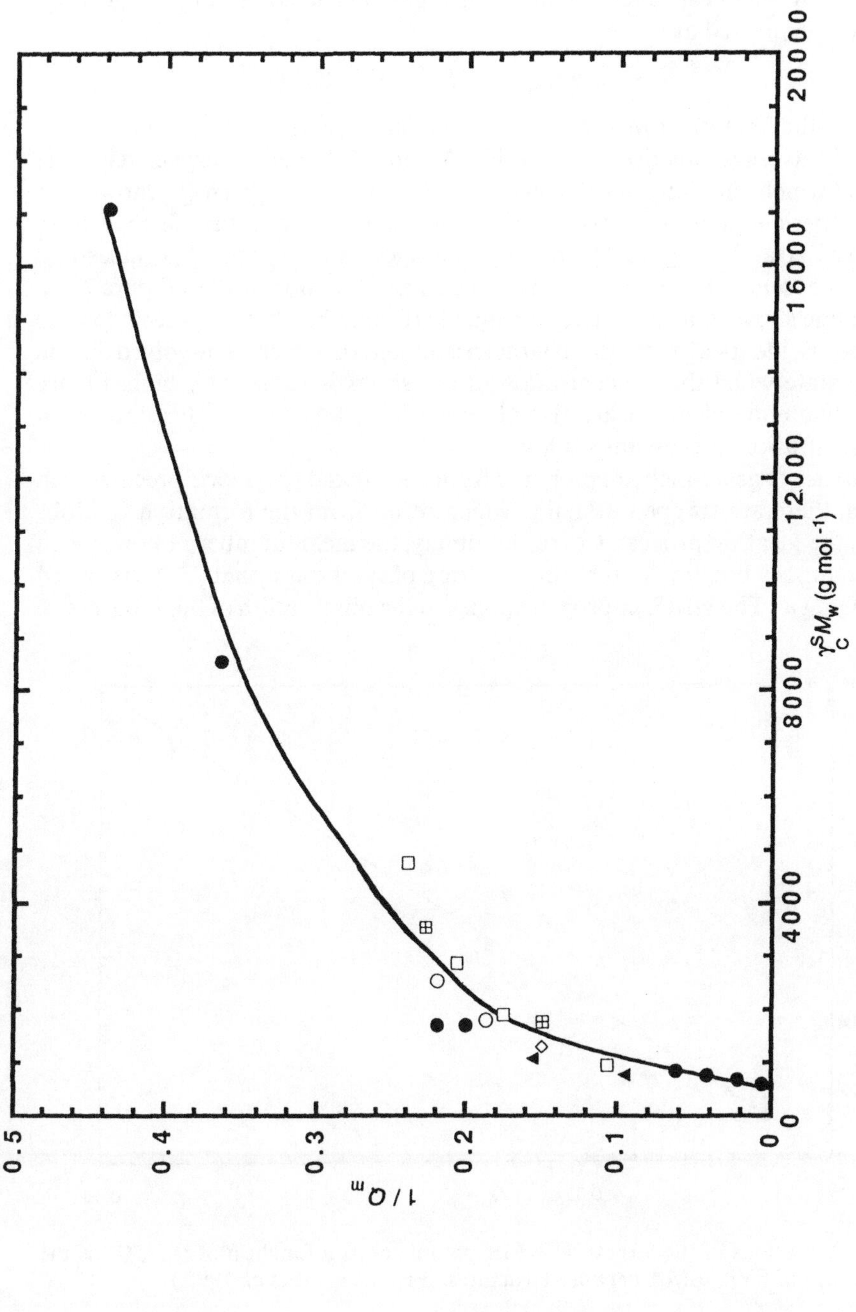

Figure 5. (A) Variations of the structural NMR parameter as a function of Q_m^{-2}; Q_m is the swelling ratio. (B) Variations of the inverse of the swelling ratio as a function of the reduced variable $\gamma_c^s M_w$ yield the value of the threshold of gelation ≈ 500 g mol^{-1}. (From Cohen Addad and Soyez, 1992. Reprinted with permission from *Macromolecules* **25**, 6855, copyright 1992, American Chemical Society)

EVA and EMA copolymers were 53.5 and 94.3 kg mol^{-1} respectively (Viallat *et al*, 1995). Then, polymer chains were crosslinked at random. Considering that VA and MA chemical functions obey a stoichiometric condition, the gelation variable is expressed as

$$\varepsilon = \phi^2(f_{MA} - 1)(f_{VA} - 1) - 1$$

where ϕ is the fraction of MA functions which have reacted and f_{MA} and f_{VA} are the weight average functionalities of EMA and EVA chains respectively. The exponent which characterizes the measured dependence of χ_c^Q on Q_m shows that this polymeric system does not behave like a network structure formed from end-linked strands (Figure 6); the swelling agent was xylene. The measured structural parameter, in a dry state, is reported as a function of χ_c^Q in Figure 7; the straight line shows that the characteristic NMR length, which is detected in the dry state, is identical with the characteristic length which is involved in the swelling state when the concentration of crosslinks is sufficiently high. Figure 7 shows that entanglements are also observed from NMR in the dry state when the concentration of crosslinks is low.

In the case of gels which are prepared from end-linked polymeric precursors in solution, there are trapped domains which result from the formation of blobs prior to the gelation process. Correspondingly, the mean number of bonds n, in one domain, is a function of the concentration of synthesis v_c: $n \propto v_c^{-5/4}$; n is varied by changing v_c. The NMR approach applies to the observation of the progressive

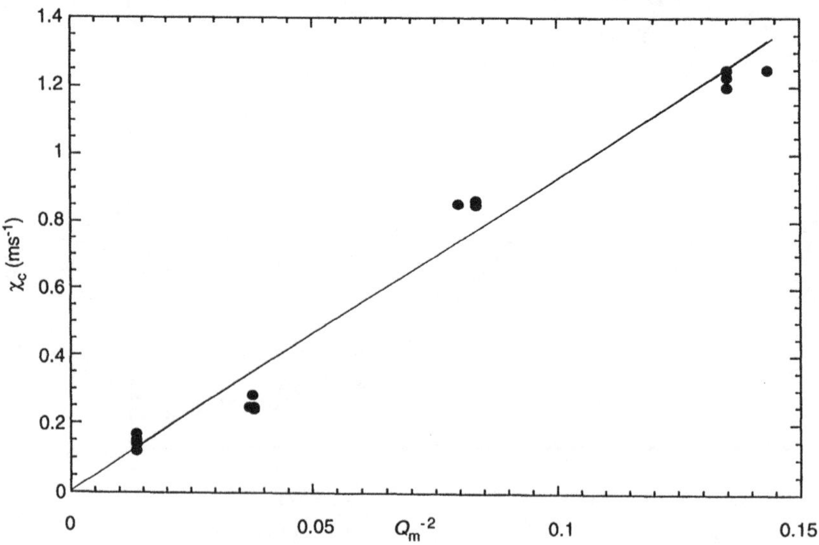

Figure 6. Variations of the structural NMR parameter as a function of Q_m^{-2}; Q_m is the swelling ratio of EVA–EMA network structures. (From Viallat *et al*, 1995.)

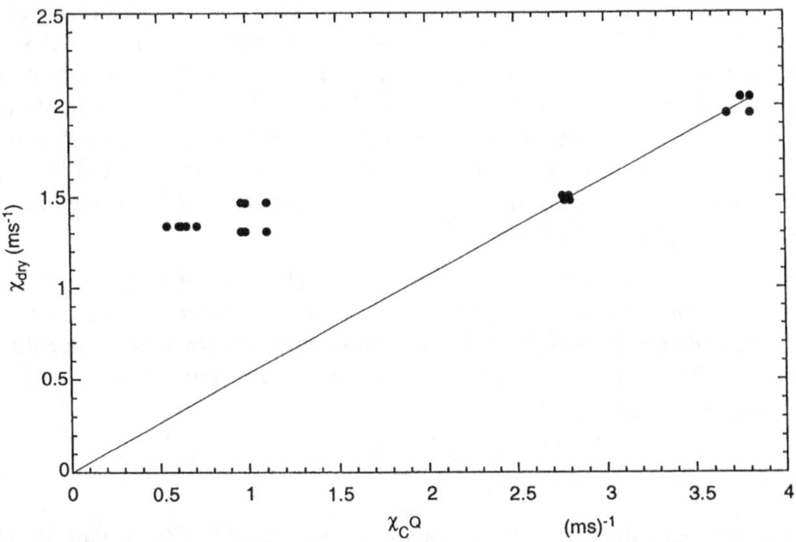

Figure 7. Relationship between structural parameters obtained from dry and swollen EVA–EMA network structures

gel swelling, induced by adding successively small amounts of solvent. Two main features are derived from observations.

1. Starting from dry gels, the relaxation rate of protons is found to exhibit a minimum when the swelling ratio Q is about equal to 2 or 3. In the range $1 \leqslant Q \leqslant 3$ the relaxation rate reflects the disengagement of chain segments from one another, then the relaxation rate increases because end-separation vectors between crosslinks are lengthened by the effect of swelling. The polymer volume fraction $1/Q^{**}$ (≈ 0.33 or 0.5) which is associated with the minimum of the relaxation rate is supposed to correspond to the crossover concentration c^{**} which separates semi-dilute from concentrated solutions; Q^{**} reflects a Gaussian state. The swelling ratio Q^{**} corresponds to the maximum value of the swelling ratio which is measured when a θ solvent is used to swell a polymeric gel.
2. Within the range $Q^{**} \leqslant Q \leqslant Q_m$, there is a transition from this Gaussian state to the state where the excluded volume effect is not screened at all. In this range the swelling ratio Q corresponds to an external condition induced by a given amount of solvent in the gel; the partially swollen gel is supposed to consist of domains which obey a packing property. The simple expression $\delta\alpha\langle \mathbf{r}\rangle^2/(\xi^2)^2$ for the proton relaxation rate is used; it is expressed as $\delta\alpha\langle \mathbf{r}\rangle^2/n^2$ for Gaussian domains which contain n skeletal bonds. Then $\delta\alpha(Q)^{2/3}/n^{4/3}$, with $\langle \mathbf{r}\rangle^2 \approx (Qn)^{2/3}$; δ is found to vary experimentally as $(Q)^{2/3}$ for a given

concentration v_c, i.e. a constant value of n. The reduced variable which is used when both the swelling ratio Q and the concentration of synthesis v_c are varied, is $(Q)^{2/3}(v_c)^{5/3}$; it corresponds to $n\alpha v_c^{-5/4}$. Finally, for the minimum, the relaxation rate is written as $\delta^{**} \approx n^{-1}$ and $Q^{**} \approx n^{1/2}$; experimentally the product $\delta(Q^{**})^2$ is found to be a constant when the concentration of synthesis is varied. In the state of maximum swelling, the relaxation rate δ is proportional to $Q_m^{-1.5}$. There is a smooth statistical transition which is detected from NMR when Q is varied from Q^{**} to Q_m.

It must be noted that there is a broad distribution of residual spin–spin interactions along one chain segment because skeletal bonds which are located near the centre of one segment experience must more isotropic random rotations than bonds which are near crosslinks. This effect is considerably enhanced when a gel is in a swelling state.

3.4.3 EFFECT OF STRETCHING

It is now supposed that an external constraint is applied to dry polymeric gels to observe the deformation of network chain segments which results from the presence of this constraint. It is well known that the interpretation of neutron scattering measurements performed on gels necessitates the introduction of a description based on statistical properties of network chain segments (Pearson, 1977). Similarly, NMR is used to probe the statistical polymeric gel structure over a semi-local space scale. This scale is determined by the average end-to-end vector $\langle \mathbf{r}_{ij} \rangle$ defined from the mesh size of domains. The analysis is based on the assumption that fluctuations which originate NMR properties are also responsible for the property of elasticity. The anisotropy of orientational diffusion of skeletal bonds which exists in any polymeric gel in a dry state is enhanced by the uniaxial extension of this system. This enhancement induces an additional residual energy of spin–spin interactions, related to the deformation of the network structure.

3.4.3.1 Modulus of elasticity

The effect of stretching is described according to the following expression for the retractile force per unit area (Cohen Addad, 1989):

$$F(\lambda) = \sum_{ij} \sum_{lp} \sum_{\alpha\beta} \langle \mathbf{r}_{ij}^{\alpha}(\lambda) \rangle \frac{kT}{\langle \Delta r_{lp}^{\alpha} \Delta r_{ij}^{\beta} \rangle} \frac{\partial \langle r_{lp}^{\beta} \rangle}{\partial \lambda} \tag{44}$$

Fluctuations of nodes which occur in the gel system are defined from the expression $\langle \Delta r_{lp}^{\alpha} \Delta r_{ij}^{\beta} \rangle$, where α and β represent any two of the three components x, y and z of $\langle r_{ij}(\lambda) \rangle$ vectors; Δr_{lp}^{α} represents the difference $r_{lp}^{\alpha} - \langle r_{lp}^{\alpha} \rangle$. Amplitudes of fluctuations are not easily estimated. They are supposed to be independent of

λ in the case of small elongations. As was discussed in section 3.2, the rate of time fluctuations of nodes has a crucial importance in NMR observations. The dependence of internal deformations upon the amplitude λ of the macroscopic elongation is taken into consideration in the last derivative of Eq. (44); this equation shows simply that the effect of elasticity is increased when the stretching of end-to-end vectors is increased and/or when amplitudes of fluctuations are reduced. When the stretching force is applied along the Z-axis direction, the deformation is usually described by the well-known tensor of affine transformation of $\langle \mathbf{r}_{ij} \rangle$ vectors (Flory, 1979)

$$\langle x_{ij}(\lambda) \rangle = \langle x_{ij} \rangle / \sqrt{\lambda}$$
$$\langle y_{ij}(\lambda) \rangle = \langle y_{ij} \rangle / \sqrt{\lambda}$$
$$\langle z_{ij}(\lambda) \rangle = \langle z_{ij} \rangle \lambda$$

the sample volume is kept invariant. The well-observed $(\lambda - 1/\lambda^2)$ dependence of the retractive force is obtained by assuming that cross-fluctuations which concern different components of a given chain segment are stochastically independent from one another. Fluctuations which concern end-to-end vectors issuing from a given mode or from different nodes are correlated with one another. Then, the force per unit initial area, corresponding to the λ ratio of elongation is given by

$$F(\lambda) = kT\left(\lambda - \frac{1}{\lambda^2}\right)\frac{\nu_c}{V_0}\frac{\xi_c^2}{3\langle \Delta\xi_c^2 \rangle} \tag{45}$$

where

$$\sum_{ij}\sum_{lp} \langle x_{ij} \rangle \langle x_{lp} \rangle / \langle \Delta x_{ij} \Delta x_{lp} \rangle = \nu_c \xi_c^2 / 3 \langle \Delta\xi_c^2 \rangle \tag{46}$$

The summation extends over all characteristic chain segments, $\nu_c(\varepsilon)$ in number. The modulus of elasticity E is expressed as

$$E = \frac{\nu_c}{V_0}\frac{\xi_c^2}{3\langle \Delta\xi_c^2 \rangle} kT \tag{47}$$

where V_0 is the initial volume of the network structure. The $\xi_c^2/\langle \Delta\xi_c^2 \rangle$ ratio is a measure of both the state of stretching of elementary chains in the absence of any elongation of the gel and the state of fluctuations of coupling junctions. The larger the stretching of elementary chains in the initial state and correspondingly the smaller the amplitude of fluctuations, the higher the modulus of elasticity. The increase of constraints exerted on chain segments increases the modulus of elasticity, whatever the nature of these constraints. Reported experimental results show that the modulus of elasticity is not a simple linear function of the concentration of crosslinks; E may be considered as a function of the square root of $\gamma_c^s M_w - 500$).

Fluctuations of nodes $\langle \Delta r_{lp}^{\alpha} \Delta r_{ij}^{\beta} \rangle$ cannot be described simply when the macroscopic deformation is large. Amplitudes of fluctuations of nodes become dependent on λ. Fluctuations which concern different components of a given end-to-end vector $\langle \mathbf{r}_{ij} \rangle$, are correlated with one another. Finally, the law of internal deformation of end-to-end vectors is probably not in affinity with the macroscopic deformation of the gel.

Transverse magnetic relaxation functions, observed in polymeric gels which undergo an uniaxial deformation, exhibit two key properties. The first concerns the strong dependence of the relaxation function on the stretching ratio λ while the second concerns its dependence upon the angle which the axis of elongation makes with the steady magnetic field direction.

3.4.3.2 Gaussian fluctuations

The nature of fluctuations is not specified in Eqs (44) and (45). The interpretation of NMR properties is based on the assumption that fluctuations are Gaussian. A given skeletal bond n which belongs to the (i, j) network chain segment is considered; its orientation is slightly polarized by the weak elongation which characterizes any chain segment in a dry polymeric gel in the absence of any deformation. Then the uniaxial deformation of the gel induces a new polarization of every skeletal bond. Correspondingly, the ε_{ij} $(\mathbf{a}_n, \langle \mathbf{r}_{ij} \rangle)$ residual spin–spin interaction, associated with the n skeletal bond, is transformed into ε_{ij} $(\mathbf{a}_n, \langle \mathbf{r}_{ij}(\lambda) \rangle$; this quantity is a function of the $\langle \mathbf{r}_{ij}(\lambda) \rangle$ vector which is transformed from the initial $\langle \mathbf{r}_{ij} \rangle$ vector. More precisely, considering the basic angular function $Y_2^0(\mathbf{a}_n)$ involved in magnetic relaxation processes, it is assumed that conformational fluctuations of chain segments are Gaussian in the absence as well as in the presence of a small gel deformation. Therefore, the relation

$$\langle Y_2^m(\mathbf{a}_n, \lambda) \rangle_{ij} \propto \langle \mathbf{r}_{ij}(\lambda) \rangle^2 \, Y_2^m(\langle \mathbf{r}_{ij}(\lambda) \rangle)/n_{ij}^2 a^2 \tag{48}$$

applies to the reference frame $\{\mathbf{B}_0\}$, associated with the direction of the steady magnetic field. In this reference frame coordinates of the $\langle \mathbf{r}_{ij}(\lambda) \rangle$ vector are called $\langle X_{ij}(\lambda) \rangle$, $\langle Y_{ij}(\lambda) \rangle$ and $\langle Z_{ij}(\lambda) \rangle$, the Z-axis is parallel to $\mathbf{B}_0$. Equation (7) holds for any end-separation vector of network chain segments.

One transverse magnetization $m_{ij}^x[t\varepsilon_{ij}(\mathbf{a}_n, \langle (\mathbf{r}_{ij}(\lambda) \rangle)]$ is ascribed to each (i, j) network chain segment. Then, the very nature of NMR leads to the following method of interpretation: the quantity which is averaged over the whole gel is not the residual interaction $\varepsilon_{ij}(\mathbf{a}_n, \langle \mathbf{r}_{ij}(\lambda) \rangle)$, but the magnetization $m_{ij}^x[t\varepsilon_{ij}(\mathbf{a}_n, \langle \mathbf{r}_{ij}(\lambda) \rangle]$. The direct substitution of the mean residual interaction for ε_{ij}, into the expression of the nuclear magnetization $m_{ij}^x(t)$, would contradict the fundamental description of NMR properties.

3.4.3.3 Spatial distribution of anchoring junctions

Let $M_x(t, \lambda)$ denote the relaxation function observed over the whole gel, the average being found by using the Gaussian probability distribution function of end-to-end vectors after deformation (Flory, 1979).

$$W(\langle \mathbf{r}_{ij}(\lambda)\rangle) = \exp[-3((z_{ij}(\lambda)/\lambda)^2 + \lambda(y_{ij}(\lambda))^2 + \lambda(x_{ij}(\lambda))^2)/2\sigma_{ij}^2] \tag{49}$$

This function applies to an uniaxial deformation. Then the relxation function is expressed as

$$M_x(t, \lambda) = \int m_{ij}^x[t\varepsilon_{ij}(\mathbf{a}_n, \langle(\mathbf{r}_{ij}(\lambda)\rangle)]\, W(\langle \mathbf{r}_{ij}(\lambda)\rangle)\, d\langle \mathbf{r}_{ij}(\lambda)\rangle \tag{50}$$

It is assumed that there are no magnetic correlations between spins located on different chain segments, although there may exist motional correlations between these segments. The effect of gel elongation, as detected from proton relaxation, is well illustrated from one proton pair attached to one skeletal bond; according to Eq. (48) the magnetization $m_{ij}^x(t)$ is expressed as

$$m_{ij}^x(t) = \exp(it\Delta_G \langle \mathbf{r}_{ij}(\lambda)\rangle^2\, Y_2^0(\langle \mathbf{r}_{ij}(\lambda)\rangle)\Lambda/n_{ij}^2 a^2 \tag{51}$$

where $\langle Y_{ij}(\lambda)\rangle$ and $\langle Z_{ij}(\lambda)\rangle$ coordinates are replaced with

$$\langle Y_{ij}(\lambda)\rangle = \langle y_{ij}(\lambda)\rangle \cos\theta_F + \langle z_{ij}(\lambda)\rangle \sin\theta_F$$

and

$$\langle Z_{ij}(\lambda)\rangle = \langle z_{ij}(\lambda)\rangle \cos\theta_F - \langle y_{ij}(\lambda)\rangle \sin\theta_F$$

in which θ_F is the angle which the steady magnetic field makes with the direction of stretching. The relaxation function reads

$$\begin{aligned} M_x(t, \lambda) = \mathcal{R}\{(1 + it^*[(\lambda^2 - 1/\lambda)(3\cos^2\theta_F - 1)] + t^{*2}[\lambda(3\cos^2\theta_F + 1) \\ + (3\sin^2\theta_F - 1)/\lambda^2] - 2it^{*3})^{-1/2}\} \end{aligned} \tag{52}$$

where t^* is equal to $t\Delta_G\Lambda/3N_{ij}$ and Δ_G is the average dipole–dipole interaction of one proton pair embedded in a solid. The force **F** is supposed to rotate in the ZY plane. The main feature perceived from Eq. (52) is the presence of a first-order term which cancels when λ is equal to unity or when the direction of elongation makes a magic angle with respect to the magnetic field direction $\theta_F \approx 54°$. According to this theoretical description the relaxation process of protons, attached to vulcanized chains, is enhanced when the polymer system is submitted to uniaxial stretching.

The expression of $M_x(t, \lambda)$, as given by Eq. (52), accounts for characteristic properties usually observed about relaxation functions (Cohen Addad, 1993b). The elongation effect is detected in a continuous way along relaxation curves. In no case should any experimental curve be split into two distinct parts.

At short times ($t^* \leqslant 1$), the magnetic relaxation is mainly governed by the $(\lambda^2 - 1/\lambda)(3\cos^2\theta_F - 1)$ first-order imaginary term, responsible for the enhancement of the relaxation mechanism. The term induces the well-known behaviour of properties of the transverse relaxation; it exhibits both a $(\lambda^2 - 1/\lambda)$ dependence and a $(3\cos^2\theta_F - 1)$ variation, whether it is observed directly or from splittings of resonance doublets obtained from Fourier transforms.

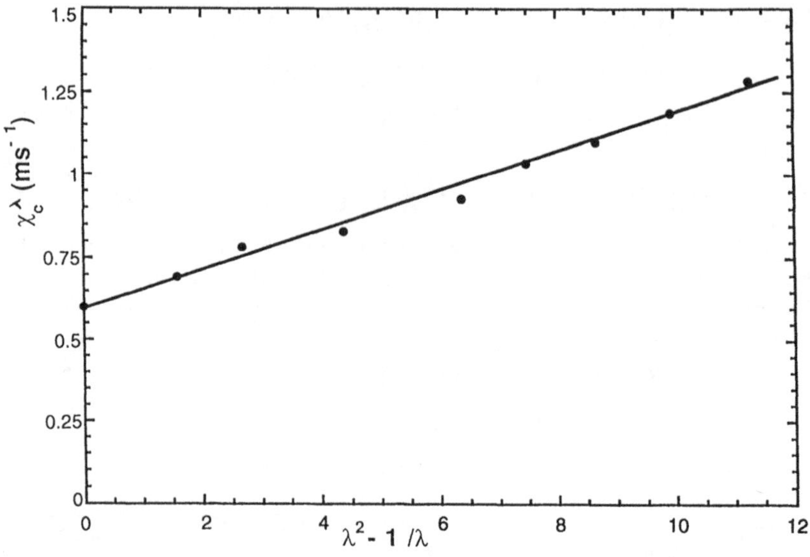

Figure 8. Variations of the structural NMR parameter as a function of the elongation function. The concentration of sulphur is $\gamma_c^s = 0.005\,\mathrm{g\,g^{-1}}$ and the polybutadiene molecular weight is $1.7 \times 10^5\,\mathrm{g\,mol^{-1}}$. (From Cohen Addad and Soyez, 1992. Reprinted with permission from *Macromolecules* **25**, 6855, copyright 1992, American Chemical Society)

At long times ($t^* \geqslant 1$), the t^{*3} dominant term is independent of the stretching ratio λ: the magnetic relaxation is less and less sensitive to the effect of elongation. This leads to the spindle-shaped pattern of relaxation curves observed as a function of λ.

The measured structural NMR parameter χ_c^λ, in a state of elongation, is very sensitive to the $[(\lambda^2 - 1/\lambda)(3\cos^2\theta_F - 1)]$ product because of the strong emphasis given to the short-time behaviour of relaxation functions in integrals (Eqs (27) and (28)). The χ_c^λ dependence on $(\lambda^2 - 1/\lambda)$ is illustrated in Figure 8. The molecular weight of vulcanized polybutadiene chains was equal to 1.7×10^5 and the χ_c^s sulphur concentration was equal to $0.005\,\mathrm{g\,g^{-1}}$. The χ_c^λ parameter is not equal to zero for $\lambda = 1$ because of the small elongation of segments in the absence of gel deformation.

The slope which is derived from the linear dependence of γ_c^λ on $\lambda^2 - 1/\lambda$ is called s_λ; its variation as a function of the variable of gelation is illustrated in Figure 9. This variation shows that the slope goes to zero near the threshold of gelation; this corresponds to an infinite NMR characteristic length ξ_n.

The dominant role played by the first-order term is experimentally illustrated by assuming that the relaxation function depends on the simple $([(\lambda^2 - 1/\lambda)(3\cos^2\theta_F - 1)]t)$ variable. Then, the effect of the rotation of the **F** direction around the X-axis can be considered; the angle that **F** makes with the X-axis is called θ_i while the angle of the Z-axis with the **F** projection is still called θ_F. Both θ_i

and θ_F are varied. Then by using a simple transformation of spherical harmonics the theoretical dependence of χ_c on θ_i and θ_F can be predicted. This is shown in Figure 10 while experimental results are reported in Figure 11.

3.4.4 KINETICS OF GEL FORMATION

The gelation process, associated with the growth of a Cayley tree, can be observed from NMR; the structural parameter is a time function: $\chi_c(\theta)$ where θ is the elapsed time since the beginning of the crosslinking reaction. The main purpose of such an approach is to compare the kinetics of formation of the network structure with the kinetics of the chemical reaction of the crosslinking agent which is characterized by the absence of polymer. The principle of the approach is first to make measurements of the NMR structural parameter as a function of the gelation variable when the gelation process is completed. The experimental curve drawn for χ_c as a function of ε serves as a reference to analyse the kinetics of formation of the network structure which is studied. Then, the kinetics of gelation are observed as a time function, at a fixed temperature, for a given concentration of crosslinking agent. Typical curves obtained from sulphur–polybutadiene systems heated at 150 °C are shown in Figure 12 (Cohen Addad *et al*, 1992). Comparing any value of $\chi_c(\theta)$ with the reference curve, it is assumed that it corresponds to a well-determined value of the gelation variable ε; then, the value of the sulphur concentrations that has been used at a time θ, is derived from ε using Eq. (41). This analysis is applied to mixtures formed from different initial sulphur concentrations and different chain molecular weights. Then, one single curve $\gamma_c^s(\theta)$ can be drawn, provided the network structure is built in an invariant way whatever the chain length or the initial concentration of sulphur.

3.4.5 FIELD OF MEAN ORIENTATIONAL ORDER

It is worth emphasizing that the transverse magnetization of protons attached to chain segments has been always found to be purely governed by residual spin–spin interactions in all polymeric network structures whether these are in a dry state or in a state of maximum swelling. This property reveals that the mean end-to-end vector $\langle \mathbf{r}_{ij} \rangle$ associated with any network chain segment is not equal to zero, in a dry gel. This effect is induced by the presence of crosslinks, these cannot move through one another and they create necessarily topological constraints. All end-to-end vectors are submitted to a residual stretching. There is a probability distribution function of such vectors; this distribution is isotropic in dry or swollen gels, in the absence of any elongation. In a dry gel, residual spin–spin interactions of protons are detected without any ambiguity by forming pseudo-solid spin echoes; however, the residual stretching of segments is too weak to give rise to a doublet of the resonance spectrum of deuterium nuclei whether these are attached to chain segments or to host molecules.

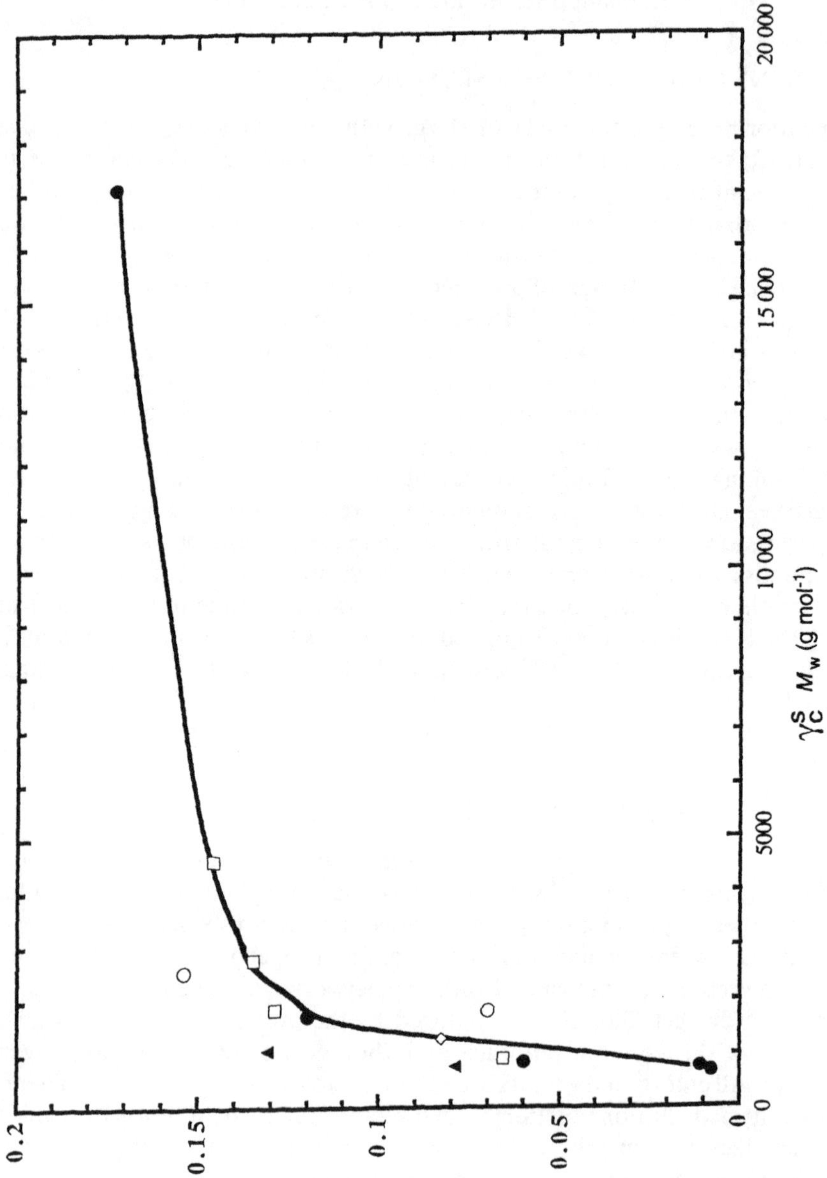
0.2
0.15
0.1
0.05
0
s_λ (ms^{-1})
0
5000
10 000
15 000
20 000
γ_c^S M_w (g mol^{-1})

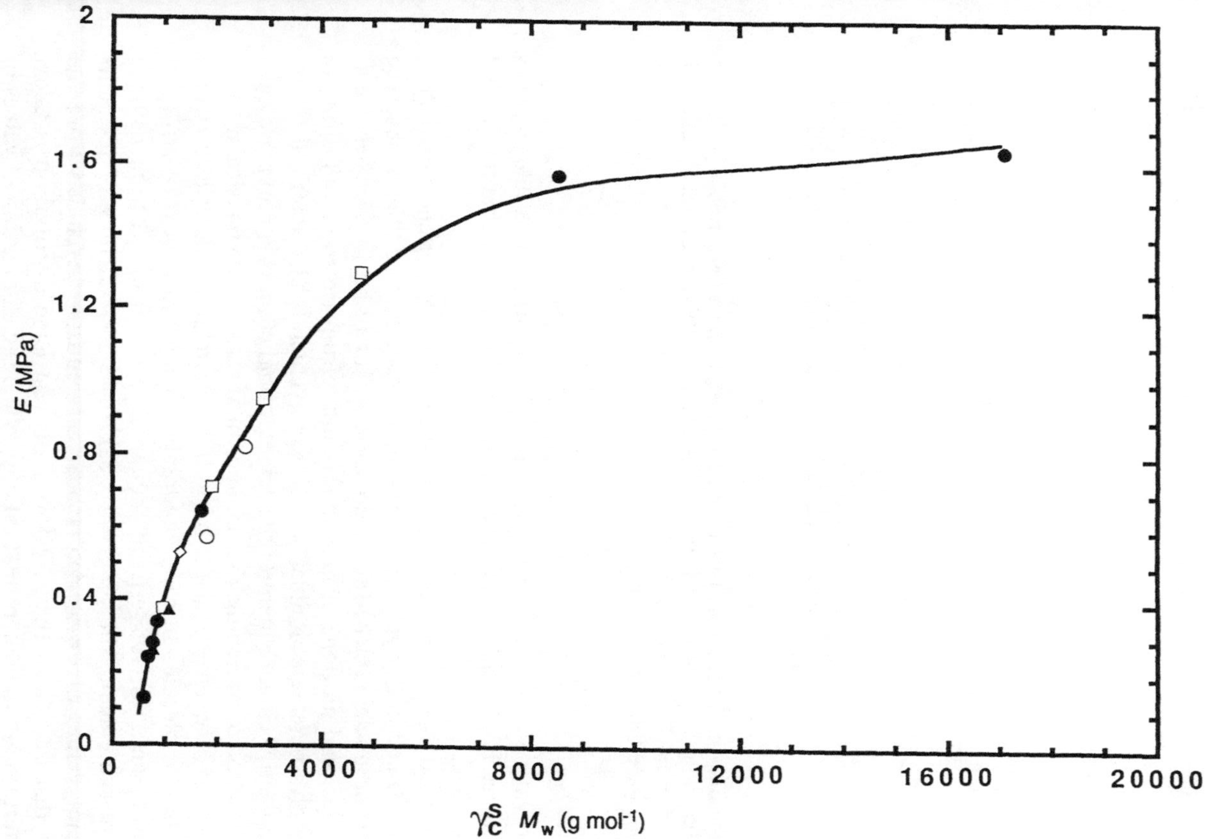

Figure 9. (A) Variations of the slope S_λ as a function of the reduced variable $\gamma_c^s M_w$: $M_w = 0.77 \times 10^5$ g mol^{-1} (▲); 1.3×10^5 g mol^{-1} (☆); 1.7×10^5 g mol^{-1} (●), 1.8×10^5 g mol^{-1} (○); 1.9×10^5 g mol^{-1} (□). (From Cohen Addad and Soyez, 1992.) (B) Variations of the modulus of elasticity E as a function of the reduced variable $\gamma_c^s M_w$; key to symbols as Figure 9A. (From Cohen Addad and Soyez, 1992. Reprinted with permission from *Macromolecules* **25**, 6855, copyright 1992, American Chemical Society)

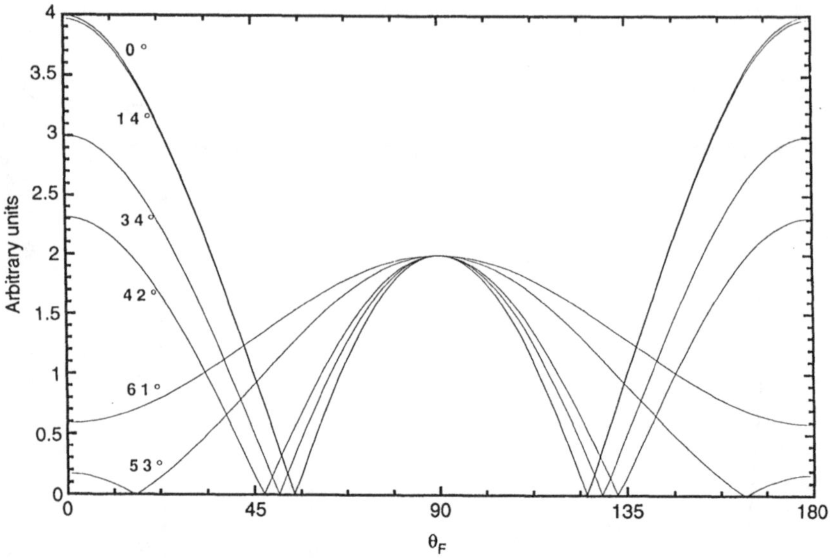

Figure 10. Simulation of the angular variation of the spin–spin interaction as a function of the rotation angle θ_F, for several inclination angles θ_i. (From Cohen Addad and Soyez, 1992. Reprinted with permission from *Macromolecules* **25**, 6855, copyright 1992, American Chemical Society)

The observation of pseudo-solid spin echoes on swollen gels is illustrated in Figure 13; these EVA–EMA gels were already evoked in this chapter (Viallat *et al*, 1995).

Under uniaxial elongation there is an enhancement of the orientational effect of monomeric units in dry gels; then a resonance spectrum doublet is observed from deuterium nuclei attached to chain segments or to host molecules. Deviations of rotational motions of monomeric units from isotropic diffusion are strong enough to create a doublet which is conveniently characterized. There may be a collective enhancement of the orientational effect of monomeric units along the direction of stretching; this enhancement results from physical interactions which exist between chemical groups. It has been given a description which applies to host molecules moving randomly through the network structure (Deloche and Samulski, 1981). Even though it is not detected from neutron scattering measurements, the observed weak residual stretching of segments indicates that values of $\langle \mathbf{r}_{ij} \rangle$ vectors near zero must be excluded from the probability distribution of these vectors. This restricted distribution is taken into consideration in the determination of χ_c while the elimination of values of $\langle \mathbf{r}_{ij} \rangle$ vectors near zero amounts to shortening the tail of curves of transverse relaxation of protons.

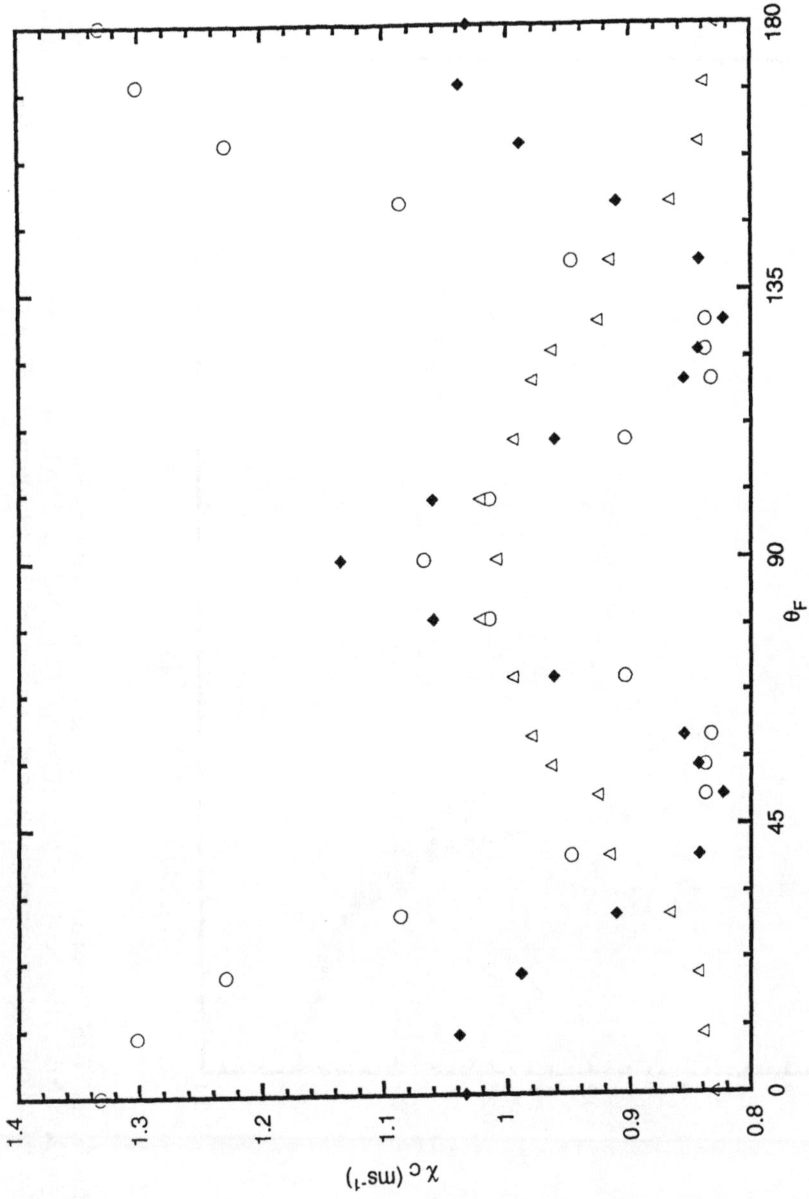

Figure 11. Experimental variation of the structural NMR parameter χ_c as a function of the rotation angle θ_F, for several inclination angles θ_i (From Cohen Addad and Soyez, 1992. Reprinted with permission from *Macromolecules* **25**, 6855, copyright 1992, American Chemical Society)

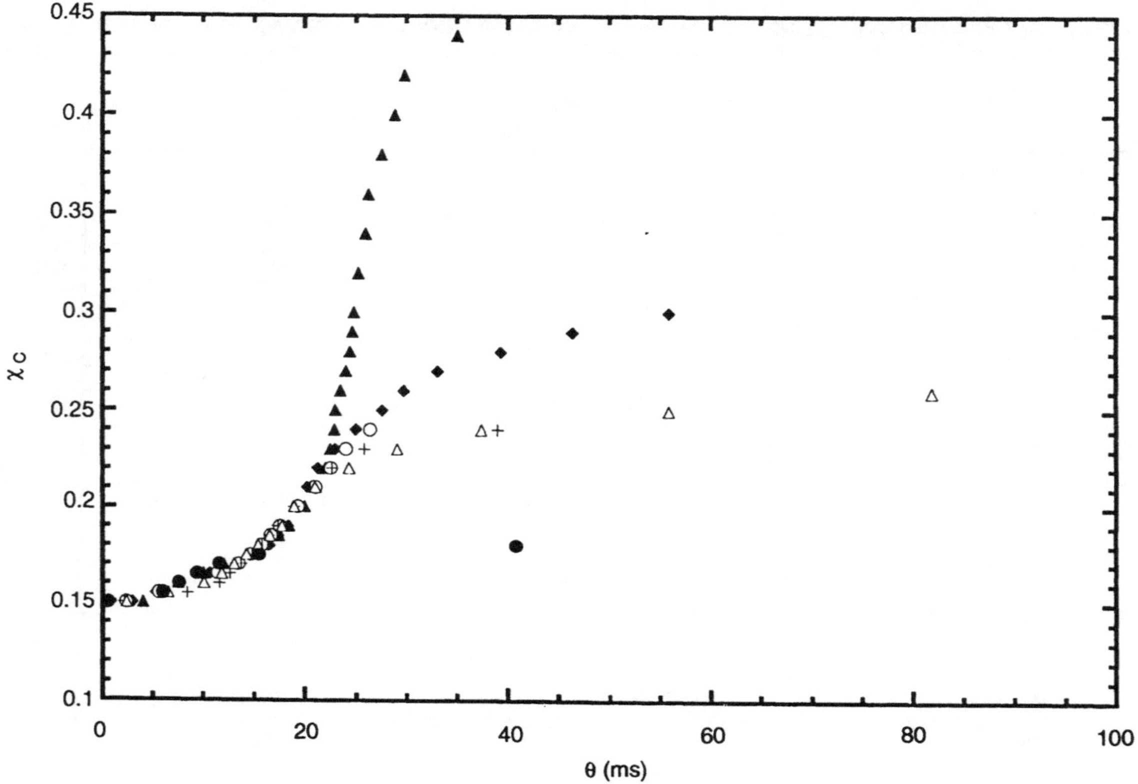

Figure 12. Kinetics of gel formation; variations of the structural NMR parameter χ_c as a function of the sulphur concentration: $\gamma_c^s = 0.01$ g/g (▲); 0.003 g/g (★); 0.0045 g g^{-1} (○); 0.004 g g^{-1} (+); 0.003 g g^{-1} (△); 0.002 g g^{-1} (●). $M_w = 2.5 \times 10^5$ g/mol (From Cohen Addad *et al*, 1992. Reprinted with permission from *Macromolecules* **25**, 1259, copyright 1992, American Chemical Society)

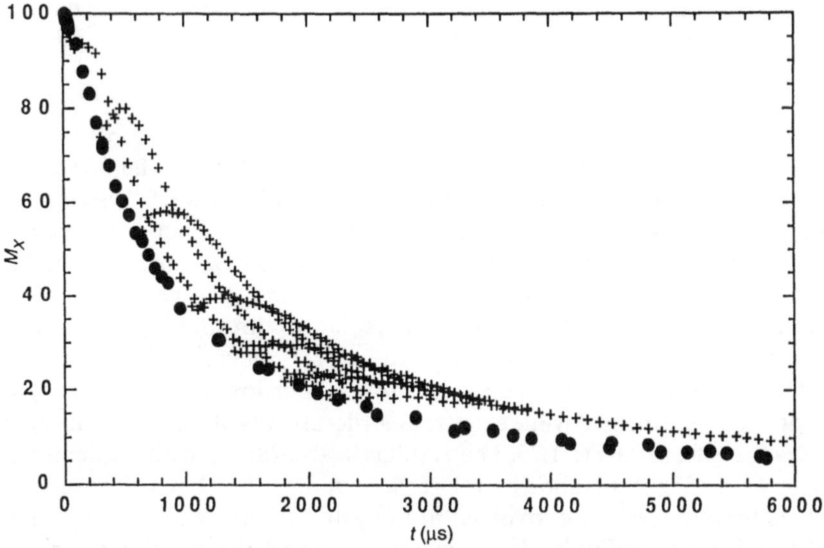

Figure 13. Typical illustration of the formation of pseudo-solid spin echoes. These echoes are specific to the existence of a residual spin–spin interaction. They are observed in swollen EVA–EMA network structures (swelling agent; xylene). (From Viallat *et al*, 1995)

3.4.6 CONCLUSION

In this section all reported experimental results clearly show that the relevant variable which can be used to describe the state of gelation of randomly crosslinked chains is the product of the concentration of crosslinks multiplied by the polymer molecular weight. This applies not only to macroscopic quantities but also to a suitably defined NMR parameter. The structural parameter χ_c is closely related to a characteristic length which is also involved in he swelling process of network structures. The relationship which exists between the characteristic lengths observed in the dry state on the one hand, and in the swelling state on the other, is clearly established.

3.5 Hydroxytelechelic polybutadiene

The three-dimensional polycondensation of hydroxytelechelic polybutadiene (PBHT) with 4,4-methylene di(phenylisocyanate) gives rise to a network structure which can be characterized from proton relaxation and swelling (the number average molecular weight was 2830 g mol^{-1} and the polydispersity index was equal to 2.3). The two basic parameters which are used to define mixtures prior to the gelation process are the weight average functionality of PBHT, hereafter called f_w and the ratio of the number of isocyanate functions over the number of hydroxyl functions; let r denote this ratio. Then the variable of gelation is determined by

$$\varepsilon = (f_w - 1)\phi_{OH}^2/r - 1 \tag{53}$$

where ϕ_{OH} is the fraction of hydroxyl functions involved in the formation of a network structure. The main feature in the study of such systems is the possibility of measuring the concentration of hydroxyl functions which have reacted during the gelation process. This measurement was made using liquid state high-resolution NMR which was applied to swollen gels (Allard, 1994). Again using the Cayley tree model, the percolation threshold is assumed to be defined by $1/(f_w - 1)$, the experimental value of f_w being 2.7. The ratio r was varied from 0.35 to 2.22; for high values ($r \geqslant 1.7$) or low values ($r \leqslant 0.65$), only finite clusters were formed while a network structure was observed when values of r were ranging from 0.65 to 1.70. This result is in accordance with the definition of ε given by Eq. (53); this is shown in the following way. For low values of r, many hydroxyl functions are not involved in the gelation process and ϕ_{OH} is nearly equal to r; correspondingly, ϕ_{OH}^2/r is equal to r and the gelation variable ε is defined by $1.7r$, the percolation threshold being given by $r \approx 0.6$. Similarly, for high values of r, many isocyanate functions are not involved in the gelation process and $\phi_{OH} \approx 1$, correspondingly, $\phi_{OH}^2/r \approx r^{-1}$ and the gelation variable is defined by $1.7/r$, the percolation threshold being given by $r \approx 1.7$.

The effect of variations of r is well illustrated from the kinetics of gelation which was observed *in situ* from NMR. The structural parameter χ_c was determined from proton transverse relaxation curves recorded during the isothermal gelation process; χ_c is reported as a function of r in Figure 14, for several time intervals after the beginning of the gelation. Considering a given time interval longer than 20 min, low values of χ_c which are associated with low values of r correspond to finite clusters. Then χ_c is increased when the relative concentration of isocyanate functions grows and it reaches a maximum value for $r \approx 1$ (stochiometry). For a further increase of r, χ_c is found to decrease; low values of χ_c which are associated with high values of r, correspond to finite clusters.

The gel fraction is reported as a function of r in Figure 15 and as a function of ϕ_{OH}^2/r in Figure 16. The swelling ratio Q_m was measured as a function of r (Figure 17); it is reported as a function of ϕ_{OH}^2/r in Figure 18. It is seen from Figures 16 and 18 that ϕ_{OH}^2/r can be considered as a reasonable reduced variable of the state of the polymeric network structure. Finally, the structural NMR parameter χ_c^Q was determined from transverse relaxation curves which were recorded from fully swollen gels, the swelling agent being cyclohexane. Variations of χ_c^Q are reported as a function of $Q_m^{-1.5}$, in Figure 19. The well-defined straight line which is drawn through experimental points, shows that these network structures behave like end-linked gels where the excluded volume effect is not screened; swollen segments are disengaged from one another.

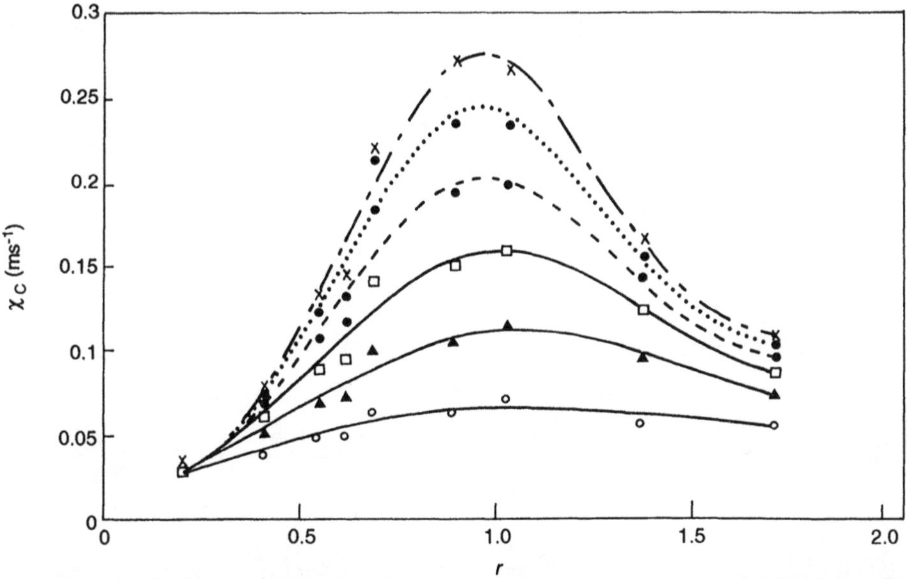

Figure 14. Kinetics of gelation observed at 50 °C as a function of *r*, for several time intervals after the beginning of the gelation: 20 mn (○); 30 mn (▲); 40 mn (□); 50 mn (●); 60 mn (★); 70 mn (×)

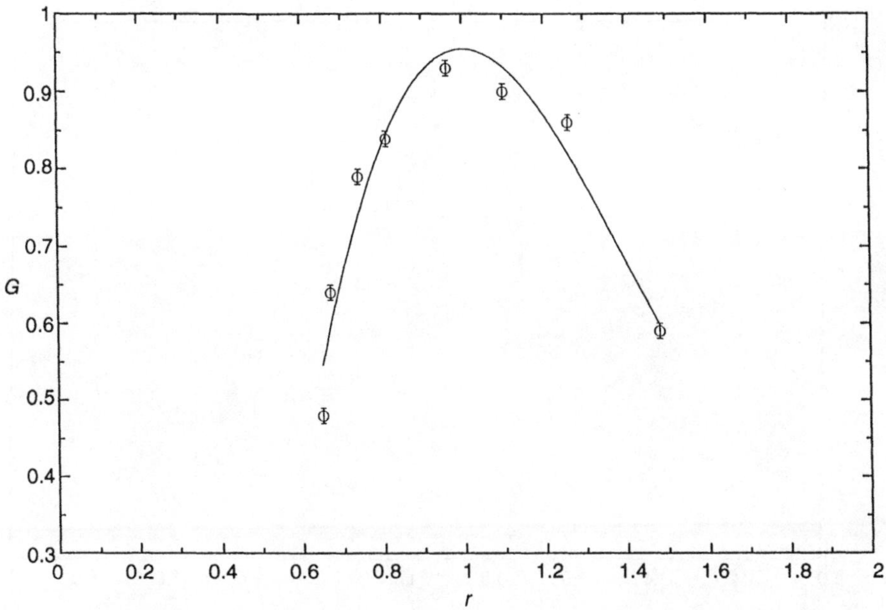

Figure 15. Gel fraction represented as a function of the ratio *r*

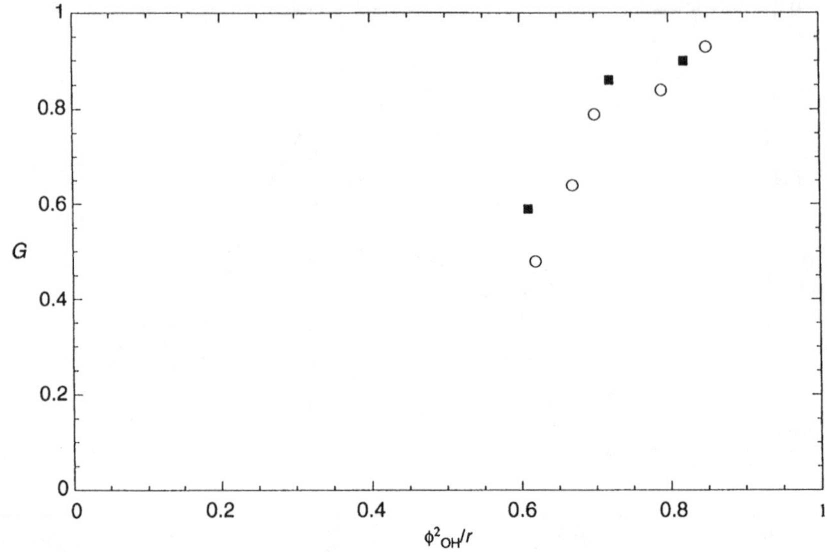

Figure 16. Gel fraction represented as a function of the reduced variable ϕ_{OH}^2/r. ○: values of r smaller than 1, ■: values of r higher than 1

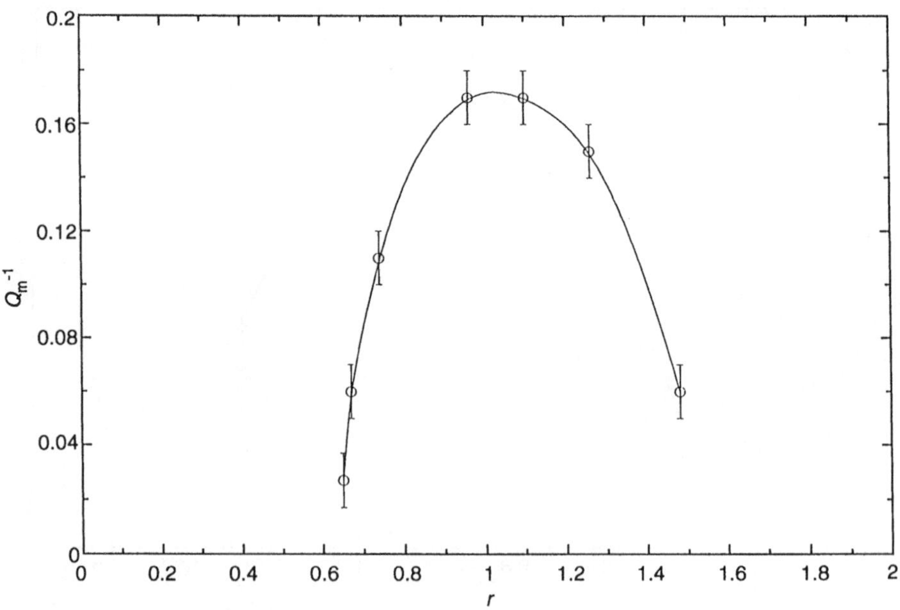

Figure 17. Swelling ratio Q_m as a function of the ratio r

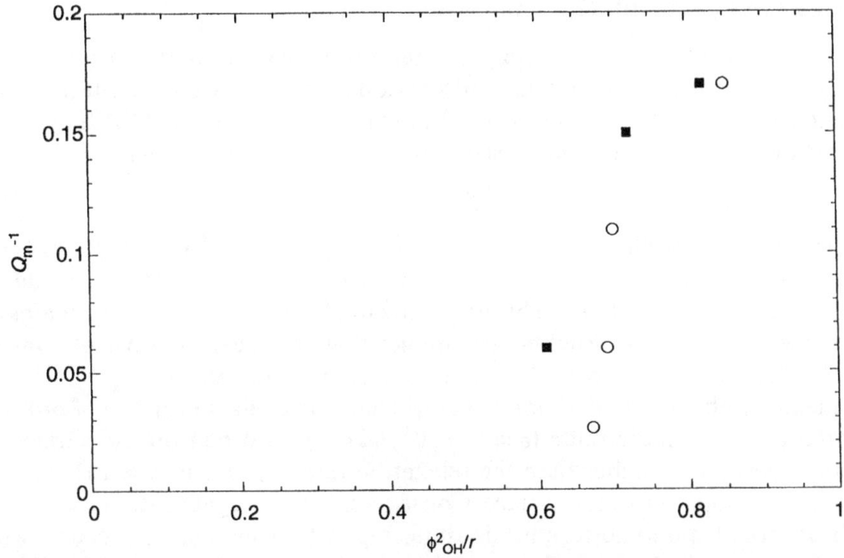

Figure 18. Swelling ratio as a function of the reduced variable ϕ^2_{OH}/r. ○: values of r smaller than 1, ■: values of r higher than 1

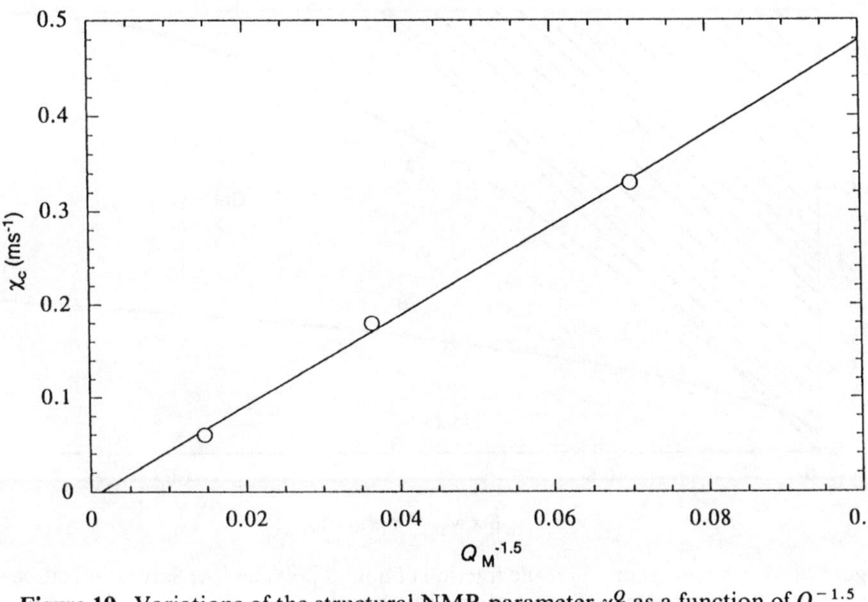

Figure 19. Variations of the structural NMR parameter χ_c^Q as a function of $Q_m^{-1.5}$

3.6 Filled polydimethylsiloxane

Observations of the transverse magnetic relaxation of protons attached to chains apply to the study of the polymer adsorption on particulate fillers. It has been recently shown that the adsorption of polydimethylsiloxane (PDMS) on silica aggregates which are fully immersed in the melt, obeys the following law:

$$Q_r^{\mathrm{m}} = \beta\sqrt{M_{\mathrm{n}}} \tag{54}$$

where Q_r^{m} is the specific amount of adsorbed polymer, $\beta = 4.3 \times 10^{-3}$ g mol$^{-1/2}$ and M_{n} is the number average molecular weight of the polymer (Cohen Addad, 1989). This law was established by weighing samples after removing free chains or from chemical analysis. Such a law implies that the mean number of contact points of one chain $\langle r_{\mathrm{c}} \rangle$ with the surface of silica is proportional to $\sqrt{N}$, N being the mean number of skeletal bonds in one chain. The relaxation rate of protons located on monomeric units ($\sigma \approx 5 \times 10^4$ rad s^{-1}) adsorbed on the surface of aggregates is much higher than the relaxation rate of protons ($\sigma \approx 10^3$ rad s^{-1}) attached to loops above the surface. Consequently the magnetization associated with adsorbed monomeric units is quite separate from that associated with polymeric loops in the adsorbed layer (Cohen Addad and Ebengou, 1992). When the law of adsorption is known, a sol–gel diagram can be built. For suitable concentrations of silica and chain lengths, an infinite cluster can be formed; it behaves like a polymeric gel (Cohen Addad, 1992). The gelation variable is

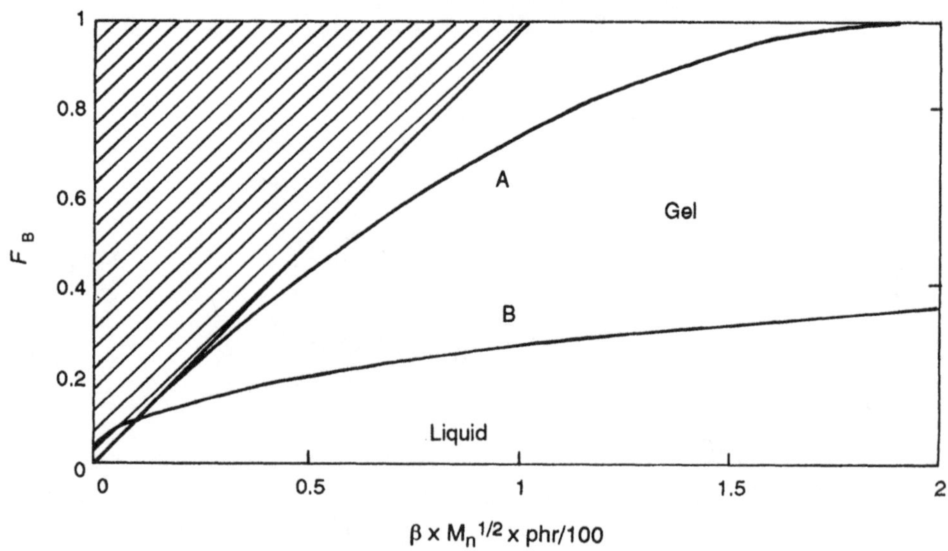

Figure 20. Sol–gel diagram; F_{B} is the fraction of bound polymer. (A) Saturation curve of silica; (B) curve of threshold of percolation. (Mn: g mol^{-1})

expressed as

$$\varepsilon = 2[1 - \sqrt{(1 - F_B)}]^2 \frac{100\mathscr{A} m^0_{Si}}{M_n phr} - 1 \tag{55}$$

where F_B is the fraction of bound polymer and m^0_{Si} is the average mass of one silica aggregate and *phr* means parts of silica per hundred parts of rubber. Such network structures can be swollen by using a good solvent, the swelling effect is also described within a mean field approximation. The phase diagram is illustrated in Figure 20; the curve of percolation threshold and the curve of surface saturation have been well observed experimentally.

3.7 Conclusion

Polymeric gels, by nature of their mechanical behaviour and ultimate properties are of interest as a fundamental domain. The first step of the description of properties of polymeric network structures concerns the search for a relevant variable of gelation which must contain the actual fractions of chemical functions which have reacted after the formation of a gel. These fractions are not always easily measured in the case of intermediate states of gelation. Then the gel fraction, the modulus of elasticity, the swelling ratio and the NMR structural parameters must be standard functions of the variable of gelation. However, the elaboration of the expressions of these functions is still in a state of refinement.

The transverse relaxation of protons attached to a polymeric gel is sensitive to a characteristic length which also governs the swelling effect induced by the osmotic pressure of a good solvent or the property of elasticity; there is a linear dependence of the modulus of elasticity on the NMR structural parameter which indicates that fluctuations which are involved in the response to a weak stretching of network chain segments are also detected from NMR in the absence of any deformation. For a network structure which is formed from end-linked polymeric precursors in bulk, the characteristic length is simply that of the precursor. Domains are trapped when the synthesis occurs in semi-dilute solutions and then the characteristic length is the segmental contour length which determines one domain. In the case of randomly crosslinked chains, the NMR structural parameter is sensitive to the number of crosslinks per chain and not to the simple mean contour length between two crosslinks along one chain.

In addition to structural properties, NMR can be used to characterize the dynamics of short chains moving randomly in a gel. The transverse relaxation of protons attached to strands can be differentiated from that of protons attached to host free chains which are used to probe the free volume of network structures.

References

Abragam, A. (1961) in *Principles of Nuclear Magnetism*, Oxford University, London
Allard, B. (1994) Thesis, Institut National des Sciences Appliquées, Lyon

Clough, P. S., Koenig, J. L. (1989). *J. Polym. Sci., Polym. Lett.* **27**, 451
Cohen Addad, J. P. (1974) *J. Chem. Phys.* **60**, 2440
Cohen Addad, J. P. (1982) *J. Physique (Paris)*, **43**, 1509
Cohen Addad, J. P. (1989) *Polymer* **60**, 1820
Cohen Addad, J. P. (1992) *Polymer* **33**, 2762
Cohen Addad, J. P. (1993a) in *NMR and Fractal Properties of Polymeric Liquids and Gels*, Progress in NMR Spectroscopy, eds Emsley, J. W., Feeney, F. and Sutcliffe, L. H., vol. 25, Pergamon Press, Oxford
Cohen Addad, J. P. (1993b) *Phys. rev. B.* **48**, 1224
Cohen Addad, J. P., Domard, M., Herz, J. (1978) *J. Chem. Phys.* **68**, 1194
Cohen-Addad, J. P., Dupeyre R (1983) *Polymer*, **24**, 7400
Cohen Addad, J. P., Ebengou, R. (1992) *Polymer* **33**, 379
Cohen Addad, J. P., Girard, O. (1992) *Macromolecules* **25**, 593
Cohen Addad, J. P., Schmit, C. (1987) *J. Polym. Sci., Polym. Lett.* **25**, 487
Cohen Addad, J. P., Soyez, E. (1992) *Macromolecules* **25**, 6855
Cohen Addad, J. P., Soyez, E., Viallat, A., Queslel, J. P. (1992) *Macromolecules* **25**, 1259
Cohen Addad, J. P., Vogin, R. (1974) *Phys. rev. Lett.* **33**, 940
Curran, S. A., Padira, A. R. (1987) *Macromolecules* **20**, 625
De Gennes, P. G. (1976) *J. Physique Lett.* **37**, L1
De Gennes, P. G. (1979) in *Scaling Concepts in Polymer Physics*, Cornell University Press, Ithaca
Deloche, B., Samulski, E. T. (1981) *Macromolecules* **14**, 575
Dickinson, L. C., Chien, J. C. W., Macknight, W. J. (1988a) *Macromolecules* **21**, 2959
Dickinson, L. C., Morganelli, P., C. W., Petrovic, Z., MacKnight, W. J., Chien, J. C. W. (1988b) *Macromolecules* **21**, 338
Doskocilova, D., Schneider, B., Jakes, J. (1978) *J. Magn. Reson.* **20**, 79
Dubault, A., Deloche, B., Herz, H. (1984) *Polymer* **25**, 1405
Feio, G., Cohen-Addad, J. P. (1988) *J. Polym. Sci.* **26**, 389
Ferry, J. D. (1983) in *Viscoelastic Properties of Polymers*, Wiley, New York Ed. J. Wiley
Flory, P. J. (1979) in *Principles of Polymer Chemistry*, Cornell University Press, Ithaca Ed Cornell University
Garrido, L., Mark, J. E., Ackerman, J. L., Kinsey, R. A. (1988) *J. Polym. Sci., Polym. Phys. Ed.* **26**, 2367
Gronski, W., Stadler, R., Jacobi, M. (1984) *Macromolecules* **17**, 741
Muthukumar, M., Edwards, S. F. (1982) *Polymer* **23**, 345
Pearson, D. S. (1977) *Macromolecules* **10**, 696
Sotta, O., Deloche, B., Herz, J. (1988) *Polymer* **29**, 1171
Sotta, P., Deloche, B., Herz, J., Lapp, A., Durand, D., Rabateno, J. C. (1987) *Macromolecules* **20**, 2769
Soyez, E. (1992) Thesis, Université Joseph Fourier, Grenoble
Stauffer, D. (1976) *J. Chem. Soc. Faraday Trans.* **II,72**, 1354
Stockmayer, W. H. (1943) *J. Chem. Phys.* **11**, 45
Vanderhart, D. L. Perez, E., McKenna, G. B. (1988) *Macromolecules* **21**, 2418
Viallat, A., Cohen Addad, J. P., Casagnau, P., Michel, A. (1995) *Polymer*, in press
Von Meerwall, E., Ferguson, R. D. (1981) *J. Polym. Sci., Polym. Phys. Ed.* **19**, 77
Yamakawa, H. (171) in *Modern Theory of Polymer Solutions*, Harper and Row, New York

4 SOL–GEL TRANSITION

M. Adam and D. Lairez
Laboratoire Léon Brillouin, Gif-sur-Yvette, France

The Physical Properties of Polymeric Gels. Edited by J. P. Cohen Addad.

4.1 Introduction

The physical properties of gels has fascinated scientists for a long time. The sol–gel transition is a compulsory stage to form these gels. This critical transition was first studied by well-known polymerists as Flory, Stockmayer and Gordon in the 1950s, but since their works the problem has not attracted the attention of researchers. Seventeen years ago the de Gennes and Stauffer theoretical challenge provoked a revival of interest: is percolation relevant to describe gelation? Since then, several teams have focused their attention on this very difficult problem which implies interdisciplinary knowledge. If gelation obeys the laws of percolation, then from a fundamental point of view gelation would be the unique physical system which allows the handling of clusters generated by this critical phenomenon.

In this chapter we first review the theoretical background of gelation from the view point of statics and dynamics of these systems. In section 4.3 we will describe the relevant experimental techniques conjointly with what it is expected to measure and the experimental results obtained so far. This review does not pretend to be exhaustive, but rather to quote the most significant results obtained during the last decade.

4.2 Theory

4.2.1 STATIC PROPERTIES OF BRANCHED POLYMERS NEAR THE SOL–GEL TRANSITION

4.2.1.1 Definitions and general purpose: the classical approach

In order to familiarize the reader with the gelation phenomenon, we present in this section basic concepts starting from the classical approach. Let us consider a population of molecules (polymers) consisting of identical elementary units (monomers) having a mass equal to one. This population is characterized by the mass distribution law of molecules which is the curve corresponding to the number n_i of molecules having a mass M_i. The various moments of this distribution law have different meanings. The expected mass of a randomly chosen molecule is equal to the ratio of the first-order moment to the zero-order moment.

$$M_{\mathrm{n}} = \frac{\sum_i n_i M_i}{\sum_i n_i} \tag{1}$$

where M_{n} is the number average molecular mass of molecules. On the other hand, if a monomer, rather than a molecule, is randomly chosen among the population, the expected mass M_{w} of the molecule, to which this monomer belongs, differs. The monomer fraction belonging to molecules having a mass M_i is equal to $n_i M_i / \sum n_i M_i$, thus one obtains

$$M_{\mathrm{w}} = \frac{\sum_i n_i M_i^2}{\sum_i n_i M_i} \tag{2}$$

where M_{w} is the weight average molecular mass of the molecules.

These two quantities M_{n} and M_{w} are linked to the statistical variance of the mass distribution law of the molecules. Indeed, polymerists usually express the width of the mass distribution law around the average value by the polydispersity index I, which is equal to 1 in the case of a zero variance:

$$I = \frac{M_{\mathrm{w}}}{M_{\mathrm{n}}} \tag{3}$$

Let us now imagine a chemical system which is made at time zero by monomer units which are able to link themselves, covalently and randomly, to f neighboring molecules: f is the functionality of monomers. As the chemical reaction continues, the mass distribution law of molecules in the medium changes with the apparition of dimers, trimers, etc. For example, this process is encountered in the case of the polycondensation leading to macromolecules.

The classical theory of Flory and Stockmayer (Flory, 1953; Stockmayer, 1943, 1944) allows us first to understand some particularities of such a system. This approach supposes that the probability p for one chemical function to have reacted is the same whatever the molecule to which is belongs. It also considers a tree-like structure for the macromolecules within which loops are not allowed. Consequently, a new created chemical link causes the disappearing of one molecule, i.e. two molecules join together to build a larger one. These two starting assumptions allow the average quantities M_{n} and M_{w} to be expressed as a function of p and f. The total number of molecules at a given time is equal to the number of monomers (which corresponds to the number of molecules in the medium at time zero) minus the number of chemical links so far created. One obtains:

$$\sum n_i = \sum n_i M_i - \tfrac{1}{2} pf \sum n_i M_i = \left(1 - \frac{pf}{2}\right) \sum n_i M_i \tag{4}$$

which leads to the expression of M_n in the classical theory framework:

$$M_n(p) = \frac{1}{1 - pf/2} \tag{5}$$

In order to establish the expression of M_w, one has to consider one monomer and the expected number of monomers which belong to the same molecule. This monomer has f chemical functions and among them pf have reacted. These pf chemical functions lead to the same number of monomers each of them having $(f-1)$ chemical function leading to $p(f-1)$ new monomers, etc. (see Figure 1).

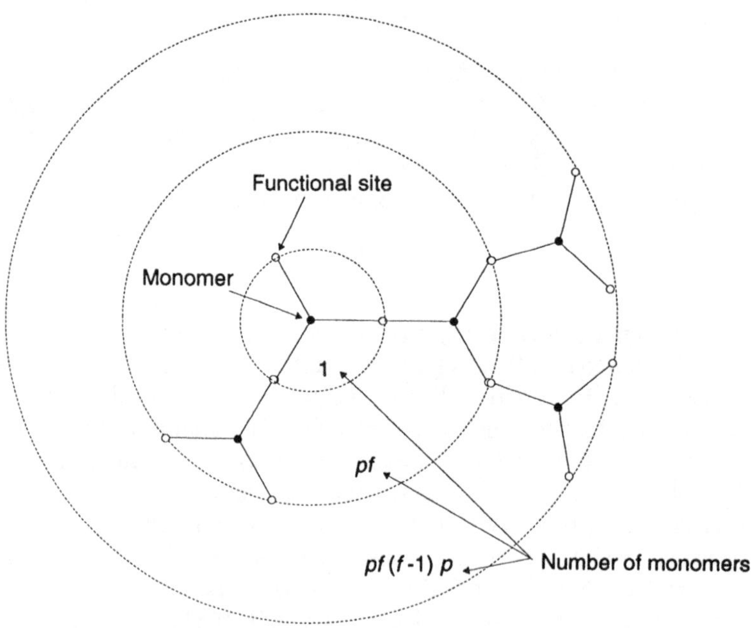

Figure 1. Tree-like branched molecules, p is the degree of connectivity and f the number of neighbors a monomer is allowed to connect with. The weight average molecular mass M_w of a population of such molecules is the total number of monomers connected to a randomly chosen origin: $M_w(p) = 1 + pf[1 + (f-1)p + (f-1)^2p^2 \ldots]$

The total number of monomers so connected to the origin is the expectation value for which we are looking:

$$M_w(p) = 1 + pf[1 + (f-1)p + (f-1)^2p^2 \cdots] \tag{6}$$

One obtains

$$M_w(p) = \frac{1+p}{1-p(f-1)} \tag{7}$$

The quantities $M_n(p)$ and $M_w(p)$ have a vertical asymptote for $p = 2/f$ and $p = 1/(f-1)$ respectively. Depending on f, different situations occur:

$f < 2$: there is no formation of macromolecules
$f = 2$: the two asymptotes correspond to $p = 1$ and macromolecules are linear, with $I = 2$ at $p = 1$.
$f > 2$: M_w diverges before M_n and the polydispersity becomes infinite for $p = p_c = \dfrac{1}{f-1} < 1$

This last case is essentially different from the others. In fact, the divergence of the polydispersity means that one or more infinite molecules appear in the medium. Such molecules have an infinite probability of connecting to each other and then forming only one infinite molecule. This molecule is called a gel and its appearance marks a qualitative change of the system—the gelation. So, gelation is a transition between a totally soluble system (made of molecules having a finite size) and a heterogeneous system made of two phases: one insoluble (the gel) and one soluble (all the other molecules: the sol). Besides the insolubility, gelation comes with important changes of physical properties: below the gelation threshold, the system is a liquid the viscosity of which increases as one approaches the critical value $p = p_c$; above the threshold the material does not flow, the gel conferring a permanent elastic modulus to the system.

The first theoretical approach which attempts to describe the sol–gel transition was due to Flory and Stockmayer. Their theory starts from three following assumptions (Flory, 1953, 1961). The first is that the aggregation process of monomers leads to clusters having a tree-like structure, i.e. there is no intramolecular chemical reaction for molecules having a finite size. Secondly, all functionalities are supposed to have the same reactivity, whatever the size of molecules to which they belong. The third assumption is that the probability for a functionality to react is simply proportional to the mean concentration of reactant. The first two assumptions have undergone a lot of refinements and are not as important as they seem at first sight (Dusek, Ilavsky and Lunak, 1975; Durand and Bruneau, 1983; Rolando and Macosko, 1987). On the contrary, the latter assumption has implications which are of great importance. A reaction probability which is proportional to the concentration amounts to supposing that virtually all the connections are possible, whatever the distance between reactants. In other words, molecules are considered as immaterial points, infinitely movable, which are able to be all together at the same time in the same place. Excluded volume and topological constraints are not considered. The Flory and Stockmayer theory is a mean field approach: one replaces the local neighborhood of one monomer by an average concentration over the medium. This causes a failure of the classical theory in the description of the behavior near the critical transition. However, the Flory and Stockmayer approach allows a good approximation of the critical threshold value p_c and of the behavior of molecular parameters from the initial conditions to the final stage of the chemical reaction (Dusek, 1979). This is why this theory is so often used for the description of various chemical systems having an industrial or practical interest. In addition, the Flory and Stockmayer theory supplies an exact solution to the problem of random connections on a particular lattice, i.e. percolation on Bethe lattice or Cayley tree (Stauffer, 1985). Unfortunately, as will be emphasized later, the Flory and Stockmayer theory does not provide a correct description of the critical behavior of gelation.

4.2.1.2 The classical approach: a particular case of percolation

Let us now look at the critical behavior of the Flory and Stockmayer model which has an exact solution, as mentioned above. Mathematical derivations of this model were carried out using either cascade theory (Gordon, 1962; Good, 1963; Dobson and Gordon, 1964), a recursive method (Macosko and Miller, 1976) or graph theory (Durand and Bruneau, 1982) in order to describe the whole course of the chemical reaction, but here we focus only on the behavior near the threshold using basic considerations. The fundamental quantity is the probability that a randomly chosen chemical function leads to a molecular sub-tree having a finite size. This probability π results from two contributions. The first is equal to the probability $(1-p)$ that the chemical function has not reacted. The second $p\pi^{f-1}$ expresses the probability that the chemical function has reacted with a monomer having its $(f-1)$ remaining chemical functions leading to molecular sub-trees of finite sizes:

$$\pi = (1-p) + p\pi^{f-1} \tag{8}$$

For simplification, let us consider the case $f = 3$. Equation (8) has two solutions: 1 and $(1-p)/p$, each one being relevant for $p < p_c$ and $p > p_c$ respectively:

$$\pi_{p<p_c} = 1 \qquad \pi_{p>p_c} = \frac{1-p}{p} \tag{9}$$

By definition, the sol phase only contains molecules of finite size. Then, the fraction of monomers belonging to these molecules is equal to π^3 and the fraction of monomers belonging to the gel phase is equal to $1-\pi^3$. Above the threshold, the gel fraction, x_{gel} increases with the reduced variable $\varepsilon = |p-p_c|/p_c$ which is called the relative distance to the threshold. Equation (9) leads to

$$\text{for } \varepsilon \to 0 \qquad x_{\text{gel}} \sim \varepsilon \tag{10}$$

and equation [7] to

$$\text{for } \varepsilon \to 0 \qquad M_{\text{w}} \sim \varepsilon^{-1} \tag{11}$$

At the gelation threshold, the mass distribution law of molecules is continuous and infinitely wide. It can be expressed by a power law:

$$\text{for } \varepsilon = 0 \qquad n_i \sim M_i^{-\tau} \tag{12}$$

However, just below the threshold, this power law behavior is valid up to the mass M^* of the largest polymer. The cut-off function is a decreasing exponential of the form $\exp(-M_i\varepsilon^2)$ (Stauffer, 1985) leading to the following expression of the polymer mass distribution:

$$n_i \sim M_i^{-\tau} \exp(-M_i/M^*) \tag{13}$$

with the mass M^* diverging as the system approaches the gelation threshold:

$$M^* \sim \varepsilon^{-2} \tag{14}$$

In order to establish the value of the exponent τ in the Flory and Stockmayer theory, one can note that in the expression of M_{w} given by Eq. (2), the denominator does not depend on the distance to the threshold. Replacing the discrete summation by the corresponding integral, leads to $M_{\text{w}} \sim \int n_i M_i^2 \, dM_i \sim \varepsilon^{2\tau-6}$. With Eq. (11) one obtains

$$\tau = 5/2 \tag{15}$$

The power law behavior of the quantities examined here will be generalized in the next section.

4.2.1.3 Critical phenomena, scaling laws and universality classes

In the particular case of the Flory and Stockmayer model, three critical exponent values were established. One may consider other exponents, each of them describing the variation of a particular quantity near the threshold. However, due to the self-similarity of polymer clusters, there exist scaling laws which link exponents between them, as for other critical phenomena (Stanley, 1971). The theory of critical phenomenon allows us to deal with processes belonging to different fields of physics, such as polymers, magnetism and semiconductors, using a common approach. For example, in the case of gelation, the weight average molecular mass of molecules below the gel point and the gel fraction above the gel point have analog quantities in a ferromagnet which are susceptibility and magnetization below and above the Curie point respectively. Such a unifying concept allows us firstly to take advantage of research advancements in other fields of physics, and secondly leads to a better global understanding of disordered systems. These scaling laws imply that only three exponents are free. Thus, it is possible to define universality classes, which regroup together physical processes having the same exponent values and being explainable by the same fundamental model. This is why, near the critical threshold, it is much more important to determine the exponent value rather than the prefactor of a power law behavior of the form $A = A_0 \varepsilon^{-x}$. The exponent reflects the nature of the involved mechanism while the prefactor depends on the local particularities of the system. So, the exponent value is a good check of the validity of a theoretical model. In a first approximation, such a model has to be very simple in order to be just representative of its universality class, and can be refined afterwards to fit more closely local particularities of one given system. The theoretical model which now seems most appropriate to describe static properties of branched polymers near the gelation threshold is the percolation model which will be detailed later.

Let us now generalize the power law behaviors obtained for a tree-like growth process. Equations (10), (11), (13) and (14) can be rewritten using the exponents β, γ, τ and σ respectively:

$$x_{\text{gel}} \sim \varepsilon^{\beta} \tag{16}$$

$$M_{\text{w}} = \varepsilon^{-\gamma} \tag{17}$$

$$n_i = M_i^{-\tau} \exp(-M_i/M^*) \tag{18}$$

and

$$M^* \sim \varepsilon^{-1/\sigma} \tag{19}$$

The classical values of these exponents are: $\beta = 1$, $\gamma = 1$, $\tau = 5/2$ and $\sigma = 1/2$. Without any a priori exponent values, starting from Eqs (18) and (19), one can calculate the kth moment of the mass distribution law:

$$M_{k>1} \sim \varepsilon^{(\tau-k-1)/\sigma}\Gamma(k - \tau + 1) \tag{20}$$

where $\Gamma(x) = \int z^{x-1} \exp(-z)\mathrm{d}z$ (with $z = M_i/M^*$) is a finite number for $x > 0$. Common values of the exponent τ are greater than 2 and therefore one has to calculate the first-order moment using a different method (Stauffer, 1985). Equation (18) leads to the expression

$$\mathrm{d}M_1/\mathrm{d}(M^{*-1}) = \mathrm{d}(\textstyle\sum n_i M_i)/\mathrm{d}(M^{*-1}) = -M_2$$

Using Eqs (19) and (20) one obtains

$$M_1 \sim \varepsilon^{(\tau-2)/\sigma}\Gamma(3 - \tau) + \text{const.} \tag{21}$$

Scaling laws can be obtained using these last two expressions. As one approaches the gelation threshold, M_{w} diverges as the second moment of the mass distribution law. Equations (17) and (20) for $k = 2$ leads to

$$\gamma = \frac{3 - \tau}{\sigma} \tag{22}$$

After the gelation threshold, one given monomer belongs either to the gel or to the sol fraction. That means that, $x_{\text{gel}} = 1 - \sum_{\text{sol}} x_i M_i$, where x_i is the number fraction of clusters having a size M_i: $x_i = n_i / \sum n_i M_i$. Consequently, the gel fraction will have the same power law behavior as the first moment of the distribution law. In the same way as for the exponent γ, one obtains the following scaling law:

$$\beta = \frac{\tau - 2}{\sigma} \tag{23}$$

Another set of exponents concerns the conformation of branched polymers and their connectivity properties. The relevant quantity is the pair correlation function $g(r)$ which is the probability that two monomers at a distance r from one another belong to the same polymer cluster. In other words, $g(r)$ represents the probability of finding a monomer belonging to the same molecule as an arbitrary origin and at a distance r from it. So we have the relation

$$\int_0^{\xi} g(r) r^2 \, \mathrm{d}r = M_{\mathrm{w}} \tag{24}$$

where ξ is the radius of the largest cluster. For $r > \xi$, the probability $g(r)$ is equal to zero which means that no monomer is at a distance larger than ξ from the origin. The size ξ is called the correlation length and diverges as one approaches the gelation threshold. Let us define a new exponent ν which takes into account the power law behavior of ξ:

$$\xi \sim \varepsilon^{-\nu} \tag{25}$$

The largest cluster having a mass M^* and a radius ξ, is self-similar and comes under a description in terms of the fractal geometry. The largest cluster obeys the scaling behavior

$$M^* \sim \xi^D \tag{26}$$

The latter expression is nothing but a generalization of the relation between mass and volume of Euclidean objects. The exponent D is the fractal dimension of branched polymers and can be fractional. Equations (19), (25) and (26) lead to the scaling law:

$$D = \frac{1}{\sigma \nu} \tag{27}$$

Zimm and Stockmayer (1949) have shown that in the mean field framework, D is equal to 4, the classical value of the exponent ν being equal to 1/2. This value of the fractal dimension of polymer clusters shows that the classical approach does not lead to exponent values consistent with the three-dimensional space in which gelation takes place. Let us focus now on the percolation theory which introduces the space dimension in the description of clusters.

4.2.1.4 Percolation model

Percolation theory was first introduced by Broadbent and Hammersley (1957) in order to describe diffusion in disordered media. More precisely, it concerned the permeation threshold. Nevertheless, the general character and simplicity of the model allowed the concept to be applied to phase transition physics and later to the gelation problem (de Gennes, 1976a; Stauffer, 1976). For details about percolation theory the reader is referred to Stauffer (1979, 1985) and Essam (1980).

In a space of any dimension, let us consider a lattice having nodes randomly occupied with a probability p or empty with probability $(1-p)$. For a given p, a cluster is a group of neighboring occupied sites. The percolation threshold p_c is the value of p at which the size of the largest cluster tends to infinity. In this case, infinity means that the largest cluster extends from one side of the lattice to the other whatever its actual size. This is the site percolation problem. One may also consider a network having all its nodes occupied by sites which form links at random with their neighbors. Above a given percent p_c of such bonds, there exists a continuous path of bonds which connects the two opposite sides of the lattice. This is the bond percolation model. Percolation is a critical phenomenon of connectivity: a continuous increasing quantity, p, induces an abrupt qualitative change. For example, if bonds are conductors, below the percolation threshold, the lattice is an insulator while above the threshold it is a conductor. The percolation threshold p_c depends on the kind of percolation: site or bond percolation, and will also depend on the lattice geometry in which percolation takes place. While the critical behavior of all quantities which are affected by the transition will only depend on the space dimension (Dean and Bird, 1967; Clerc *et al*, 1983). In this way the percolation model is universal. In the percolation/gelation analogy, percolation clusters correspond to branched polymers and the probability p corresponds to the degree of connectivity and is defined as the ratio between the number of chemical bonds between monomers and the total number of potential bonds.

How can universality be explained? Let us imagine a chemical system consisting, at zero time, of single monomers. During the first stage of the chemical reaction, the chemical nature of monomers will be of great importance, but the system will rapidly contain not only monomers but also dimers, trimers, etc. These oligomers connect with one another and the particular nature of the monomers becomes less and less important. Elementary units making up the system are renormalized and the percolation model becomes relevant to describe the system behavior near the transition. The early stages of the gelation process are governed by the laws of chemistry, thus the advancement of the chemical reaction at the threshold and the prefactors of the power laws, and hence all non-universal quantities. While the nature of the critical transition and the corresponding universality class will be described using a very simple model, neglecting all the details of chemistry and system particularities.

As it will be seen in the experimental section 4.3.1.1, gelation processes can be divided into two main categories: chainwise and stepwise polymerization. Each of them corresponds to particular chemical rules of monomer connections. Starting from percolation framework and Monte Carlo method, computer simulations are reported in the literature, which attempt to establish which of these rules are relevant from the point of view of the universality class. It emerges from their results that chainwise gelation may belong to a universality class different from that of the original model (Herrmann and Landau, 1982; Pandey and Stauffer, 1983; Bansil, Herrmann and Stauffer, 1984, 1985). However, it is not the case for stepwise gelation which corresponds at once to a limited bond percolation model (Gaunt, Guttmann and Whittington, 1979; Kertész, Chakrabarti and Duarte, 1982) and to a combination of site and bond percolation problems (Lairez, Durand and Emery, 1991).

While the percolation concept is simple, exact calculation of the corresponding critical exponents is possible only in one dimension or within the mean field framework (on a Bethe lattice). Thus the model is strongly dependent on computer simulation results provided by the Monte Carlo method. Table 1 gives a comparison of exponent values obtained in three-dimensional percolation simulations with analytical mean field results (Stauffer, 1985). As shown in Table 1, discrepancies in exponent values are sufficient to decide experimentally between theoretical models. This view will be emphasized in section 4.3, which will show that three-dimensional percolation seems the best model from the point of view of static properties.

Table 1. Theoretical exponent values in mean field and three-dimensional percolation approaches

Exponent	Mean field	3D percolation	Quantity
γ	1	1.74	$M_w \sim \varepsilon^{-\gamma}$
β	1	0.44	$x_{gel} \sim \varepsilon^{\beta}$
σ	1/2	0.46	$M^* \sim \varepsilon^{-1/\sigma}$
ν	1/2	0.88	$\xi \sim \varepsilon^{-\nu}$
D	4	2.52	$M^* \sim \xi^D$
τ	5/2	2.2	$n_i \sim M_i^{-\tau}$

The percolation model has very strong implications on the static properties of branched polymers, both on their fractal geometry and on their mass distribution law. Moreover, if one considers branched polymers in their native state, percolation predicts that these features are related. This relation is what theoreticians call the 'hyperscaling law' and will form the subject of section 4.2.1.5.

4.2.1.5 Excluded volume effects, critical dimension and the hyperscaling law

At the percolation threshold, the mass distribution of clusters is continuous and infinitely wide. This distribution curve is expressed by a power law having an exponent τ. The main particularity of such a law is its self-similarity: if one takes out the subset of clusters having a mass inferior to a given mass, the renormalized mass distribution of clusters in this subset is superimposable on the initial one. The conformation of each percolation cluster examined individually is also self-similar, this property being expressed by their fractal dimension D. These two self-similarities of branched polymers have to be consistent with the space dimension in which gelation takes place.

If one tries to introduce the space dimension d within the mean field approach, one comes up against difficulties. In a three-dimensional space the direct implication of the fractal dimension, $D = 4$, predicted by the Flory and Stockmayer theory, is a cluster density which increases with the length scale. Excluded volume effects prevent polymer clusters behaving in this way. A mean field approach allows the calculation of the critical space dimension above which this excluded volume effect can be neglected (Flory, 1953; de Gennes, 1980; Isaacson and Lubensky, 1980). Classically, the free energy and the resulting conformation of one polymer result from the competition between an entropic or elastic term and monomer–monomer repulsive interactions. The entropic terms is equivalent to a spring constant which tends to return the polymer to its Gaussian conformation. It diverges as one approaches to the gel point. The excluded volume interactions are proportional to the number of monomer pairs per unit volume divided by a screening factor which is the weight average number of monomers per screen molecule. One obtains for the largest cluster

$$F_{\text{excluded volume}} \sim \frac{M^{*2}}{M_w \xi^d} \tag{28}$$

Compared to the entropic term, this contribution will be negligible if it scales as a constant, i.e. for a space dimension d_c for which $M^{*2}/M_w \xi^{d_c} \sim 1$. Using Eqs (17), (25) and (26) one obtains the expression

$$d_c = 2D - \frac{\gamma}{\nu} \tag{29}$$

The exponent values obtained from the Flory and Stockmayer theory (see

Table 1) lead to the critical dimension $d_c = 6$ above which excluded volume interactions between monomers can be neglected. Computer simulations by Monte Carlo methods show that this is actually the case and that percolation exponent values tend to those predicted by mean field for space dimensions larger than 6. In this sense, mean field corresponds to a particular case of the percolation model.

Equation (29), obtained from polymer physics concepts, is known in critical phenomena physics as the 'hyperscaling law' and is assumed to be valid whatever the space dimension d. This relation reduces the number of free exponents required to specify a universality class. In the case of gelation, it is convenient to express this equation in terms of the exponents τ and D, which describe the self-similarity of the polymers in their native state from the point of view of their mass distribution and of their conformation respectively. Using Eqs (22) and (27) in d dimensions, the hyperscaling law becomes

$$\frac{d}{D} = \tau - 1 \tag{30}$$

The meaning of hyperscaling can be emphasized by considering the probability w_i to take out a monomer belonging to a cluster of mass M_i and radius R_i. This probability is proportional to the product $n_i M_i$. Using the cluster mass distribution law and the fractal dimension D of clusters, one obtains $w_i \sim R_i^{-D(\tau-1)}$. Finally the hyperscaling law leads to

$$w_i \sim R_i^{-d} \tag{31}$$

The probability w_i is proportional to the inverse of the volume of the sphere containing the cluster. In three dimensions, this implies that there is only one cluster of mass M_i per volume R_i^3. In other words, clusters having the same mass do not interpenetrate each other, they are at the overlap concentration C^* (Daoud, 1989). Moreover, percolation clusters in a three-dimensional space have a fractal dimension less than 3 and are therefore porous objects with pore sizes increasing with the radius of the cluster. The C^* situation of clusters having a mass M_i implies that the only free space available to smaller clusters is inside these pores. This occurs whatever the mass M_i of the distribution curve might be. Clusters arrange themselves as shown schematically in Figure 2, in which polymers are represented by circles of different diameters according to their masses.

4.2.1.6 *Width of the critical region in the gelation problem*

As in other critical phenomenoa, one may ask what is the width of the critical region inside which the percolation model is valid and outside which the mean field approximation becomes relevant. In order to discuss this problem, let us consider the gelation of a linear polymer melt. This is the so-called 'vulcanization' process, in which crosslinks occur between monomers belonging to different

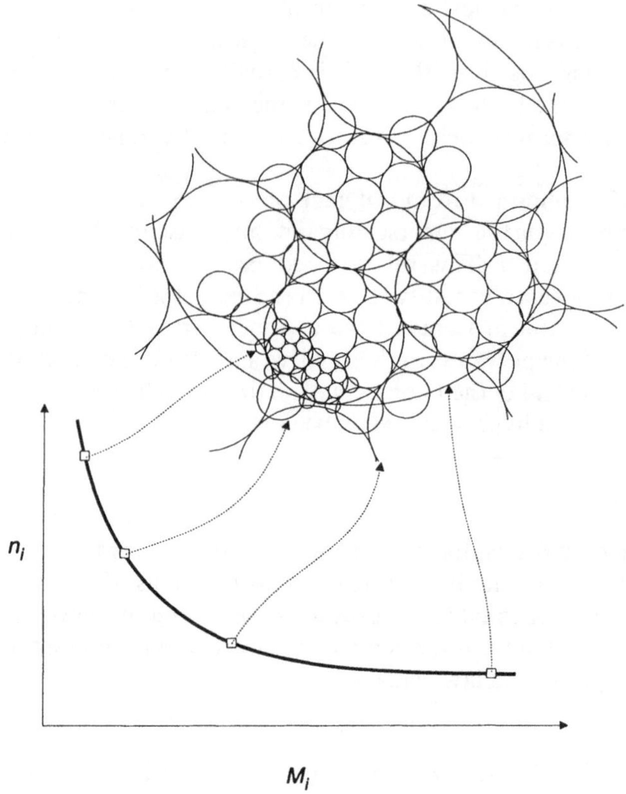

Figure 2. The hyperscaling law links the two self-similar features of branched polymers in their native state, i.e. the conformation of each polymer cluster: $M_i \sim R_i^D$, and the mass distribution curve of the overall population: $n_i \sim (M_i^{-\tau}/M^*)f(M_i/M^*)$. Clusters having the same mass are in a C^* situation and contain smaller clusters; this occurs at all length scales. In this schematic representation, only four classes of clusters and part of them are drawn as circles of different diameters according to their masses

polymer chains through the reaction of small molecules. In this case, the elementary units playing a role in the growth process are polymer chains having a mass M_0 and a radius ξ_0 rather than monomers having a mass equal to unity. Because polymer chains in a melt adopt a Gaussian conformation one has $M_0 \sim \xi_0^2$, Eqs (17) and (25) can be rewritten taking into account these prefactors:

$$M_w = M_0 \varepsilon^{-\gamma} \tag{32}$$

$$\frac{\xi}{a} = M_0^{1/2} \varepsilon^{-\nu} \tag{33}$$

The size a as well as the mass of one monomer will be taken as unit. In order to decide if the mean field approach is relevant to describe a critical phenomenon, one has to determine if fluctuations of the order parameter in a correlation volume, ξ^d, are negligible. This is the Ginsburg argument arising from phase transition physics (de Gennes, 1977). In our problem the order parameter corresponds to the number n_{gel} of monomers belonging to the gel phase. Above the gel point, monomers belong either to the gel or to the sol phase, and thus the sum of the corresponding monomer numbers is constant and equal to the total number of monomers in the volume ξ^d;

$$n_{\mathrm{gel}} + n_{\mathrm{sol}} = \xi^d \tag{34}$$

Consequently, fluctuations of the gel and of the sol phase balance each other: $\Delta n^2_{\mathrm{gel}} = \Delta n^2_{\mathrm{sol}}$. When a finite size cluster links to the gel, M_{w} monomers are lost by the sol fraction, and the fluctuation term is thus equal to $\Delta n^2_{\mathrm{sol}} = n_{\mathrm{sol}} M_{\mathrm{w}}$. Moreover, very close to the gel point, the gel fraction can be neglected compared to the sol fraction and $n_{\mathrm{sol}} \cong \xi^d$. This leads to

$$\Delta n^2_{\mathrm{sol}} = \xi^d M_{\mathrm{w}} \tag{35}$$

One has to compare this fluctuation term to the number of monomers belonging to the gel inside a correlation volume ξ^d. Introducing the gel fraction x_{gel}, one obtains

$$n^2_{\mathrm{gel}} = x^2_{\mathrm{gel}} \xi^{2d} \tag{36}$$

Finally, fluctuations can be neglected at least for a ratio, $\chi = \Delta n^2_{\mathrm{gel}}/n^2_{\mathrm{gel}}$ equal to a constant. Equations (34)–(36) lead to the following expression of this ratio:

$$\chi = \frac{M_{\mathrm{w}}}{x^2_{\mathrm{gel}} \xi^d} = \varepsilon^{\nu d - \gamma - 2\beta} M_0^{1 - d/2} \tag{37}$$

Introducing the mean field exponent values ($\gamma = 1$, $\beta = 1$, $\nu = 1/2$) leads to an estimation of the relative distance to the threshold ε^* above which fluctuations can be neglected and mean field becomes relevant:

$$\varepsilon^* = M_0^{(d-2)/(d-6)} \tag{38}$$

When $d = 3$, it becomes

$$\varepsilon^* = M_0^{-1/3} \tag{39}$$

This means that percolation describes the critical behavior close to the gel point inside a critical region ε^* which is maximum for $M_0 = 1$, i.e. the case we have considered in the previous sections. As M_0 increases, the critical region becomes narrower and mean field becomes a good approximation for the exponent values. Moreover, from an experimental point of view, the critical region is all the more narrow in absolute value because the critical threshold is known to be very small and equal to $1/M_0$ (de Gennes, 1975). Outside the critical region ε^*, polymer chains of size ξ_0 interpenetrate each other, allowing cluster fractal dimension of 4 even in space dimension $d = 3$. One can note that Eq. (37) allows the recovery of the critical dimension $d_c = 6$ above which mean field theory is pertinent.

In the case of vulcanization of polymers in semi-dilute solution, the number M_0 has to be replaced by the number of blobs per chain which is a power law of the reduced concentration C/C^*, where C^* is the overlap concentration of linear polymers in the initial solution (Daoud, 1979). The case $C = C^*$ corresponds to $M_0 = 1$, while increasing concentration amounts to reducing the width of the critical region. This could be a good way to observe a crossover from percolation to mean field behavior.

4.2.2 DYNAMIC PROPERTIES OF BRANCHED POLYMERS NEAR THE THRESHOLD

The most outstanding manifestation of the sol–gel transition concerns the change in the transport properties. The material passes from a liquid to a solid state. In order to account for this behavior, several theoretical approaches are used. This section is concerned with the description of the critical transition, and thus what is happening far from the threshold will not be considered. For example, the effective medium theory, which seems to be appropriate in the description of dynamical properties far from the gel point, will not be considered here and the reader is referred to other papers (Kirkpatrick, 1973; Allain, Limat and Salomé, 1991).

4.2.2.1 Complexity of dynamic properties

The knowledge of static properties of branched polymers, i.e. their description in terms of fractals, is necessary but not sufficient for the understanding of dynamical properties of these systems. The first reason is that transport properties depend strongly on connectivity. Fractal dimension describe how polymers fill the space; however, it gives no indication about connectivity. For example, linear polymers have a fractal dimension equal to 5/3 or 2 depending on whether or not they are swollen. However, in both the cases, connectivity remains unaffected: a walk along the chemical bonds takes place in a one-dimensional space. Inversely, two polymers having different connectivity can be in such a situation that they have the same fractal dimension, for example branched polymers in good solvent and linear polymers in the melt. The difference of connectivity affects the polymer dynamics, polymer motion is isotropic for branched polymers but anisotropic for linear polymers in the melt. Thus, an additional dimension is needed to take into account this connectivity: the spectral dimension. In the case of linear polymers the spectral dimension is equal to 1, but connectivity can be fractal and the corresponding dimension can be a fractional number such as that of branched polymers and gels. Spectral dimension reflects the network connectivity, i.e. loops, dangling ends and so on, all parameters which governs diffusion and transport on the network. Like fractals, the concept of spectral dimension comes from an extension of the physics of the ordered matter.

Elastic vibrations on a d dimensional crystal are usually characterized by a number of vibrating modes in the frequency interval $[\omega, \omega + \mathrm{d}\omega]$ equal to the corresponding volume in the reciprocal space. One defines this number to be equal to the product $\boldsymbol{D}(\omega)\mathrm{d}\omega$, where $\boldsymbol{D}(\omega)$ is the density of state. One has $\boldsymbol{D}(\omega)\mathrm{d}\omega = \int \mathrm{d}^d k$, k being the wave vector and is equal to ω/c (where c is the sound velocity). One obtains $\boldsymbol{D}(\omega)\mathrm{d}\omega = \omega^d$, and finally the phonon density of state scales as

$$\boldsymbol{D}(\omega) = \omega^{d-1} \tag{40}$$

This way of counting vibrating modes amounts to using the property of a regular network to be unchanged by translation. One can fill the reciprocal space by unit volume, whereas this is not the case for fractal networks which are only unchanged by dilation. This later property leads to a more general expression for the density of state which can be established by studying diffusion on such a network (Alexander and Orbach, 1982; Rammal and Toulouse, 1982, 1983). Diffusion on fractal network is anomalous (Gefen, Aharony and Alexander, 1983; Havlin and Ben-Avraham, 1987).

Let us imagine a particle doing a random walk along a fractal network of dimension D. This is the de Gennes problem of the ant in a labyrinth (de Gennes, 1976b). One defines the fractal dimension d_w of the walk by the relation between the radius R of the visited space and the duration t of the walk, i.e. the number of steps

$$t \sim R^{d_w} \tag{41}$$

The diffusion coefficient $\mathscr{D}$ is therefore equal to

$$\mathscr{D} = \frac{d\langle R^2 \rangle}{\mathrm{d}t} \sim t^{2/d_w^{-1}} \sim R^{2-d_w} \tag{42}$$

In the case of an Euclidean network, this diffusion coefficient is a constant whatever the space dimension, the random walk being equivalent to a Brownian motion with $d_w = 2$. In the case of a fractal network, steps of the random walk lead to dead ends, thus obliging the ant to go back. The visited space grows slower than for a free random walk, thus in a general way, d_w is larger than 2. With such a random walk, what is the behavior of the number S of new network sites visited as a function of the time? To calculate $S(t)$, one only counts each site once, this gives the same fractal dimensions for S as the network itself (Orbach, 1985), thus $S \sim R^D$. Using Eq. (41) one obtains

$$S(t) \sim t^{D/d_w} \tag{43}$$

In fact, $S(t)$ governs the autocorrelation of the random walk and the probability, $P_0(t)$, that the walker returns to its starting point after an elapsed time t, $P_0(t)$ is the stationary walk probability and is proportional to the inverse of $S(t)$: $P_0(t) \sim 1/S(t)$. Taking the Laplace transform gives the number of such stationary waves in the frequency interval $[\omega, \omega + d\omega]$, one obtains for the density of state (Alexander, Bernasconi and Orbach, 1978):

$$D(\omega) \sim \omega^{(2D/d_w)-1} \tag{44}$$

The ratio $2D/d_w$ represents the spectral dimension d_s governing the density of vibrational modes of a fractal network. These modes are called fractons by analogy with phonons. One has

$$d_s = 2D/d_w \tag{45}$$

In the case of well-ordered systems (crystals), a random walk is a Brownian motion and $d_w = 2$, moreover the fractal dimension D of these systems is equal to the space dimension d. Therefore, such systems are particular cases for which $d_s = D = d$, but in general

$$d_s \leqslant D \leqslant d \tag{46}$$

A few years ago the universal value $d_s = 4/3$ was proposed by Alexander and Orbach (1982) for percolation networks in any dimension between 1 and 6. This was a first attempt to link static (D and d_s) to dynamic properties (d_w).

As discussed above, the dynamics of branched polymers near the gelation threshold depends on the connectivity properties of polymers. Once more this is not sufficient, and one needs an assumption about the kind of hydrodynamic interactions between polymers. Information about these interactions are not contained in either the static or the connectivity properties. Hydrodynamic assumptions will have implications on the way energy transport is modeled. This problem will be discussed in the following sections.

4.2.2.2 *Scaling of rheological properties near the gel point*

One of the manifestations of the sol–gel transition is an extensive modification of the viscoelastic properties of the material. At zero frequency, the viscosity η_s diverges as one approaches the threshold, while the elastic modulus G_s emerges as one goes further off the gel point. As polymer clusters increase in size, the system passes from being a more and more viscous liquid to being a viscoelastic solid as an infinite cluster appears. As for other quantities which depend on the relative distance from the threshold, η_s and G_s have a power law behavior:

$$\text{for } p < p_c \qquad \eta_s = \eta_0 \varepsilon^{-s} \tag{47}$$

$$\text{for } p > p_c \qquad G_s = G_0 \varepsilon^{t} \tag{48}$$

where η_0 and G_0 are linked to the microscopic properties of the medium.

At the beginning of the chemical reaction, the system is liquid-like and made up of small molecules. Frequency dependences of the real and imaginary parts of the complex elastic modulus $G^* = G' + iG''$ are those of an ordinary liquid at low frequency:

$$G' \sim \omega^2 \quad \text{and} \quad G'' \sim \omega^1 \tag{49}$$

However, this liquid behavior is only valid below a cutoff frequency ω_0 which corresponds to the glass transition frequency in the case of polymers. Here, we will always assume that we are far enough below ω_0. At the final stage of the chemical reaction, when the polymer network tends to have a regular topology, the rheological behavior of the gel will be the one of a solid:

$$G' \sim \omega^0 \quad \text{and} \quad G'' \sim \omega^1 \tag{50}$$

Near the gelation threshold, the growing clusters may be described in terms of the fractal geometry at the length scales between the monomer size and the correlation length ξ. These length scales correspond respectively to two limits of a viscoelastic behavior: the very high frequency ω_0 mentioned above and a crossover frequency ω^* which corresponds to the inverse of the longest relaxation time of the largest cluster. It is not as easy as with scattering experiments to link the wavelength of a mechanical stress with the size of molecules or parts of molecules moving with this stress. Nevertheless, one can propose a scaling relation between the longest relaxation time of a polymer and its mass. Without any assumption about the exponent value, one can write, $\omega^* \sim \xi^{-\tilde{z}}$ or, using Eq. (25),

$$\omega^* = \omega_0 \varepsilon^z \quad \text{with} \quad z = \nu \tilde{z} \tag{51}$$

At length scale smaller than ξ, the self-similarity of clusters causes a similar relation between relaxation frequencies and the size of the polymer part which is moving. Thus, one expects at frequencies higher than ω^* a power law behavior of the complex elastic modulus of the form:

$$G^* \sim (i\omega/\omega_0)^u \tag{52}$$

which can be rewritten as

$$G^* \sim (\omega/\omega_0)^u \, e^{i\delta} \quad \text{with} \quad \delta = \arctan(G''/G') = u\pi/2 \tag{53}$$

This implies a similar power law behavior for the real and imaginary parts of the elastic modulus. The correlation length tending to infinity at the gel point, the frequency ω^* tends to zero and thus the rheological behavior of the system is like neither liquid nor solid and does not present any characteristic time.

The dependence of ω^*, η_s and G_s on the relative distance from the threshold (Eqs (47), (48), (51)) leads to the following expressions for the viscosity η_s and elastic modulus G_s measured at zero frequency as a function of ω^*:

$$\eta_s \sim (\omega^*)^{-s/z} \quad \text{and} \quad G_s \sim (\omega^*)^{t/z} \tag{54}$$

At the frequency ω^* one has $G' = G_s$ and $G'' = \omega\eta_s$, moreover G' and G'' are proportional whatever the frequency $\omega > \omega^*$, these remarks lead to

$$z = s + t \tag{55}$$

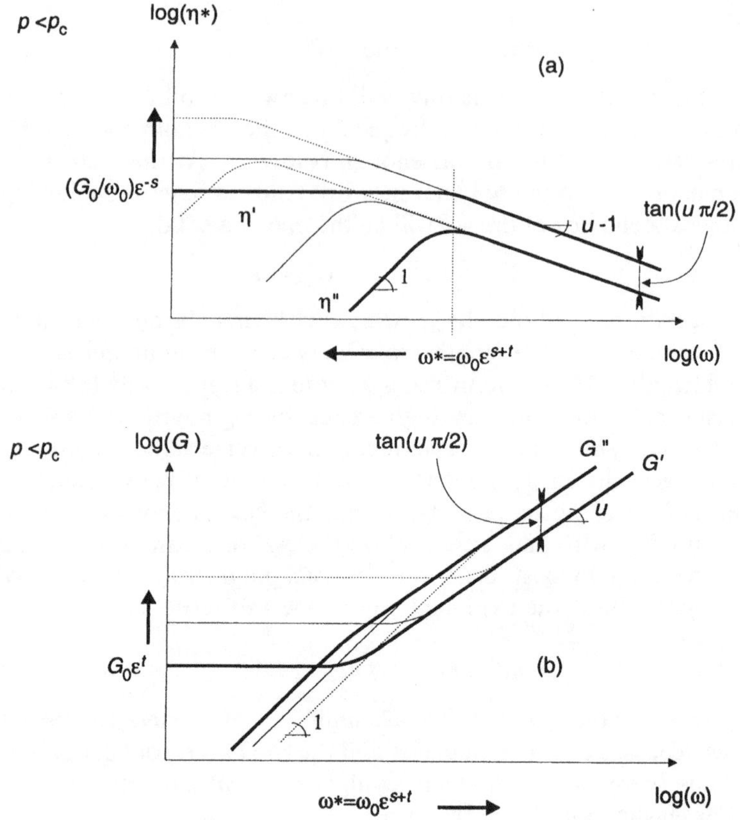

Figure 3. Schematic representation of the scaling behavior of the rheological properties, below (a) and above (b) the gelation threshold. The complex viscosity $\eta^* = G^*/i\omega$ or the complex modulus is plotted below or above the threshold respectively. The arrow in bold indicate the evolution of the curves for increasing p

and

$$u = \frac{t}{s+t} \tag{56}$$

In addition, one can note that above the threshold, the two parts of the complex elastic modulus are equal at the frequency ω^*:

$$G'(\omega^*) = G''(\omega^*) \tag{57}$$

The expected rheological behavior over all the frequency range can be summarized by a schematic diagram, shown in Figure 3, and by the general expression

$$G^* = G_0 \varepsilon^t f(i\omega/\omega^*) \quad \text{with} \quad \omega^* = \omega_0 \varepsilon^{s+t} \tag{58}$$

for $\omega < \omega^*$	$p < p_c$	$f(i\omega/\omega^*) =$ $i\omega/\omega^* + (\omega/\omega^*)^2 + \ldots$	thus	$G^* =$ $(G_0/\omega_0^2)\varepsilon^{-2s+t}\omega^2 + i(G_0/\omega_0)\varepsilon^{-s}\omega$ or $\eta^* = G^*/i\omega =$ $(G_0/\omega_0)\varepsilon^{-s} - i(G_0/\omega_0^2)\varepsilon^{-2s-t}\omega$
	$p > p_c$	$f(i\omega/\omega^*) = 1 + i\omega/\omega^* + \ldots$	thus	$G^* = G_0\varepsilon^t + i(G_0/\omega_0)\varepsilon^{-s}\omega$
for $\omega^* < \omega$	$\forall p$	$f(i\omega/\omega^*) = (i\omega/\omega^*)^{t/(s+t)}$	thus	$G^* = (G_0/\omega_0^u)\omega^u e^{i\delta}$ with $u = t/(s+t)$ and $\delta = u\pi/2$

This scaling expression of the elastic modulus comes from the study of electrical properties of resistor–capacitor random mixtures (Efros and Shklovskii, 1976) and from the analogy between percolation and gelation (de Gennes, 1976a). This expected rheological behavior follows directly from scaling concepts which can be applied due to the fractal nature of the polymer clusters and of their mass distribution. Once scaling concepts are accepted the results reported in this section are independent of the theoretical model. In fact, depending on the type of hydrodynamic interactions we assume between polymer clusters, different exponent values can be found within this framework.

4.2.2.3 Percolation dynamics: electrical analogy, Rouse and Mean field model

At present two main theoretical approaches claim to describe the dynamics of branched polymers near the gelation threshold. One is known as the electrical analogy and was first proposed by de Gennes, while the second refers to a well-known limit of the polymer dynamics, the Rouse model, which assumes no hydrodynamic interaction between polymers. In this section, we deliberately deal with these two theoretical models starting from the problem of diffusion on percolation networks (Kirkpatrick, 1973). Then, both the electrical analogy and the Rouse model will be derived from a common approach and will be thus more easily compared.

To treat the emergence of the elastic modulus after the gel point, one has to consider a percolating conductor network in an insulating medium (insulator/conductor transition (IC)), while the viscosity problem is solved by an analogy with percolation of supraconductor clusters in a conductor medium (conductor/supraconductor transition (CS)). In both the cases the conductivity Σ is expected to obey the same power law as the corresponding rheological property:

$$\text{IC: } \Sigma \sim \varepsilon^{t} \sim \xi^{-\tilde{t}} \quad \text{with} \quad \tilde{t} = t/\nu \tag{59}$$

$$\text{CS: } \Sigma \sim \varepsilon^{-s} \sim \xi^{\tilde{s}} \quad \text{with} \quad \tilde{s} = s/\nu \tag{60}$$

Conduction along a fractal network is related to a diffusion problem through the Einstein relation linking the conductivity Σ of a medium to the density ρ of charge carriers and their diffusion coefficient $\mathscr{D}$ (Gefen, Aharony and Alexander, 1983):

$$\Sigma = \mathscr{D}\rho \tag{61}$$

Let us focus on the IC problem. Above the percolation threshold the system contains a conductive network of mesh size ξ. At length scale larger than ξ the network can be considered as Euclidean, i.e. a regular three-dimensional network, while at smaller length scales dangling ends and loops have to be considered; fractal geometry becomes relevant. At the threshold, the correlation length ξ tends to infinity, thus the network is fractal at all the length scales and conductivity will depend on the anomalous diffusion on such a fractal (one ant in a labyrinth (de Gennes, 1976b)). In this case $\mathscr{D}$ scales as, ξ^{2-d_w} (diffusion is slowed down due to dead ends and loops, see Eq. (42)) and the charge carrier density is proportional to $M^*/\xi^d \sim \xi^{D-d}$. Thus for the electrical analogy, one obtains:

$$\text{electrical analogy: } \tilde{t} = d_w - 2 + d - D \tag{62}$$

Alexander and Orbach's conjecture $D/d_w = 2/3$ and percolation exponent values obtained in three dimensions leads to

$$\text{electrical analogy: } t \cong 2.0 \tag{63}$$

The Rouse model, which considers no hydrodynamic interactions between monomers, amounts to neglecting the contribution of dangling ends to the energy transport, only crosslinks of the network are efficient. In terms of the conductivity of the network near the threshold, the Rouse model consists in renormalizing the network with units of size ξ. Thus, on the renormalized network, diffusion is not anomalous: $\mathscr{D}$ is a constant, and one has one charge carrier per mesh volume ξ^d. This leads to

$$\text{Rouse model: } \tilde{t} = d \tag{64}$$

and

$$t \cong 2.7 \tag{65}$$

With the CS problem, particles which diffuse have an infinite velocity inside supraconductor clusters and spend all their time at the periphery of clusters. This is the termite problem of de Gennes (1980). In this case, diffusion is governed by the time which is needed to jump from one cluster to another. This time is proportional to the ratio of the total number of monomers in the cluster to the number of peripheral monomers, i.e. ξ^{D-d_u} where d_u is the fractal dimension of the cluster periphery. The diffusion coefficient $\mathscr{D}$ is then equal to, ξ^{2-D+d_u}, while the density of the charge carrier remains the same as in the ant problem: $\rho \sim \xi^{D-d}$. This leads to

$$\tilde{s} = 2 - d + d_u \tag{66}$$

In the electrical analogy model, the periphery, from which the particle emerges or to which particles reach a new cluster, does not represent the total perimeter of the cluster but rather is an 'unscreened perimeter' (Hentschel, 1984; Coniglio and Stanley, 1984; Coniglio, 1985; Stanley *et al*, 1985). Indeed, percolation clusters being fractal, its periphery shows up fjords. A particle emerging from a bottom of a fjord will be probably snapped up again by the same cluster and will not contribute to energy transport from one cluster to another. On the other hand, Brownian charge carrier which approaches a cluster will be snapped up before reaching the bottom of fjords. Therefore, only a fraction of the cluster perimeter is viewed by charge carriers. This is the unscreened perimeter which has a fractal dimension smaller than the cluster itself. It has been shown that for percolation clusters $d_u = [(d + D)/2] - 1$, using Eq. (66) this leads to

$$\text{electrical analogy:}\ \tilde{s} = 1 + \frac{D - d}{2} \tag{67}$$

It is important to note that such an approach considers only 'inter modes' of energy transport between clusters having the same size. In the Rouse approach, each monomer belonging to a cluster behaves independently from its neighbor. Monomer/monomer friction, which causes energy loss, occurs equally between monomers belonging to different clusters and between monomers belonging to the same cluster. In terms of the conductance of a system near the CS transition, this amounts to taking into account not only the 'inter modes' of energy transfer but also the 'intra modes'. Thus, all the cluster's perimeter is viewed by charge carriers. In addition, mass and perimeter fractal dimensions are identical for percolation clusters (Pfeifer and Obert, 1989). In the Rouse model, the fractal dimension d_u is then equal to the fractal dimension: $d_u = D$. One obtains

$$\text{Rouse model:}\ \tilde{s} = 2 - d + D \tag{68}$$

The fractal dimension of percolation clusters in three dimensions and Eqs (66)–(68) lead to the exponent values

$$\text{electrical analogy:}\ s = 0.66 \tag{69}$$

$$\text{Rouse model:}\ s = 1.32 \tag{70}$$

The values found for the exponents s and t allow us to deduce the exponent u which describes the frequency behavior of the complex modulus in the viscoelastic frequency range using Eq. (56):

$$\text{electrical analogy:}\ u = 0.75 \tag{71}$$

$$\text{Rouse model:}\ u = 0.66 \tag{72}$$

For the electrical analogy, numerical calculations of exponent values using transfer-matrix methods give: $s = 0.75 \pm 0.04$ and $t = 1.9 \pm 0.1$, leading to $u = 0.72 \pm 0.02$ (Herrmann, Derrida and Vannimenus, 1984; Derrida *et al*, 1983;

Laugier and Luck, 1987). In addition, the value of the exponent s derived from the diffusion approach differs slightly from that obtained by the study of random resistor/capacitor networks (Clerc *et al*, 1990). This can be due to the 'intra modes' which are not considered at all in the diffusion approach: because clusters are superconductor, 'intra modes' are short-circuited. Is this approximation relevant for the viscosity problem (Daoud, 1988)? As for the Rouse exponent values, they can also be derived using more conventional polymer physics methods (de Gennes, 1978; Rubinstein, Colby and Gillmor, 1989; Martin, Adolf and Wilcoxon, 1988; Martin, Adolf and Wilcoxon, 1989a, b). Finally, one has to emphasize that both these theoretical predictions, i.e. electrical analogy and the Rouse model, start from the assumption that the elasticity is a scalar; nevertheless, theoretical approaches attempt to take a possible vectorial feature of elasticity into account (Bergman and Kantor, 1984; Webman and Kantor, 1984; Kantor and Webman, 1984; Roux and Hansen, 1988). These approaches lead to values of the elasticity exponent t which is higher than that of the conductivity.

One may ask if there exists a space dimension at which the electrical analogy and the Rouse model are equivalent. Equality between Eqs (67) and (68) leads to $d - D = 2$, while equality between Eqs (62) and (64) gives $d_w - D = 2$. These equalities are verified in six dimensions. We have seen that in six dimensions excluded volume interactions can be neglected and that percolation is equivalent to the mean field, thus it is not surprising that in such conditions hydrodynamic interactions do not act. In this case, we recover the non-diverging viscosity underlain within the mean field approach (Dobson and Gordon, 1964).

4.3 Experiment

4.3.1 EXPERIMENTAL TECHNIQUES

4.3.1.1 Chemical systems

Synthesis of branched polymers is usually carried out by random chemical processes. In this section we will consider only chemical networks where monomers are linked by covalent bonds, i.e. permanent and irreversible chemical bonds. Such networks can be obtained through different chemical reactions which can be regrouped into three main classes: stepwise polymerization, chainwise polymerization and vulcanization. We will illustrate each of these processes by an example. For more details about chemistry and polymers synthesis the reader is referred to Elias (1984).

Stepwise polymerization is due to a chemical reaction between two different species of monomers, one of which has a functionality greater than two. A typical example is the synthesis of polyurethane network by condensation of triol monomers (having three hydroxyl (—OH) functional groups) with diisocyanate monomers (having two isocyanate (—C≡N) functional groups). In such a growth process, condensation of one monomer to a polymer cluster is not correlated to the previous ones. This is not the case for chainwise polymerization in which the growth of polymer clusters is due to the 'propagation' of a chemical reaction. Classically, the reactive groups are double bonds (C=C) which can be opened for example by free radicals. Within this chemical way, network synthesis needs the presence of monomers having at least two C=C groups, i.e. being able to link with at least four other monomers. A typical example of chainwise polymerization is the radical copolymerization of methyl-methacrylate (bifunctional unit) with ethylene-methacrylate (tetrafunctional unit). As for the vulcanization process, the polymer network is formed by a chemical reaction between functional groups belonging to prepolymer linear chains. As an example we can quote the crosslinking of polyisoprene by sulfur or the one of polystyrene under the action of γ-rays.

In order to study the transition of connectivity and to characterize experimentally the polymer cluster population, it is necessary to quench the gelation process at a known distance from the critical threshold. This means that the experimentalist sufficiently controls the growth process to avoid it progressing further during the time necessary to perform experiments. Furthermore, one has to determine the critical value p_c of the degree of connectivity and the actual value p of the population studied. Table 2 attempts to summarize the different methods used by authors in order to satisfy these requirements.

4.3.1.2 Measured quantities: swelling and polydispersity effects

4.3.1.2.1 Scattering experiments in very dilute solution

Among the available experimental techniques for the study of branched polymers near the gelation threshold, scattering experiments (neutron or light) are of major importance (see for example Lindner and Zemb, 1991; Chu, 1991). The relevant parameter in such experiments is the transfer vector **q**:

$$\mathbf{q} = \frac{4\pi}{\lambda}\sin(\theta/2) \tag{73}$$

where λ is the effective wavelength in the medium and θ is the scattering angle. This vector represents the inverse of the length scale at which the polymer is observed. The scattered intensity $I(q)$ is the result of constructive interferences between waves scattered by each monomer. So, very generally, $I(q)$ may be expressed by a summation over the monomers i and j belonging to polymers α and β:

$$I(q) \sim \left[\sum_{i,j,\alpha,\beta} \exp[-iq(r_i^\alpha - r_j^\beta)]\right]^2 \tag{74}$$

For a solution of n monodisperse polymers having a radius of gyration R, the

Table 2. Chemical systems used by the different authors

Reference	Chemical system	Quench of the growth process	Determination of p	Determination of p_c
Patton *et al* (1989) Colby *et al* (1992)	Polyester (polycondensation)	Cooling at room temperature	Titration of hydroxyl end groups	Non-linear regression of data using the general form: $\ln(M_w) = -\gamma \ln(p_c - p) + \ln(M_0)$
Adam *et al* (1987, 1991) Bouchaud *et al* (1986) Kajiwara *et al* (1984)	Polyurethane (polycondensation)	Non-stoichiometric ratio, chemical reaction of the minority species completed	Initial stoichiometric ratio determined by weighting	Solubility test
Lapp *et al* (1989)	Endlinking of linear PDMS (polycondensation)	Chemical poisoning		
Collette *et al* (1988)	Friedel–Craft cross-linking of semi-dilute polystyrene solutions	Chemical poisoning of catalyst activity	Crosslinking agent titration by UV spectroscopy	Solubility test
Candau, Ankrim and Munch (1985)	Radical crosslinking copolymerization of styrene with either divinylbenzene or ethylene dimethacrylate or diisopropenyl-benzene	Free radical poisonning	The reaction time corresponds to a given chemical reaction extent, the correspondence being determined by gas chromatography	Solubility test
Adolf, Martin and Wilcoxon (1990) Adolf and Martin (1990)	Anionic polymerization of epoxy	Cooling and dissolution for statics	Chemical reaction extent determined by Fourier Transform Infrared	Solubility test
Martin, Adolf and Wilkoxon (1988)		Measurements as the reaction proceeds: $\varepsilon = \|t - t_c\|/t_c$, for dynamics		Linearization of $\log(\eta)$ versus $\log(\varepsilon)$

Schosseler and Leibler (1984, 1985) Schosseler, Daoud and Leibler (1990) Schosseler *et al* (1989)	γ-rays crosslinking of polystyrene semi-dilute solutions	Stopping the irradiation		
Bauer *et al* (1991) Bauer and Burchard (1992)	Polycyanurate (polycondensation)	Mixture of mono functional and tri-functional monomers chemical reaction completed	Initial ratio of tri-functional to total number of monomers	Linearization of $\log(M_w)$ versus $\log(\varepsilon)$
Trappe, Richtering and Burchard (1992)	Anionic polymerization of epoxy			
Wu, Bauer and Su (1989)	Epoxy being crosslinked by either diamine or triamine (polycondensation)	Cooling and dissolving		
Adam, Delsanti and Durand (1985)	Polyurethane (polycondensation)	Measurements as the reaction proceeds: $\varepsilon = \lvert t - t_c \rvert / t_c$	Taken as proportional to reaction time t	Simultaneous linearization of $\eta^{-1/s}$ and $G^{1/t}$ as a function of time t
Adam *et al* (1981)	Radical copolymerisation styrene (40%) and divinylbenzene (1%) in benzene	Measurements as the reaction proceeds: $\varepsilon = \lvert t - t_c \rvert / t_c$	Taken as proportional to reaction time t	Simultaneous linearization of $\eta^{-1/s}$ and $G^{1/t}$ as a function of time t

scattered intensity can be rewritten

$$I(q) \sim n\left[\sum_{r=0}^{R} e^{-iqr}\right]^2 \tag{75}$$

The factor of proportionality in this later relation depends on the geometry of the experiment and on the contrast between polymers and their surrounding. In the case of light scattering, the contrast is due to a difference of refractive index, i.e. a difference of molecular polarizability, and is obtained by dilution of polymers in a solvent belonging to another chemical species than the polymers. In the case of neutron scattering the contrast is related to the difference of scattering length of polymers and their surroundings, i.e. difference of the mass between nuclei of scatterers. This difference is obtained by using different isotope species (hydrogen and deuterium, for example) for the chemical synthesis of the polymers and their surroundings. The advantage of creating contrast without using another chemical species other than the polymer itself is that the use of a solvent is not required.

Scattered intensity expressed by the summation of Eq. (75) can be factorized into two terms reflecting the correlations between monomers belonging either to the same polymer or to different polymers respectively. In the case of polymer solutions at a concentration tending to zero, interactions between polymers can be neglected and the scattered intensity is only due to spatial correlations between monomer belonging to the same polymer. The form factor $P(q)$ of one single polymer cluster is measured. Moreover, at length scales q^{-1} tending to infinity, all the M monomers belonging to a given polymer scatter coherently. In order to compute the scattered intensity one has to add scattered amplitude over all the M^2 pairs of monomers $\{i, j\}$ belonging to one polymer and to sum the scattered intensity contributions due to each polymer. Assuming point-like polymers, the form factor is constant, the intensity is simply proportional to the square M^2 of the number of monomers per molecule multiplied by the number n of molecules in the solution. At length scale q^{-1} between the size of monomers and the radius of gyration, R, monomers at a distance larger than q^{-1} from each other do not scatter coherently. The scattered intensity is only due to local spatial correlations between monomers in a volume q^{-3} and thus is proportional to the square, $M(q)^2$, of the number of such monomers multiplied by the number $n(q)$ of volumes q^{-3} in the total volume. The concentration C represents the number of monomers per unit volume and is simply equal to $n(q)M(q)$. Thus, one obtains the general form for the scattered intensity per monomer:

$$I(q)/C \sim M(q) \tag{76}$$

with

$$\begin{array}{ll} \text{for } q \to 0 & M(q) = M \\ \text{for } qR > 1 & M(q) = q^{-D} \end{array}$$

The q^{-D} dependence of $M(q)$ for $qR > 1$ is the direct expression of the fractal nature of polymers. Equation (76) implies that at short length scales ($qR > 1$), the scattered intensity per monomer is not sensitive to the overall mass nor to the size of the polymer, but only to its fractal conformation. Nevertheless, we have to keep in mind that characterization of branched polymers by scattering experiments is often performed in solution, which implies a different conformation of polymers than in their native state, i.e. without any solvent.

The excluded volume effect apart (two monomers cannot be at the same place), units constituting percolation clusters do not interact. The fractal conformation of percolation clusters is the only result of random connection. On the other hand, fractal conformation of branched polymers results from the sum of interactions between monomers belonging to the same polymer and those between monomers belonging to different polymers. If we consider branched polymers in their native state, these interactions cancel and only topological constraints remain, leading to the same conformation as percolation clusters. But now if such a system is diluted in order to be characterized, the fractal conformation of branched polymers in this state will depend on the new interactions balance. If one considers a single branched polymer in good solvent ($C = 0$ limit), monomer–monomer repulsions lead to a swollen conformation. In this case, simple mean field considerations predict a fractal dimension D_s lower than the one, D_p, of percolation clusters in their native state (Isaacson and Lubensky, 1980; Daoud and Joanny, 1981):

$$D_s = 2.00 \tag{77}$$

One has to outline that this fractal dimension value in the swollen state is predicted to be the same whatever the theoretical framework: percolation or mean field.

Apart from this swelling effect, scattering experiments are also sensitive to an important effect linked to the polydispersity of branched polymers. We have seen that the scattered intensity per monomer is the expression of a mass, i.e. the number of monomers which scatter coherently with the origin. In fact, at $qR < 1$ polymers are not point-like, their form factors tend to the expression obtained for spherical as well as for Gaussian particles, where the radius of gyration plays a role:

$$(I/C)_{C \to 0, qR < 1} \sim M\left(1 - \frac{q^2R^2}{3}\right) \tag{78}$$

Because of the polydispersity of branched polymer near the gelation threshold, intensity scattered per monomer is obtained by summing, over all the mass distribution curve, the contribution of each class of polymer having a mass M_i

and a radius R_i:

$$(I/C)_{qR<1} \sim \frac{\int_0^{M^*} n_i M_i^2 \left(1 - \frac{q^2 R_i^2}{3}\right) \mathrm{d}M_i}{\int_0^{M^*} n_i M_i \mathrm{d}M_i}$$

$$\sim \frac{\int_0^{M^*} n_i M_i^2 \mathrm{d}M_i}{\int_0^{M^*} n_i M_i \mathrm{d}M_i} \left[1 - \frac{q^2}{3} \frac{\int_0^{M^*} n_i M_i^2 R_i^2 \mathrm{d}M_i}{\int_0^{M^*} n_i M_i^2 \mathrm{d}M_i} \right] \sim M_\mathrm{w} \left[1 - \frac{q^2 \langle R^2 \rangle_z}{3} \right] \tag{79}$$

with

$$M_\mathrm{w} = \frac{\int_0^{M^*} n_i M_i^2 \mathrm{d}M_i}{\int_0^{M^*} n_i M_i \mathrm{d}M_i}$$

and

$$\langle R^2 \rangle_z = \frac{\int_0^{M^*} n_i M_i^2 R_i^2 \mathrm{d}M_i}{\int_0^{M^*} n_i M_i^2 \mathrm{d}M_i}$$

The first consequence of the polydispersity is that the scattered intensity is proportional to the weight average molecular mass M_w of polymer clusters in solution. The second is that experiments are sensitive to the z-average square radius of gyration $\langle R^2 \rangle_z$.

At smaller length scales, i.e. at $qR_z > 1$ with $R_z = \langle R^2 \rangle_z^{1/2}$, the scattered intensity is also sensitive to the polydispersity and is proportional to the square number of monomers belonging to the same cluster as a randomly chosen origin and closer than q^{-1} from it. In other words, scattered intensity per monomer depends on the weight average molecular mass $\langle M(q) \rangle_\mathrm{w}$ of clusters at the length scale q^{-1} i.e. the subset of clusters for which $qR_i < 1$. Thus, instead of Eq. (76) which is valid for a monodisperse sample, one obtains

$$I/C \sim \langle M(q) \rangle_\mathrm{w} \tag{80}$$

with

$$\text{for } qR_z < 1 \qquad \langle M(q) \rangle_\mathrm{w} = M_\mathrm{w}$$

$$\text{for } qR_z > 1 \qquad \langle M(q) \rangle_\mathrm{w} = \frac{\int_0^{M(q)} n_i M_i^2 \mathrm{d}M_i}{\int_0^{M(q)} n_i M_i \mathrm{d}M_i} \sim q^{-(3-\tau)D}$$

It is important to outline again that the fractal dimension D in this latter equation, depends on the polymer state: native state ($D = D_p$) or dilute solution ($D = D_s$). Therefore, different exponent values are expected for the q dependence of the scattered intensity.

As one approaches the gelation threshold, one can show that, in the reaction bath, the z-average radius of gyration, $R_z = \langle R^2 \rangle_z^{1/2}$ has the same scaling behavior as the radius of the largest cluster, i.e. the correlation length ξ. However, for experiments performed in dilute solution, one expects to measure a size R_z larger than the correlation length ξ in the reaction bath. The mass of polymers being unaffected by dilution one has $M^* \sim \xi^{D_p} \sim R_z^{D_s}$ and therefore the effective exponent ν', which governs the critical behavior of R_z is expected to be larger than ν:

$$\nu' = (D_p/D_s)\nu \tag{81}$$

One way to circumvent the difficulty in locating precisely the gelation threshold is to compare the variation of one measured quantity as a function of another quantity on samples quenched at different distances to the threshold. Then only exponent ratios are determined. This is the case with light scattering data which allow the experimentalist to plot the weight average molecular mass M_w as a function of the z-average radius of gyration R_z of various samples. From a theoretical point of view, Eqs (17) and (25) show that in the reaction bath M_w and the correlation length ξ are linked by $M_w \sim \xi^{\gamma/\nu}$. In solution, replacing the exponent ν by ν' and using scaling law (Eqs (22) and (27)) this relation becomes

$$M_w \sim R_z^{(3-\tau)D_s} \tag{82}$$

where D_s is the fractal dimension of diluted branched polymers and τ the exponent which characterize the polymer mass distribution (Daoud, Family and Jannink, 1984). It is interesting to note that Eqs (80) and (82) are two different expressions of the scaling properties of branched polymers. Equation (80) considers one sample at a given distance from the threshold, while Eq. (82) is concerned with different samples.

4.3.1.2.2 Weakly interacting polymers solutions: scattering and intrinsic viscosity measurements

At non-zero concentration, another relevant quantity which is of interest is the osmotic pressure of the polymer solution. Classically, the osmotic pressure π of n interacting particles per volume unit is expressed by the virial expansion: $kTn(1 + a_2 n)$, where k is Boltzmann's constant, T the temperature and a_2 the second virial coefficient. In the case of monodisperse polymers in good solvent having a mass M and a concentration C, one has $n = C/M$ and $a_2 = R^3$. One can introduce the polymer internal concentration called the overlap concentration

$C^* = M/R^3$ in the expression of π, which can be rewritten

$$\pi = \pi_0[1 + C/C^* + \ldots] \tag{83}$$

with $C^* = 1/(MA_2)$, $A_2 = a_2/M^2$ and $\pi_0 = kTC/M$.

At the $q = 0$ limit the scattered intensity reflects the macroscopic properties of the material. Scattering is due to statistical heterogeneities, i.e. variation of refractive index in the material for light scattering. In the case of polymer solutions these heterogeneities are mainly concentration fluctuations and the scattered intensity per monomer I/C is proportional to the osmotic compressibility, $(\mathrm{d}\pi/\mathrm{d}C)^{-1}$, of the solution. Using Eq. (83) for $C/C^* \ll 1$, one obtains

$$\left[\frac{I}{C}\right]_{q\to 0} \sim M(1 - 2C/C^*) \tag{84}$$

In the case of polydisperse branched polymers in solution, the overlap concentration C^* corresponds to the situation where all the smaller clusters have gone out of the larger ones. At C^* all the polymers are in closed packing condition. Then, the summation over all the volumes occupied by each cluster is equal to the total volume V of the solution:

$$\int_0^{M^*} R_i^3 n_i \,\mathrm{d}M_i = V = \frac{\int_0^{M^*} n_i M_i \,\mathrm{d}M_i}{C^*} \tag{85}$$

The first-order moment of the mass distribution curve being constant and equal to the total number of monomers in the solution, one obtains

$$C^* = \frac{1}{\int_0^{M^*} R_i^3 n_i \,\mathrm{d}M_i} \tag{86}$$

Finally, introducing the fractal dimension D_s of polymer clusters in solution, the exponent τ of the mass distribution and using the hyperscaling law (Eq. (30)) one obtains (Daoud and Leibler, 1988)

$$C^* = M^{*3\left(\frac{1}{D_p} - \frac{1}{D_s}\right)} = M_{\mathrm{w}}^{-\frac{3}{3-\tau}\left(\frac{1}{D_s} - \frac{1}{D_p}\right)} \tag{87}$$

This concentration C^* corresponds to the inverse of the second virial coefficient which is measured in scattering experiment (Adam *et al*, 1987). One has to note that direct measurements of the osmotic pressure using osmometry techniques lead to number averaged molecular masses which do not diverge as one approaches the threshold.

The overlap concentration can also be determined by viscosity measurements of polymer solutions in good solvent. The quantity which is of interest is the relative increase of the viscosity due to the presence of the polymer:

$$\frac{\eta - \eta_0}{\eta_0} = [\eta]C(1 + \kappa[\eta]C) \tag{88}$$

where η and η_0 are the solution and solvent viscosity respectively, and κ is the Huggins constant. The quantity $[\eta]$ is called the intrinsic viscosity and has the dimension of the inverse of a concentration. In solution, hydrodynamic interactions are important, i.e. polymer behavior corresponds to the Zimm limit. In addition, if one considers a zero shear flow, no deformation of the polymer occurs during the experiment. These two remarks allow us to apply the Einstein relation which links the viscosity η of a solution of hard spheres occupying a volume fraction Φ: $\eta = \eta_0(1 + \frac{5}{2}\Phi)$. From Eq. (88), $[\eta] \sim \Phi/C$ which is the inverse of the overlap concentration C^* we have defined above ($\Phi \sim nR^3$ and $C \sim nM$ leading to $\Phi/C \sim R^3/M = 1/C^*$):

$$[\eta] \sim \frac{1}{C^*} \tag{89}$$

One has to outline that the second virial coefficient and intrinsic viscosity measurements can be a good check of the branched polymer swelling under dilution. In fact, without any swelling, Eq. (87) clearly shows that the second virial coefficient and the intrinsic viscosity will be independent of the weight average molecular mass M_w of the sample.

4.3.1.2.3 Size exclusion chromatography

A size exclusion chromatography (SEC) apparatus is composed of a series of columns of porous media having different pore sizes, from 10^3 to 10^6 Å. Dilute solutions of polymers are injected into the columns. Molecules with small hydrodynamic volume explore a greater fraction of the porous voids than larger ones, and therefore will need a long time to flow through the column. The larger the hydrodynamic volume of the molecule, the shorter the time needed to flow through the columns (for details see for example Yau, Kirkland and Bly, 1979).

Taking the outflow from the columns at constant time interval corresponds to selecting from the solution a fraction of polymers having a given hydrodynamic volume. This outflow is analyzed by a low angle light scattering (LALS) device which allows the measurement of the scattered intensity at very low scattering angle ($\theta = 6°$) and very low concentration, and then the determination of the weight average molecular mass of each fraction of the polymer solution (elution volume). The polymer concentration of each fraction of the solution is measured using a differential refractometer. Such an experiment amounts to measuring the weight fraction $\Phi(M_i) \sim n_i M_i$ of monomers belonging to molecules having a mass

M_i, which is expected to behave as

$$\Phi(M_i) \sim M_i^{1-\tau} \tag{90}$$

This can provide the experimentalist with a direct determination of the exponent τ.

The maximum observed in the light scattering chromatograms corresponds to the mass M^* of the largest polymer cluster present in the solution (Schosseler *et al*, 1989), while the second moment of the mass distribution curve obtained by the whole chromatogram gives the weight average molecular mass M_w of the polymer population. These two quantities are expected to behave as $M_w \sim M^{*\sigma\gamma}$ or, using the scaling law (Eq. (22)),

$$M_w \sim M^{*(3-\tau)} \tag{91}$$

The calibration curve of the SEC device performed on linear and branched polymers allows the determination of the ratio of the fractal dimension linking mass and size. If D_{lin} and D_s are the fractal dimensions in diluted solution of linear and branched polymers respectively, it is easy to show that the slopes A of the curves representing the mass versus the elution volume in log–log scale behave as $A_{branched}/A_{lin} = D_s/D_{lin}$ which is expected to be equal to $D_s/D_{lin} = 2/1.7 = 1.18$.

Besides this analytical use, size exclusion chromatography can be employed to prepare narrow distributed polymer samples. Branched polymers obtained near the gel point can be thus fractionated in order to be studied with other experimental techniques, being free of the polydispersity effects.

4.3.1.3 Dynamical properties measurements

4.3.1.3.1 Rheological measurements

In order to measure the viscoelastic properties of a material, one may measure the stress, ϕ, needed to impose an external strain, ψ, to the medium (Eirich, 1960). When this strain is sufficiently low, linear relations exist between stress and either strain or strain rate for solid or liquid materials, respectively:

$$\text{solid:}\quad \phi = E\psi \tag{92}$$

$$\text{liquid:}\quad \phi = \eta\, \mathrm{d}\psi/\mathrm{d}t \tag{93}$$

where E is the elastic modulus and η is the viscosity. Classically, in order to measure the elastic modulus or the viscosity of one sample, this latter is sandwiched between two parallel plates, the upper being able to rotate and the lower allowing measurement of the strain. This situation corresponds to a shear experiment which allows the shear modulus G to be measured. Usually gelling media are simultaneously viscous and elastic; this means that strain and stress are neither in phase nor in quadrature. Imposing a given strain of frequency ω to the medium, one can measure parts of the stress which are in-phase and out-of-phase with the strain, as a function of ω. These correspond to real and imaginary parts of the complex modulus $G^*(\omega)$ respectively. Very generally one has

$$G^*(\omega) = \frac{\int \phi(t)\,\mathrm{e}^{-i\omega t}\,\mathrm{d}t}{\int \psi(t)\,\mathrm{e}^{-i\omega t}\,\mathrm{d}t} \tag{94}$$

in terms of the Fourier transforms of the imposed strain and measured stress.

4.3.1.3.2 Quasi-elastic light scattering measurements

Quasi-elastic light scattering device allows the time-dependent fluctuations of the refractive index to be analyzed by computing the autocorrelation function of the scattered intensity $I(t)$. The relaxation function of these fluctuations leads to the dynamical structure factor $S(q, t)$:

$$S(q,t) = \frac{\langle I(q,0)I(q,t)\rangle - \langle I(q,t)\rangle^2}{\langle I(q,t)^2\rangle - \langle I(q,t)\rangle^2} \tag{95}$$

In the frequency domain the dynamical structure factor $S(q, \omega)$ is proportional to the imaginary part of the dynamical longitudinal compliance $J_{\mathrm{L}}^*(\omega) = 1/M_{\mathrm{L}}^*(\omega)$, where $M_{\mathrm{L}}(\omega)$ is the longitudinal elastic modulus (Wang and Fischer, 1985; Meier, 1987):

$$S(q,\omega) = \frac{nkT}{\omega} J''(q,\omega) \tag{96}$$

with n the number of particles per volume unit. The longitudinal elastic modulus is a sum of the bulk modulus K and of the shear modulus G: $M_{\mathrm{L}}(\omega) = K(\omega) + 4/3G(\omega)$. Different cases can be considered:

1. In the case of liquids at zero frequency (static measurements) only the bulk modulus remains. In addition, for dilute polymer solutions, density fluctuations can be neglected compared to concentration fluctuations, thus the bulk modulus corresponds to the osmotic modulus $K = C(\mathrm{d}\pi/\mathrm{d}C)$. This allows those results obtained in sections 4.3.1.2.1 and 4.3.1.2.2 with a virial expansion of K to be recovered.
2. At higher frequency for dilute polymer solutions at $qR < 1$, the bulk osmotic relaxation shows a single characteristic time $\tau_c = 1/(\mathcal{D}q^2)$ which is the time needed for the concentration fluctuation to cover a distance q^{-1}, $\mathcal{D}$ being the diffusion coefficient of polymers. In this case, the dynamical structure factor has a Lorentzian shape leading to a simple exponential decay for the autocorrelation function of the scattered intensity.

3. Nevertheless, in a general manner, it is necessary to take into account the respective contributions of bulk and shear to the longitudinal modulus. In most of the cases, the bulk contribution dominates the measured relaxation function ($K > G$) and quasi-elastic light scattering does not reflect the viscoelastic properties of the material, e.g. good solvent semi-dilute solutions of linear polymers (Candau, Bastide and Delsanti, 1982) for which $K/G \gg 1$ (Adam, Lairez and Raspaud, 1992; Raspaud, Lairez and Adam, 1995). However, in some cases, shear and osmotic bulk moduli are of the same order of magnitude. This is case for semi-dilute solutions of linear polymers in a theta solvent (Brochard, 1983; Adam and Delsanti, 1985, 1989) within which the two elastic moduli present a single relaxation time corresponding to a cooperative diffusion for the bulk and to a q independent relaxation for the shear contribution. Thus, depending on the scattering angle, the autocorrelation function is complex and especially at high q, it reflects the viscoelastic properties of the solution.

The case of branched polymer near the gelation threshold needs particular attention. The bulk modulus of polymers in the melt is so high that experimentalists need to increase the scattered intensity by adding a small percentage of solvent. This tends to hide density fluctuations by concentration fluctuations governed by the osmotic modulus. In this situation the polydisperse population of branched polymers is divided into two subsets. The first contains the smaller clusters which are in dilute situation and mainly contribute to the osmotic modulus, while the second contains larger clusters which remain unaffected by diluent addition. In this condition the longitudinal elastic modulus is expected to be a sum of the form

$$M_{\mathrm{L}}^*(\omega) = K(1 + i\omega\tau_0) + 4/3G^*(\omega)$$

where τ_0 corresponds to the diffusive mode of diluted small clusters and $G^*(\omega)$ is given by Eq. (58). In the time domain this leads to a diffusive exponential decay at short time with a characteristic time τ_0, while at long time one expects a power law decay corresponding to the Fourier transform of $1/G(\omega) \sim \omega^{-u}$, up to the longest viscoelastic relaxation time of the system $\tau^* = 1/\omega^*$ (Adam *et al*, 1988):

$$S(q,t) \sim (t/\tau_0)^{u-1} f(t/\tau^*) \tag{97}$$

with $\tau_0 \sim 1/q^2$ and $f(x \ll 1) = 1$, $f(1 \ll x) = 0$.

Besides this approach, some authors attempt to compute the dynamical structure factor in a different way (Martin and Wilcoxon, 1988). They assume that clusters are submitted to hydrodynamic interactions during their diffusive motion: the diffusion coefficient $\mathscr{D}_i$ of clusters having a mass M_i is inversely proportional to their radius R_i. The local viscosity viewed by these clusters depends also on their size and is calculated from the Rouse model, which assumes no-hydrodynamic interaction between the clusters. This approach seems not to be self-consistent. In addition, one may wonder how the diffusion of these clusters can be observed, while the total scattered intensity measurements are not sensitive to their size.

4.3.2 STATIC PROPERTIES EXPERIMENTAL RESULTS

In this section we will review experimental results reported in the literature concerning the critical behavior of the sol–gel transition. First, it is important to know that in spite of the different recipes used by each experimentalist to circumvent the inherent difficulties of gelation study, none of the results are obtained very close to the transition, as is the case for thermodynamic critical phenomenon. Consequently, the determination of exponent values is usually

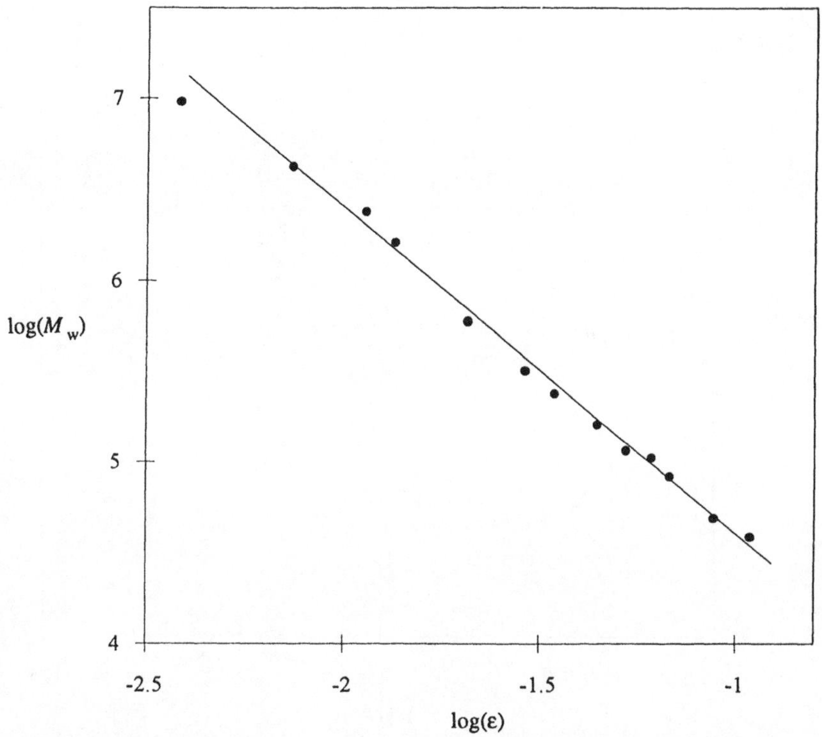

Figure 4. Variation of the weight average molecular mass, M_w, as a function of the relative distance to the gelation threshold, ε, from Adam *et al* (1987). Results were obtained by light scattering experiments performed on polyurethane samples having different initial stoichiometric ratios, r. In this case $\varepsilon = (r_c - r)/r_c$, where r_c, the critical stoichiometric ratio, was determined by solubility test with an accuracy of 5×10^{-4}. The expected power law behavior is confirmed with an exponent value $\gamma = 1.7 \pm 0.06$, the accuracy on γ being determined from that on r_c

carried out with a poor precision compared to those obtained in other fields of physics. Secondly, this review does not claim to be exhaustive, but rather intends to present the most significant results reported in the literature. For example, we will quote only the results of experimentalists who determined the critical threshold with a precision much smaller than ε_{min}, the relative distance to the threshold of the sample the closest to it. When this is not the case, we will quote only exponent ratios or exponents which are independent of ε, for example τ and D. In Table 3, experimental results for the exponents values are listed, some of them being plotted in Figures 4–10.

Before discussing of the results reported in Table 3, some remarks are needed concerning results reported in the literature but not quoted here.

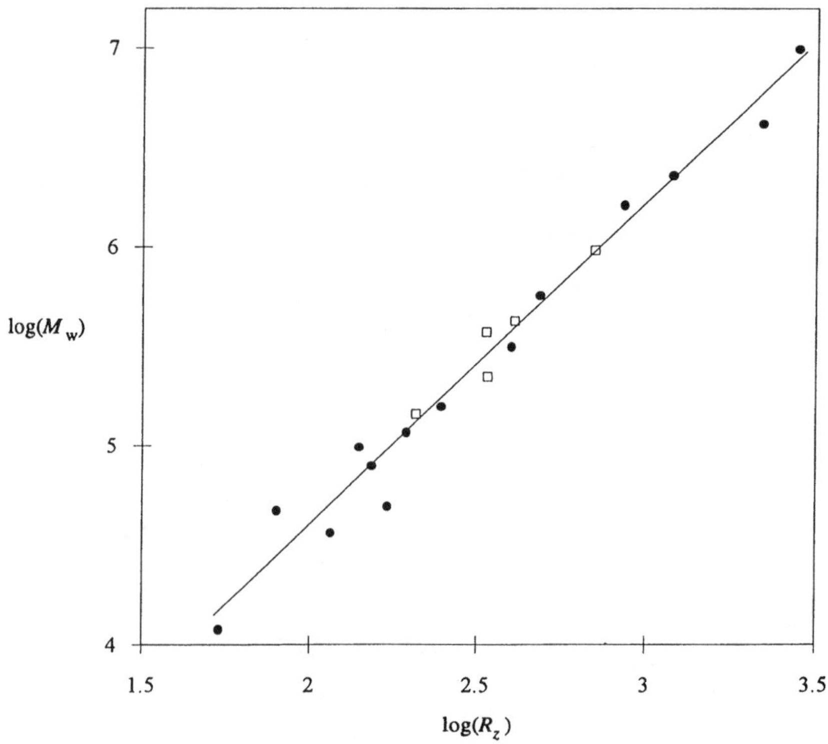

Figure 5. Variation of the weight average molecular mass (M_w in g mol^{-1}) as a function of the z-average radius of gyration (R_z in Å) from two independent light scattering experiments on polyurethane samples obtained in the melt by Adam *et al* (●) (1987) and Kajiwara *et al* (□) (1984). In spite of the analysis of Kajiwara *et al*, one can see that the two sets of experimental data fit on a single straight line. The corresponding exponent ratio $\gamma/\nu' = (3-\tau)D_s$ is found to be equal to 1.61 ± 0.07, in good agreement with percolation predictions, while from the mean field one expects $(3-\tau)D_s = 1$

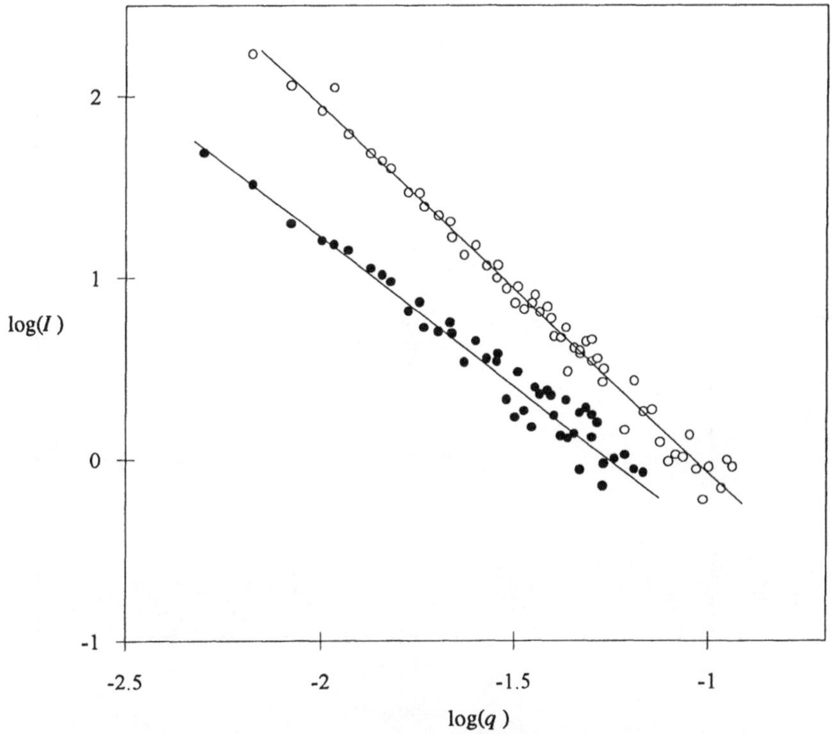

Figure 6. Variation of neutron scattered intensity I (a.u.) versus transfer vector $\mathbf{q}$ (Å^{-1}) obtained on fractionated (○) (with $qR > 1$) and unfractionated (●) (with $qR_z > 1$) polyurethane samples in dilute solutions in deuterated tetrahydrofuran, from Bouchaud *et al* (1986). For both samples, a power law behavior $I \sim q^{-x}$ is found. For fractionated samples $x = 1.98 \pm 0.03$ and corresponds to the expected fractal dimension D_s of swollen branched polymers, while for unfractionated $x = 1.59 \pm 0.05$ and corresponds to the product $(3 - \tau)D_s$. The ratio of these two exponent values leads to $\tau = 2.20 \pm 0.04$, in good agreement with percolation

Sometimes, the gel point is approached by varying the fraction x_c of crosslinkers taken to be the degree of connectivity (Fang, Brown and Konak, 1991). When a chemical reaction takes place indiscriminately between difunctional monomers or crosslinkers, changing their fraction amounts to varying the power law prefactors, i.e. the network unit size ξ_0, and its mass M_0. Introducing these prefactors leads to Eqs (32) and (33), in which ξ has to be replaced by R_z and the exponent ν by ν', as experiments are performed in dilute solutions. To a first approximation, the mass M_0 between two crosslinkers varies as $1/x_c$. This behavior leads to a major error in the determination of the exponent γ, while fortuitously the exponent ratio $\gamma/\nu' = (3 - \tau)D_s$ remains practically unaffected. In

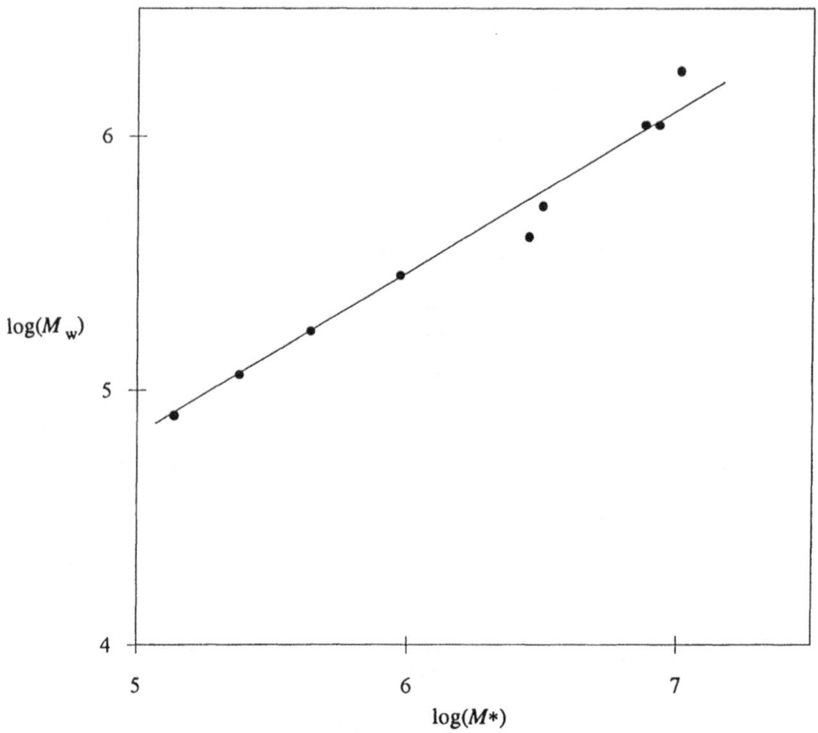

Figure 7. Variation of the weight average molecular mass (M_w in g mol^{-1}) as a function of the mass of the largest cluster (M^* in g mol^{-1}). Experiments were performed on γ-irradiated polystyrene samples by Schosseler *et al.* (1989) using size exclusion chromatography. M^* is proportional to the maximum in the light scattering chromatogram. The straight line corresponds to the law $M_w \sim M^{*(3-\tau)}$ with $\tau = 2.3 \pm 0.1$

fact, taking into account prefactors, one obtains: $M_w = M_0(R_z/R_0)^{(3-\tau)D_s}$. The mass M_0 corresponds to a linear portion of the polymer having a fractal dimension $D_1 = 1.70$. According to the three-dimensional percolation exponent values one obtains $M_w = M_0^{1-(3-\tau)D_s/D_1} R_z^{(3-\tau)D_s} = M_0^{0.06} R_z^{(3-\tau)D_s}$. This remark may also explain the results of Schmidt and Burchard (1981) and Whitney and Burchard (1980). Obviously, these systems undergo a sol–gel transition, but the reduced variables describing the growth of connectivity are ill defined.

Several teams have reported interesting experimental studies on silica gels in solutions (see for example: Schaeffer and Keefer, 1984; Cabane, Dubois and Duplessix, 1987; Dubois and Cabane, 1989; Martin and Wilcoxon, 1989; Martin and Odinek, 1990). These systems undergo two growth processes as the chemical reaction progresses. At the earlier stages of the reaction, cluster growth

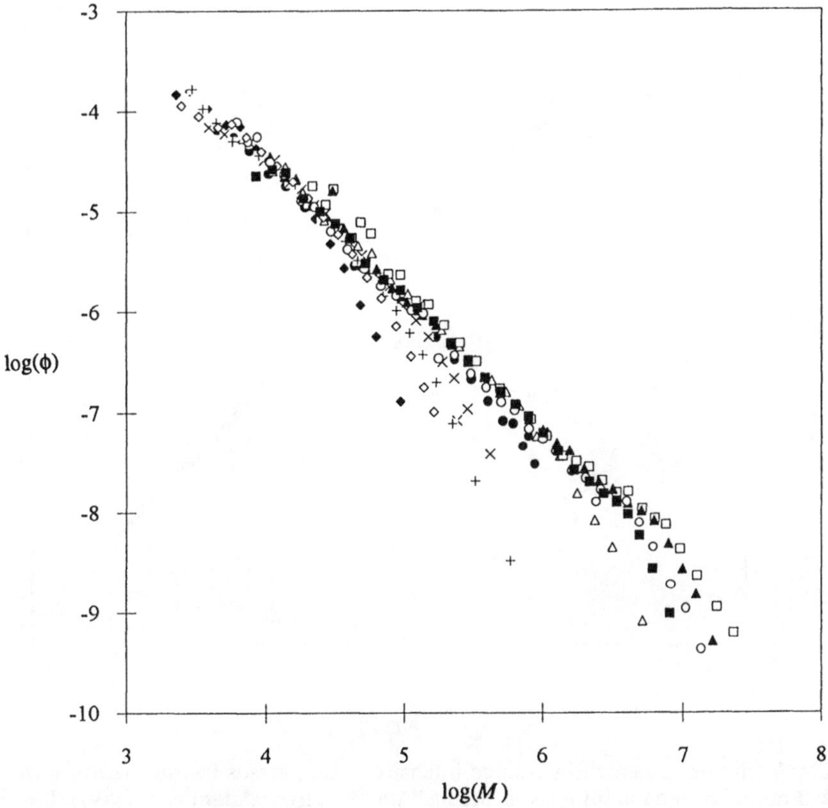

Figure 8. Scaling of the reduced mass distribution curve: variation of the weight fraction of polymers Φ versus the mass M. Results are obtained on polyester samples synthesized by polycondensation (Patton *et al*, 1989). The different symbols used correspond to samples prepared at different distances to the gelation threshold. For $M < M^*$, the weight is independent of the sample, all data points lie on a single line and one recovers and expected power law behavior with an exponent $\tau - 1 = 1.29 \pm 0.03$, while for $M > M^*$, Φ depends on the distance to the threshold

seems to be described by the reaction-limited aggregation model, while at latter stages these aggregates percolate. One recovers this complex growth behavior within static properties of clusters which show two fractal dimensions depending on the length scale. As in the vulcanization problem (see section 4.2.1.6) elementary units constituting the percolation network are very large, yielding to a narrow critical domain. The study of the crossover from aggregation to percolation behavior would be of great interest (Gimel *et al*, 1995).

To our knowledge, only one team has studied the crossover from mean field to percolation behavior in the vulcanization problem (Colby *et al*, 1992). Authors

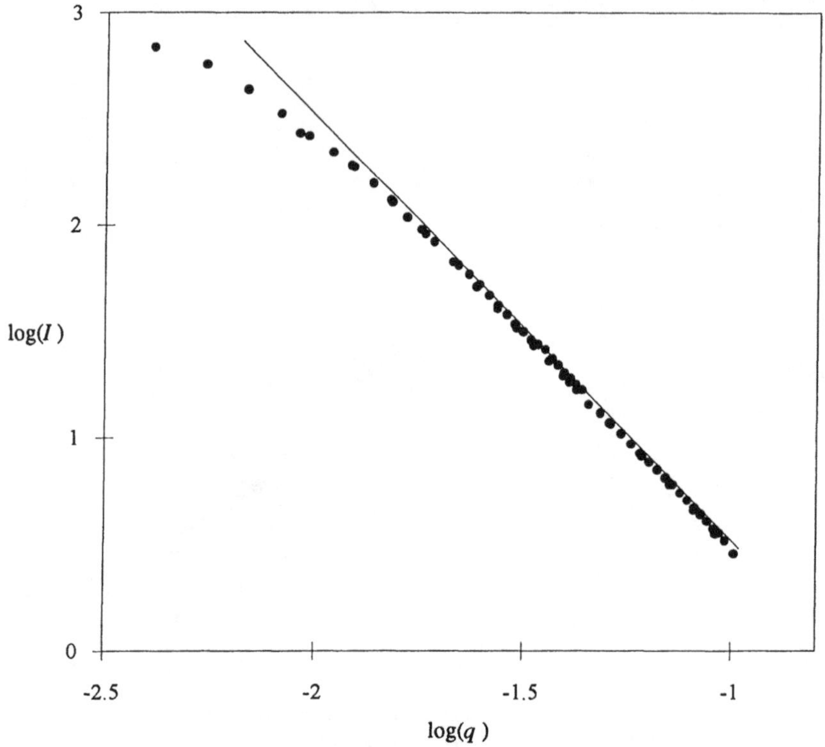

Figure 9. Variation of neutron scattered intensity I (a.u.) versus transfer vector $\mathbf{q}$ ($Å^{-1}$) obtained on polyurethane samples in the melt ($q\xi > 1$), from Adam *et al* (1991). Results were obtained by mixing two pregel samples quenched at the same relative distance to the threshold, one of them being deuterated. The slope of the straight line corresponds to the exponent $(3-\tau)D_p = 2.00 \pm 0.05$. This value has to be compared to $(3-\tau)D_s = 1.59 \pm 0.05$ obtained on the same system in dilute solutions (see Figure 6). These two sets of experimental results allow us to verify the hyperscaling relation (Eq. (30)) which can be rewritten as $\frac{d}{D_p(3-\tau)} = \frac{\tau - 1}{3-\tau}$. In fact, in the melt one finds $d/D_p(3-\tau) = 1.50 \pm 0.04$ and in diluted solutions $(\tau-1)/(3-\tau) = 1.50 \pm 0.12$. Hyperscaling and experimental results lead to $D_p = 2.50 \pm 0.06$

show that the variation of the gel fraction as a function of the weight average molecular mass of the sol is of the form

$$x_{gel} \sim M_w^{-1/4}(1 + (M_w/M_x)^{-3/4}$$

where M_x is the weight average mass corresponding to the center of the crossover. With such a law, one recovers two limiting power laws: $x_{gel} \sim M_w^{-\beta/\gamma}$ with $\beta/\gamma = 1/4$ and $\beta/\gamma = 1$ for the percolation $(M_w/M_x > 1)$ and for mean field $(M_w/M_x < 1)$ limits respectively (see Figure 10).

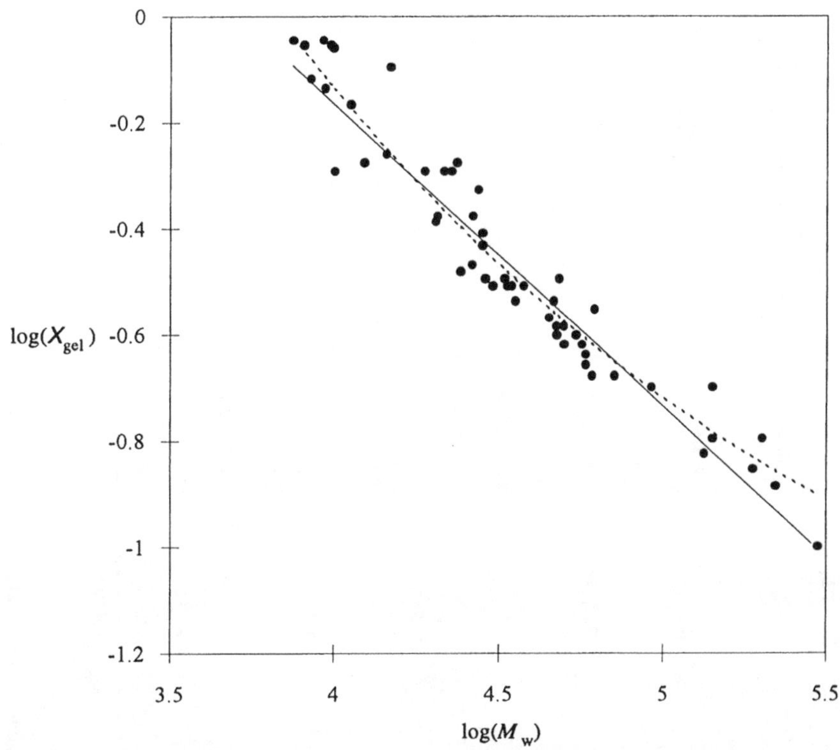

Figure 10. Variation of the gel fraction x_{gel} versus the weight average molecular mass, M_w, of the extracted sol. Results were obtained on polyester samples quenched above the gel point by Colby *et al* (1992). The straight line has a slope which corresponds to an effective exponent $\beta/\gamma = 0.56 \pm 0.04$. The dotted line represents the crossover function $x_{gel} \sim M_w^{-1/4}(1 + (M_w/M_x)^{-3/4})$, which allows the recovery of the exponent values corresponding to percolation ($\beta/\gamma = 1/4$) and to mean field ($\beta/\gamma = 1$) for high and low M_w values respectively. For $M_w = M_x$, mean field and percolation contribution to x_{gel} are equal ($M_x = 2.3 \times 10^4\,\mathrm{g\,mol^{-1}}$)

One can note that the most frequently measured exponent ratio is $\gamma/\nu' = (3 - \tau)D_s$, the mean value of which is 1.7 ± 0.1. Curiously, results obtained on epoxy resins deviate from this mean value. In Table 3, one can see from Adolf, Martin and Wilcoxon (1990) that one obtains $(3 - \tau)D_s = 1.2 \pm 0.15$. On the other hand, Wu, Baues and Su (1989) find $(3 - \tau)D_s$ equals either 1.2 or 1.65, with diamine or triamine as crosslinker, respectively. This can be due to a peculiar growth process leading to non-self-similar objects, as found by Chu *et al* (1987) and Chu and Wu (1988) who report non-power law spectra of X-rays scattered intensity for $qR_z > 1$.

Table 3. Experimental determination of static exponent values. Numerical values in Bold referred to exponent values theoretically expected

References	Fig.	γ $M_w = f(\varepsilon)$	$\nu' = \nu(D_p/D_s)$ $R_z = f(\varepsilon)$	$(3-\tau)D_s$ (a) $M_w = f(R_z)$ (b) $I = f(q)$	$(3-\tau)D_p$ $I = f(q)$	D_s (a) SEC (b) $I = f(q)$ (c) $M = f(R)$	τ (a) $M_w = f(M^*)$ (b) $n_i M_i = f(M_i)$ (c) $(3-\tau)D_s/D_s$	γ/β $M_w = f(x_{gel})$	$[3/(3-\tau)](1/D_s - 1/D_p)$ $C^* = f(M_w)$
Mean field		**1**	**1**	**1**	**2**	**2**	**52**	**1**	**3/2**
Percolation		**1.74**	**1.10**	**1.60**	**2.00**	**2.00**	**2.20**	**0.25**	**0.37**
Patton *et al* (1989)	8	1.8 ± 0.3		(a) 1.52 ± 0.07		(a) 2.05 ± 0.07[1]	(a) 2.30 ± 0.04 (b) 2.30 ± 0.02		0.28 ± 0.03[3]
Colby *et al* (1992)	10						(a) 2.33 ± 0.04[3]		0.19 ± 0.02[3]
Adam *et al* (1987)	4, 5	1.71 ± 0.06	1.01 ± 0.05	(a) 1.61 ± 0.07					0.35 ± 0.05
Bouchaud *et al* (1986)	6			(b) 1.59 ± 0.05		(b) 1.98 ± 0.03	(c) 2.20 ± 0.04		
Adam *et al* (1991)	9				2.00 ± 0.05[4]				
Kajiwara *et al* (1984)	5			(a) 1.6 ± 0.2[5] (a) 1.7 ± 0.1[5]					
Lapp *et al* (1989)							(a) 2.27 ± 0.04		
Collette *et al* (1988)		1.65 ± 0.10		(a) 1.8 ± 0.2					
Candau, Ankrim and Munch (1985)		1.65 ± 0.08[6]		(a) 1.75 ± 0.15					0.35 ± 0.05
Adolf, Martin and Wilkoxon (1990)		1.7 ± 0.1	1.4 ± 0.1						
Schosseler *et al* (1989)	7					(a) 2.04 ± 0.06	(a) 2.3 ± 0.1		
Schosseler and Leibler (1984, 1985)				(a) 1.7 ± 0.2					
Schosseler, Daoud and Leibler (1990)				(a) 1.67 ± 0.04 (b) 1.55 ± 0.15		(b) 2.06 ± 0.1 (c) 2.04 ± 0.1			
Bauer *et al* (1991)				(a) 1.9 ± 0.1			(a) 2.18 ± 0.05		
Bauer and Burchard (1992)							(b) 2.22 ± 0.02		
Trappe, Richtering and Burchard (1992)		1.75 ± 0.4	1.0 ± 0.2						
Wu, Bauer and Su (1989)				(a) 1.67; (b) 1.6[7] (a) 1.15; (b) 1.2[8]					

[1] This value is deduced using the theoretical fractal dimension of linear polymers ($D_1 = 1.70$) in Eq. (32) of Patton *et al* (1989).
[2] $p < p_c$.
[3] $p > p_c$.
[4] Neutron scattering experiments performed in the melt, part of clusters being labeled.
[5] Table 2 of Kajiwara *et al* (1984) reports data obtained on samples prepared either in the melt or in the presence of solvent. Exponent values presented here are obtained by fitting separately each set of data.
[6] Average over five exponent values measured on samples obtained with different comonomers.
[7] Triamine crosslinker (which can be connected with four epoxides).
[8] Diamine crosslinker (which can be connected with six epoxides).

Results obtained by size exclusion chromatography give a direct evidence of the scaling behavior of the clusters growth near the gelation threshold. Actually, Figure 7 shows the master curve $\Phi(M/M^*)$ obtained for one chemical system at different relative distances from the threshold (Patton *et al*, 1989). It has been shown that results obtained on different chemical systems scale on a single universal curve (Lapp *et al*, 1989).

Experimental results concerning statics of the sol–gel transition obtained during the last decade can be summarized as follows:

1. Scaling of the branched polymers mass distribution curve was verified.
2. At least two exponents, γ and τ, were measured and found to be in good agreement with computer simulations following three-dimensional percolation model.
3. In one case, the exponent β of the gel fraction was found to be compatible with the one deduced from γ and τ experimental determinations and scaling laws (Durand *et al*, 1989).
4. The hyperscaling law was verified from two independent neutron scattering experiments (Bouchaud *et al*, 1986; Adam *et al*, 1991).
5. Two sets of experimental results confirm the swelling of polymer clusters under dilution:
   (a) The fractal dimension of branched polymers in dilute solution D_s was measured directly and found to be smaller than D_p, the fractal dimension of polymers in their native state.
   (b) The second virial coefficient deduced from scattering experiments and intrinsic viscosity measurements is found to be dependent on the weight average molecular mass. For polydisperse systems this indicates that D_s is smaller than D_p.

All these results indicate clearly that gelation is a critical phenomenon of connectivity which can be described using scaling arguments and which belongs to the same universality class as three-dimensional percolation. However, one can note that very few experiments are performed above the gel point. It should be interesting to measure the weight average molecular mass of the sol fraction at $p > p_c$ and to compare it to M_w of the sol fraction at $p < p_c$. The percolation model predicts that: $M_w \sim C^- \varepsilon^{-\gamma}$ for $p < p_c$, $M_w \sim C^+ \varepsilon^{-\gamma}$ for $p > p_c$ and a ratio of amplitudes C^-/C^+ equal to 10 (Nakanishi and Stanley, 1980). There is no experimental determination of this ratio on polymeric systems near the gel point, but computer simulations (Herrmann, Landau and Stauffer, 1986) seem to indicate that this ratio is much smaller for radical copolymerization than for percolation.

While static properties of branched polymers near the gelation threshold seems to be well understood, we will see in the next section that dynamics of these systems is not so clear.

4.3.3 DYNAMICAL PROPERTIES EXPERIMENTAL RESULTS

The signature of the sol–gel transition is the passage from a viscous liquid to an elastic medium as the chemical reaction proceeds. Scaling of the dynamical properties has been detailed in Section 4.2.2.2. This scaling behavior has strong implications which have to be verified whatever the theoretical model invoked. Before reviewing the experimental results reported in the literature (see Table 4), let us recall what one expects to observe near the gelation threshold. From our point of view, experimental evidence of the scaling behavior of rheological properties passes inevitably through three main points:

1. Scaling of the reduced rheological curves: $G(\omega)/G(\omega^*)$ versus ω/ω^* obtained on samples at different relative distances to the threshold lie on a single curve. This has direct corolaries:
   (a) same power law behavior at high frequency ($\omega^* < \omega$) for the real and imaginary parts of the complex elastic modulus: $G' \sim G'' \sim \omega^u$;
   (b) a high-frequency ($\omega^* < \omega$) particular loss angle δ, which is ε and ω independent and related to the exponent u: $\tan(\delta) = G''/G' = \tan(u\pi/2)$.
2. ε dependence power law of zero frequency rheological properties, η_s and G_s, and of the characteristic frequency ω^*: $\eta_s = (G_0/\omega_0)\varepsilon^{-s}$, $G_s = G_0\varepsilon^t$ and $\omega^* = \omega_0\varepsilon^z$.
3. Scaling relation between exponents: $z = s + t$ and $u = t/s + t$, the direct consequence being a high-frequency complex modulus ($\omega^* < \omega$) independent of ε.

The first point is the counterpart in dynamics of the scaling observed in statics, for example the reduced mass distribution curve (Figure 8). It reflects the self-similarity of both the conformation and the mass distribution of clusters. Only two teams report experimental results obtained on chemical gels which were analyzed in term of scaling law. Hodgson and Amis (1990, 1991) working on silica gels in solutions built a rheological master curve valid in the overall pregel domain. Nevertheless, their results has to be considered, keeping in mind that gelation of these systems implies two growth processes as yet mentioned and shown by rheological study (Devreux *et al*, 1993). Adolf and Martin (1990) have obtained pre- and post-gel rheological master curves by phenomenological x–y translation instead of using the theoretical reduced variables. However, several teams have observed a high-frequency power law behavior of the complex elastic modulus, the exponent u so determined being equal to 0.695 ± 0.005, while only one team reports a loss angle compatible with the value of the exponent u and independent of ε in the near vicinity of the threshold (Durand *et al*, 1987, 1988). It is important to note that the crossover frequency range in which the material passes from a simple liquid- or solid-like behavior (see Eq. (58)) to a viscoelastic behavior is very wide—between two and three frequency decades (Durand *et al*,

Table 4. Experimental determination of dynamic exponent values. Numerical values in Bold referred to exponent values theoretically expected

	s (a) $\eta_s = f(\varepsilon)$ (b) $\eta_s = f(M_w)$[1]	t (a) $G_s = f(\varepsilon)$ (b) $G_s = f(M_w)$[1] (c) Scaling	u (a) $G^* = f9\omega)$ (b) $u = \delta 2/\pi$ (c) $G(\omega^*) = f(\omega^*)$	z $\omega^* = f(\varepsilon)$	Comments
Electrical analogy	**0.75**	**1,9**	**0.72**	**2.65**	
Rouse model	**1.32**	**2.7**	**0.66**	**4**	
Rubinstein, Colby and Gillmor (1989)			(a) 0.69 ± 0.02		
Colby, Gillmor and Rubinstein (1993)	(b) 1.8 ± 0.3	(b) 4 ± 1.7			
Durand *et al* (1987, 1988)			(a) 0.70 ± 0.02 (b) 0.69 ± 0.04		G independent of ε at $\omega = 5$ Hz for $5 \times 10^{-4} < \varepsilon < 7 \times 10^{-3}$
Lairez *et al* (1992)	(b) $1.4 < \ldots < 1.7$				An apparent exponent s is measured which dependent on the temperature and diluent concentration
Adam, Delsanti and Durand (1985)	(a) 0.8 ± 0.1	(a) 3.2 ± 0.5			
Adam *et al* (1981)	(a) 0.85 ± 0.06	(a) 2.1 ± 0.3			
Martin, Adolf and Wilkoxon (1988)	(a) 1.4 ± 0.2		(a) 0.70 ± 0.05		
Adolf and Martin (1990)		2.8 ± 0.2		3.9 ± 0.2	Master curve obtained by x, y translation
			Physical gels[2]		
Allain and Salomé (1987a, b, 1990)	(a) 0.9 ± 0.2	(a) 1.9 ± 0.1			Semi-dilute solutions of polyacrylamide crosslinked by Cr(III) complexation
Axelos and Kolb (1990)	(a) 0.82 ± 0.05	(a) 1.93 ± 0.08	(b) 0.70 ± 0.01 (c) 0.71 ± 0.02	2.66 ± 0.12	Pectin biopolymers Master curve obtained using reduced variables: $G(\omega)/G(\omega^*) \pm f(\omega/\omega^*)$
Adam and Aimé (1991)	(a) 1.0 ± 0.1	(a) 1.9 ± 0.1			Semi-dilute solutions of polydiacetylene in the red phase

[1] Measurements lead to the exponent ratios s/γ and t/γ, s and t values are deduced from γ values reported in Table 3.
[2] Gelation threshold is determined by simultaneous linearisation $G^{1/t}$ and $\eta^{-1/s}$ imposing a common x-axis intercept corresponding to the threshold.

1988; Adolf and Martin, 1990). Thus, examination of the loss angle as a function of ω is indispensable to determine the frequency range corresponding to the power law behavior and inside which the exponent u value can be measured. This surprisingly wide crossover frequency range could explain the scattered values of the exponent u sometimes reported (Winter, Morganelli and Chambon, 1988; Muller *et al*, 1991).

Very few experimental determinations of the values of the exponents s and t were published and no direct determination of z is reported on chemical gels. As the longest relaxation time of the polymeric system diverges at the gel point, and as it is necessary to wait until the system has relaxed to perform measurements, no rheological measurements have been carried out in the near vicinity of the critical point. Because of this, determination of the exponent s and t is particularly difficult when measurements are performed in the reaction bath as the chemical reaction is proceeding. One would prefer measurements on chemically stable systems quenched at different distances from the threshold (e.g. non-stochiometric systems). In this case measurement time limitation is due only to the patience of the experimenter. In addition, quenched systems present the advantage of allowing independent static measurements and to determine exponent ratios s/γ and t/γ, getting rid of the ε inaccuracy (Lairez *et al*, 1992; Colby, Gillmor and Rubinstein, 1993). Measurements on samples prepared in this way clarify the role of power laws prefactors (high-frequency limit of the viscoelastic behavior) which may depend on the degree of connectivity (Lairez *et al*, 1992). In this case one measures an apparent exponent, and this outlines the necessity to use the reduced variable G^*/G_0 in order to determine real exponent values. The study of the very-high-frequency part of the viscoelastic spectra implies very specific experimental devices and quenched samples on which these measurements would be performed independently; unfortunately these measurements were never carried out. Prefactors of power laws reflect the local properties of the material; in most cases when gelation takes place in the melt they are functions of the degree of connectivity, p, rather than ε. This p dependence cannot be neglected when exponents are determined far from the threshold, as it is usually the case. Table 4 shows how scattered are the values of exponent s and t, while on the contrary the values of the exponent u reported by the different teams are very close together. This can be due to a wrong choice of reduced variables for the determination of s and t, while the exponent u may be determined from the high-frequency behavior of one sample at a given ε and consequently does not present this disadvantage.

Despite the scattered results on s and t values, one can see that all the teams verify the scaling relation $u = t/(s + t)$ within narrow error bars. This is not so surprising as it seems at first sight. In fact, the two limiting exponent values provided by the Rouse model and the electrical analogy (see section 4.2.2.3) leads to average values of $s = 1.0 \pm 0.3$ and $t = 2.3 \pm 0.3$, using these values one gets $u = 0.7 \pm 0.1$. In fact the exponent u so determined benefits from cancellation of errors. For the other scaling relation: $z = s + t$, one has to note that if the high-frequency limit of the viscoelastic behavior ω_0 decreases as the growth process proceeds (due to increase of the glass transition temperature for example), apparent exponents z and s will be both overvalued by the same amount, leaving the scaling relation unaffected. Finally, it is clear that the verification of these two relations between experimental exponent values cannot constitute a proof of the scaling behavior of rheological properties. This is not the case for the high-frequency modulus, which is expected to be ε independent. The scaling relation $u = t/(s + t)$ implies that the high-frequency power law behavior of the modulus of one sample at a given ε is recovered considering different samples at different ε at the frequency ω^*: $G(\omega^*) = \omega^{*u}$. In other words, using reduced variables amounts to shifting each rheological curve obtained at different ε by a vector having the same slope as the high-frequency part of each spectra. This implies that as the viscoelastic medium is probed at a sufficiently high frequency $\omega > \omega^*$, measurements are not sensitive to the growth process. The complex elastic modulus has to be ε independent in these conditions and this will constitute the major piece of evidence for the scaling relation between rheological exponents. If it is not the case, equality between experimental exponent values u and $t/(s + t)$ is fortuitous and measured exponents have to be considered as apparent exponents. Unfortunately, to our knowledge, this is only the case in the work reported by Durand *et al* (1988).

As yet, we have focused only on chemical gel dynamics. One can see that, considering the state of the art, the problem consists in deciding on the validity of the scaling approach rather than in determining exponent values. This was done on one physical gel in which gelation is a reversible phenomenon due to non-covalent crosslinks between polymers in semi-dilute solutions. The nature of their crosslinks prevents these systems being diluted in order to characterize their static properties. This is the reason why they have not been discussed until now. Nevertheless, scaling of rheological properties was fully verified on one system (Axelos and Kolb, 1990). Authors have verified the scaling of the reduced modulus $G(\omega)/G(\omega^*)$ versus the reduced frequency ω/ω^* for samples prepared at a different distance from the threshold, below and above the gel point (Figure 11). They report a loss angle independent of ε consistent with the exponent u over a frequency range of two decades (Figure 12) and, moreover, the elastic modulus $G(\omega^*)$ varies as ω^{*u} as expected (Figure 13). Within this framework, examination of the exponent values is meaningful. In Table 4, one can see that these exponent values are suprisingly narrowly distributed compared to those obtained on chemical gels. These values are in general agreement with three-dimensional percolation and electrical analogy predictions.

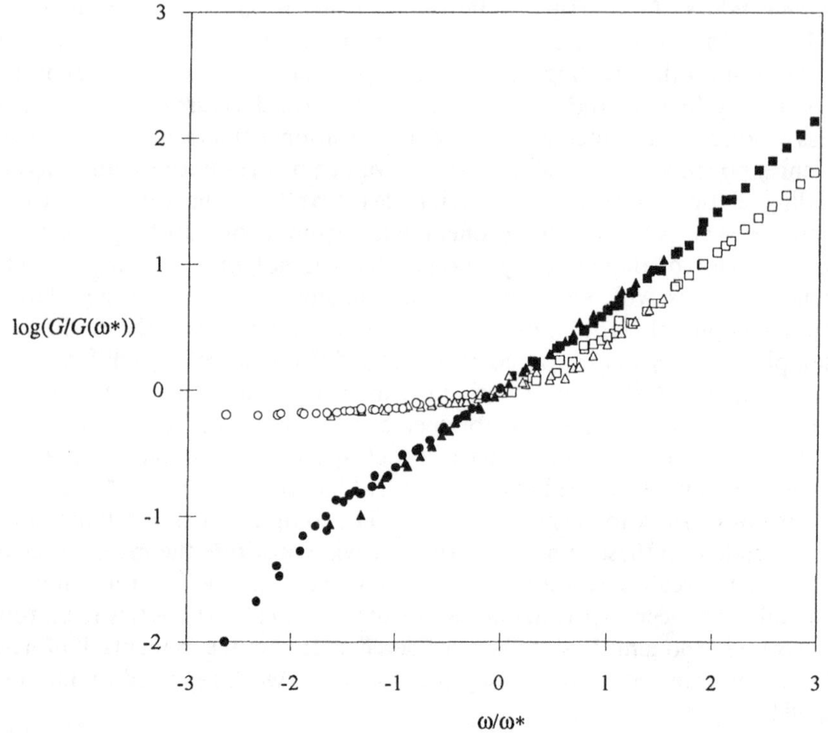

Figure 11. Scaling of the reduced modulus $G(\omega)/G(\omega^*)$ versus ω/ω^* obtained on pectin gels at different distances from the gelation threshold ($p > p_c$, $7 \times 10^{-2} < \varepsilon < 0.3$) by Axelos and Kolb (1990). The characteristic frequency ω^* is determined on each sample by G' and G'' intersection. Solid and empty symbols correspond to G'' and G' respectively. The straight line has a slope equal to $u = 0.71$. Once reduced, experimental data lie on a single curve, whatever the value of ε

Very few quasi-elastic light scattering results have been obtained so far. None of them relate to the long-time cut-off of the relaxation function in order to verify whether or not it is q independent and if it scales as $1/\omega^*$. The power law decay of the dynamical structure factor $S(q, t)$ was verified (Martin and Wilcoxon, 1988; Adam *et al*, 1988; Bauer and Burchard, 1992; Lang and Burchard, 1991). Apart from silica gels (Martin and Wilcoxon, 1988: $l - u = 0.13 \pm 0.01$), the mean

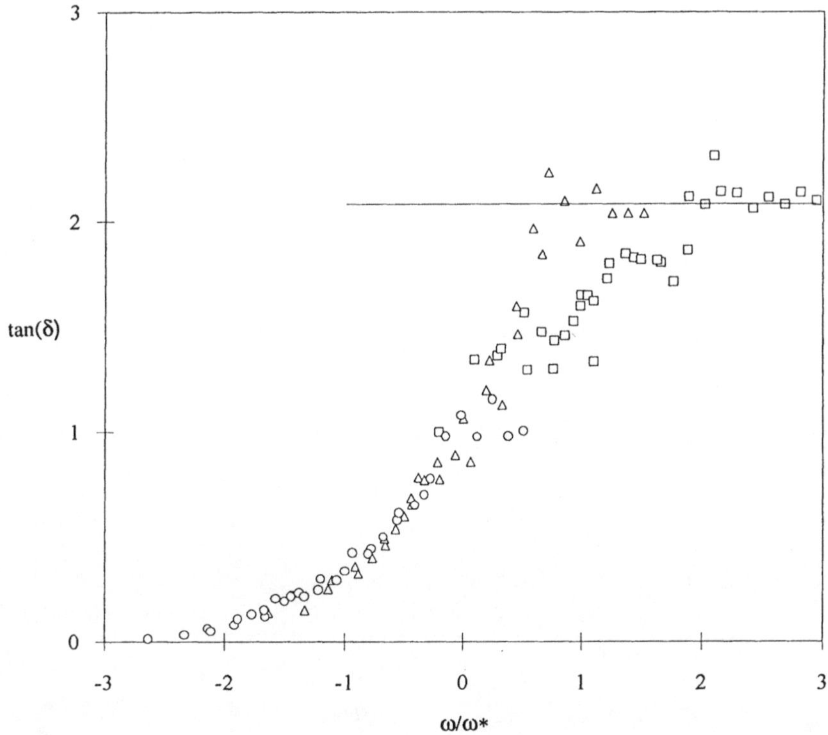

Figure 12. Ratio of the imaginary to the real part of the elastic modulus (tan(δ)) versus the reduced frequency ω/ω^* obtained on pectin gels at different distances from the gelation threshold ($p > p_c$, $7 \times 10^{-2} < \varepsilon < 0.3$) by Axelos and Kolb (1990). As expected, tan(δ) is ε independent and ω independent over two frequency decades. The horizontal line corresponds to the theoretical value $G''/G' = \tan(u\pi/2) = 2.1$ obtained with $u = 0.72$.

exponent value is found to be $l - u = 0.3 \pm 0.05$, in agreement with the value of the exponent u from rheological measurements.

One can see that a lot of work remains to be done in order to understand the dynamical properties of branched polymers near the gelation threshold. Scaling behavior in chemical gels has to be fully verified, then exponent values measured and compared to theoretical predictions. For this problem, quasi-elastic light scattering may be a powerful tool when one has firmly established what is theoretically expected to be measured.

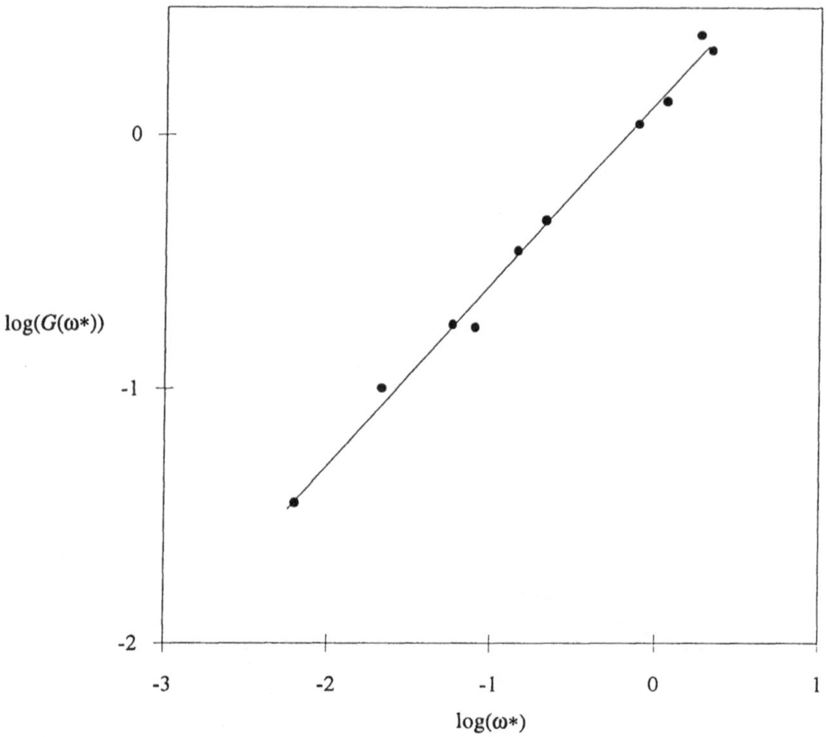

Figure 13. Variation above the gel point of the elastic modulus $G(\omega^*)$ measured at the characteristic frequency ω^* on the pectin system by Axelos and Kolb (1990). The straight-line has a slope $u = 0.71 \pm 0.02$ in good agreement with the value deduced from $\tan(\delta)$ reported in Figure 12. This clearly indicates that at high frequency ($\omega > \omega^*$) the modulus is independent of the distance from the threshold

4.4 Conclusion

Since the question asked by de Gennes and Stauffer in 1976 about the gelation universality class, experimental studies have led to a better understanding of this transition. At present, it is firmly established that static properties as well as growth of branched polymers near the gelation threshold come under the percolation model. This has very strong implications on the way to represent the conformation of the polymers and their relative arrangement. The hyperscaling law which links the fractal conformation of one polymer cluster to the self-similarity of the cluster mass distribution is of major importance. Until now, growing branched polymers which undergo a sol–gel transition constitute the unique physical system allowing us to handle percolation clusters. However, the problem of the crossover between mean field and percolation behavior is still current and provides a worthwhile objective for research. More precisely, the validity of the Ginzburg criterion has never been established either experimentally or by computer simulation.

On the other hand, the dynamics of these systems remain largely unsolved. The problem concerns much more the relevance of scaling concepts rather than discussion about exponent values. Progress on this problem suffers from the lack of experimental teams working on it. Understanding of the dynamical behavior of gelation should greatly benefit from a revival of interest.

References

Adam, M., Aimé, J. P. (1991) *J. Phys. II France* **1**, 1277
Adam, M., Delsanti, M. (1985) *Macromolecules* **18**, 1760
Adam, M., Delsanti, M. (1989) in *New Trends in Physics and Physical Chemistry of Polymers*, ed. Lieng-Huang Lee, Plenum Press, New York, p. 293
Adam, M., Delsanti, M., Durand, D. (1985) *Macromolecules* **18**, 2285
Adam, M., Delsanti, M., Durand, D., Hild, G., Munch, J. P. (1981) *Pure Appl. Chem.* **53**, 1489
Adam, M., Delsanti, M., Munch, J. P., Durand, D. (1987) *J. Physique* **48**, 1809
Adam, M., Delsanti, M., Munch, J. P., Durand, D. (1988) *Phys. Rev. Lett.* **61**, 706
Adam, M., Lairez, D., Boué, F., Busnel, J. P., Durand, D., Nicolaï, T. (1991) *Phys. Rev. Lett.* **67**, 3456
Adam, M., Lairez, D., Raspaud, E. (1992) *J. Phys. II France* **2**, 2067
Adolf, D., Martin, J. E. (1990) *Macromolecules* **23**, 3700
Adolf, D., Martin, J. E., Wilcoxon, J. P. (1990) *Macromolecules* **23**, 527
Alexander, S., Bernasconi, J., Orbach, R. (1978) *J. Physique Colloq.* **39**, C6–706
Alexander, S., Orbach, R. (1982) *J. Physique Lett.* **43**, L625
Allain, C., Limat, L., Salomé, L. (1991) *Phys. Rev. A* **43**, 5412
Allain, C., Salomé, L. (1987a) *Polymer Com.* **28**, 109
Allain, C., Salomé, L. (1987b) *Macromolecules* **20**, 2958
Allain, C., Salomé, L. (1990) *Macromolecules* **23**, 981
Axelos, M. A. V., Kolb, M. (1990) *Phys. Rev. Lett.* **64**, 1457
Bansil, R., Herrmann, H. J., Stauffer, D. (1984) *Macromolecules* **17**, 998
Bansil, R., Herrmann, H. J., Stauffer, D. (1985) *J. Polym. Sci.: Polym. Symp.* **73**, 175
Bauer, J., Burchard, W. (1992) *J. Phys. II France* **2**, 1053
Bauer, J., Lang, P., Burchard, W., Bauer, M. (1991) *Macromolecules* **24**, 2634
Bergman, D. J., Kantor, Y. (1984) *Phys. Rev. Lett.* **53**, 511
Bouchaud, E., Delsanti, M., Adam, M., Daoud, M., Durand, D. (1986) *J. Physique* **47**, 1273
Broadbent, S. R., Hammersley, J. M. (1957) *Proc. Camb. Phil. Soc.* **53**, 629
Brochard, F. (1983) *J. Physique* **44**, 39
Cabane, B., Dubois, M., Duplessix, R. (1987) *J. Physique* **48**, 2131
Candau, S. J., Ankrim, M., Munch, J. P. (1985) in *Physical Optics of Dynamic Phenomena and Processes in Macromolecular Systems*, ed. Sedlacek B., Walter de Gruyter, Berlin, p. 145
Candau, S. J., Bastide, J., Delsanti, M. (1982) *Adv. Polym. Sci.* **44**, 27
Chu, B. (1991) in *Laser Light Scattering*, 2nd edn, Academic Press, Boston
Chu, B., Wu, C. (1988) *Macromolecules* **21**, 1729
Chu, B., Wu, C., Wu, D., Phillips, J. C. (1987) *Macromolecules* **20**, 2642

Clerc, J. P., Giraud, G., Laugier, J. M., Luck, J. M. (1990) *Adv. Physics* **39**, 191
Clerc, J. P., Giraud, G., Roussenq, J., Blanc, R., Carton, J. P., Guyon, E., Ottavi, H., Stauffer, D. (1983) *Ann. Physique* **8**, 1
Colby, R. H., Gillmor, J. R., Rubinstein, M. (1993) *Phys. Rev. E.* **48**, 3712
Colby, R. H., Rubinstein, M., Gillmor, J. R., Mourey, T. H. (1992) *Macromolecules* **25**, 7180
Collette, C., Lafuma, F., Audebert, R., Leibler, L. (1988) in *Biological and Synthetic Networks*, ed. Kramer, O., Elsevier Appl. Sci., Amsterdam, p. 277
Coniglio, A. (1985) in *Physics of Finely Divided Matter*, eds Boccara, N. and Daoud, M., Springer, Berlin, p. 84
Coniglio, A., Stanley, H. E. (1984) *Phys. Rev. Lett.* **52**, 1068
Daoud, M. (1979) *J. Physique Lett.* **40**, L201
Daoud, M. (1988) *J. Phys. A: Math. Gen.* **21**, L237
Daoud, M. (1989) *Prog. Reaction Kinetics* **15**, 1
Daoud, M., Family, F., Jannink, G. (1984) *J. Physique Lett.* **45**, L199
Daoud, M., Joanny, J. F. (1981) *J. Physique Lett.* **42**, 1359
Daoud, M., Leibler, L. (1988) *Macromolecules* **21**, 1497
Dean, P., Bird, N. F. (1967) *Proc. Camb. Phil. Soc.* **63**, 477
Derrida, B., Stauffer, D., Herrmann, H. J., Vannimenus, J. (1983) *J. Physique Lett.* **44**, L701
Devreux, F., Boilot, J. P., Chaput, F., Malier, L., Axelos, M. A. V. (1993) *Phys. Rev. E* **47**, 2689
Dobson, G. R., Gordon, M. (1964) *J. Chem. Phys.* **41**, 2389
Dubois, M., Cabane, B. (1989) *Macromolecules* **22**, 2526
Durand, D., Adam, M., Delsanti, M., Munch, J. P. (1988) in *University in Condensed Matter*, vol. 32, eds Jullien, R., Peliti, L., Rammal, R. and Boccara, N., Springer, Berlin, p. 27
Durand, D., Bruneau, C. M. (1982) *Makromol. Chem.* **183**, 1007
Durand, D., Bruneau, C. M. (1983) *Polymer* **24**, 592
Durand, D., Delsanti, M., Adam, M., Luck, J. M. (1987) *Europhys. Lett.* **3**, 297
Durand, D., Naveau, F., Busnel, J. P., Delsanti, M., Adam, M. (1989) *Macromolecules* **22**, 2011
Dusek, K. (1979) *Makromol. Chem.* Suppl., **2**, 35
Dusek, K., Ilavsky, M., Lunak, S. (1975) *J. Polym. Sci.* **53**, 29
Efros, A. L., Shklovskii, B. (1976) *Phys. Status Solidi B* **76**, 47580
Eirich, F. R. (1960) in *Rheology: Theory and Applications*, vol. 3, Academic Press, New York
Elias, H.-G. (1984) in *Macromolecules 2: Synthesis, Materials and Technology*, 2nd Edn, Plenum Press, New York
Essam, J. W. (1980) *Rep. Prog. Phys.* **43**, 53
Fang, L., Brown, W., Konak, C. (1991) *Macromolecules* **24**, 6839
Flory, P. J. (1953) in *Principles of Polymer Chemistry*, Cornell University Press, Ithaca
Flory, P. J. (1961) *J. Am. Chem. Soc.* **63**, 3083
Gaunt, D. S., Guttmann, A. J., Whittington, S. G. (1979) *J. Phys. A: Math. Gen.* **12**. 75
Gefen, Y., Aharony, A., Alexander, S. (1983) *Phys. Rev. Lett.* **50**, 77
de Gennes, P. G. (1975) *J. Physique* **36**, 1049
de Gennes, P. G. (1976a) *J. Physique Lett.* **37**, L1
de Gennes, P. G. (1976b) *La Recherche* (*France*) **7**, 919
de Gennes, P. G. (1977) *J. Physique Lett.* **38**, L355
de Gennes, P. G. (1978) *C.R. Acad. Sci.* **286**, B, 131
de Gennes, P. G. (1980) *J. Physique Colloq.* **41**, C3–17
Gimel, J. C.; Durand, D.; Nicolai, T. (1995) *Phys. Rev. B.* **51**, 11348
Good, I. J. (1963) *Proc. Roy. Soc.* **A272**, 54

Gordon, M. (1962) *Proc. Roy. Soc.* **A268**, 240
Havlin, S., Ben-Avraham, D. (1987) *Adv. Phys.* **36**, 695
Hentschel, H. G. E. (1984) *Phys. Rev. Lett.* **52**, 212
Herrmann, H. J., Derrida, B., Vannimenus, J. (1984) *Phys. Rev.* **B30**, 4080
Herrmann, H. J., Landau, D. P. (1982) *Phys. Rev. Lett.* **49**, 412
Herrmann, H. J., Landau, D. P., Stauffer, D. (1986) *Phys. Rev. Lett.* **49**, 412
Hodgson, D. F., Amis, E. J. (1990) *Macromolecules* **23**, 2512
Hodgson, D. F., Amis, E. J. (1991) *J. Non-Cryst. Sol.* **131**, 913
Isaacson, J., Lubensky, T. C. (1980) *J. Physique Lett.* **41**, L469
Kajiwara, K., Burchard, W., Kowalski, M., Nerger, D., Dusek, K., Matejka, L., Tuzar, Z. (1984) *Makromol. Chem.* **185**, 2543
Kantor, Y., Webman, I. (1984) *Phys. Rev. Lett.* **52**, 1891
Kertész, J., Chakrabarti, B. K., Duarte, J. A. M. S. (1982) *J. Phys. A: Math. Gen.* **15**, L-13
Kirkpatrick, S. (1973) *Rev. Mod. Phys.* **45**, 574
Lairez, D., Adam, M., Raspaud, E., Emery, J. R., Durand, D. (1992) *Prog. Colloid Polym. Sci.* **42**, 37
Lairez, D., Durand, D., Emery, J. R. (1991) *J. Phys. II France* **1**, 977
Lang, P., Burchard, W. (1991) *Macromolecules* **24**, 814
Lapp, A., Leibler, L., Schosseler, F., Strazielle, C. (1989) *Macromolecules* **22**, 2871
Laugier, J. M., Luck, J. M. (1987) *J. Phys. A: Math. Gen.* **20**, L885
Lindner, P., Zemb, Th. (1991) in *Neutron, X-Ray and Light Scattering*, North-Holland, Amsterdam
Macosko, C. W., Miller, D. R. (1976) *Macromolecules* **9**, 199
Martin, J. E., Adolf, D., Wilcoxon, J. P. (1988) *Phys. Rev. Lett.* **61**, 2620
Martin, J. E., Adolf, D., Wilcoxon, J. P. (1988a) *Polym. Preprint (Washington)* **30**, 83
Martin, J. E., Adolf, D., Wilcoxon, J. P. (1989b) *Phys. Rev.* **A39**, 1325
Martin, J. E., Odinek, J. (1990) *Macromolecules* **23**, 3362
Martin, J. E., Wilcoxon, J. P. (1988) *Phys. Rev. Lett.* **61**, 373
Martin, J. E., Wilcoxon, J. P. (1989) *Phys. Rev.* **A39**, 252
Meier, G., Hagenah, J.-U., Wang, C. H., Fytas, G., Fischer, E. W. (1987) *Polymer* **28**, 1640
Muller, R., Gérard, E., Dugand, P., Rempp, P., Gnanou, Y. (1991) *Macromolecules* **24**, 1321
Nakanishi, H., Stanley, H. E. (1980) *Phys. Rev.* **B22**, 2466
Orbach, A. (1985) in *Scaling Phenomena in Disordered Systems*, eds Pynn, R. and Skjeltorp, A., Plenum Press, New York, p. 335
Pandey, R. B., Stauffer, D. (1983) *Phys. Lett.* **95A**, 511
Patton, E. V., Wesson, J. A., Rubinstein, M., Wilson, J. C., Oppenheimer, L. E. (1989) *Macromolecules* **22**, 1946
Pfeifer, P., Obert, M. (1989) in *The Fractal Approach to Heterogeneous Chemistry*, ed. Avnir, D., Wiley, Chichester, p. 11
Rammal, R., Toulouse, G. (1982) *Phys. Rev. Lett.* **49**, 1194
Rammal, R., Toulouse, G. (1983) *J. Physique Lett.* **44**, L13
Raspaud, E., Lairez, D., Adam, M. (1995) *Macromolecules* **28**, 927
Rolando, R. J., Macosko, C. W. (1987) *Macromolecules* **20**, 2707
Roux, S., Hansen, A. (1988) *Europhys. Lett.* **6**, 301
Rubinstein, M., Colby, R. H., Gillmor, J. R. (1989) *Space-Time Organization in Macromolecular Fluids*, eds Tanaka, F., Doi, M. and Ohta, T., Springer, Berlin, p. 66
Schaeffer, D. W., Keefer, K. D. (1984) *Mat. Res. Soc. Symp. Proc.* **32**, 1
Schmidt, M., Burchard, W. (1981) *Macromolecules* **14**, 370

Schosseler, F., Benoit, H., Grubisic-Gallot, Z., Strazielle, C., Leibler, L. (1989) *Macromolecules* **22**, 400
Schosseler, F., Daoud, M., Leibler, L. (1990) *J. Phys. France* **51**, 2373
Schosseler, F., Leibler, L. (1984) *J. Physique Lett.* **45**, L501
Schosseler, F., Leibler, L. (1985) *Macromolecules* **18**, 398
Stanley, H. E. (1971) in *Introduction to Phase Transitions and Critical Phenomena*, Clarendon Press, Oxford
Stanley, H. E., Bunde, A., Coniglio, A., Hong, D. C., Meakin, P., Witten, T. A. (1985) in *Scaling Phenomena in Disordered Systems*, eds Pynn, R. and Skjeltorp, A., Plenum Press, New York, p. 307
Stauffer, D. (1976) *J. Chem. Soc. Farad. Trans.* **II 72**, 1354
Stauffer, D. (1979) *Phys. Reps.* **54**, 1
Stauffer, D. (1985) in *Introduction to Percolation Theory*, Taylor & Francis, London
Stockmayer, W. H. (1943) *J. Chem. Phys.* **11**, 45
Stockmayer, W. H. (1944) *J. Chem. Phys.* **12**, 125
Trappe, V., Richtering, W., Burchard, W. (1992) *J. Phys. II France* **2**, 1453
Wang, C. H., Fischer, E. W. (1985) *J. Chem. Phys.* **82**, 632
Webman, I., Kantor, Y. (1984) in *Kinetics of Aggregation and Gelation*, eds Familly, F. and Landau, D., Elsevier Sci. Pub. BV, Amsterdam, p. 133
Whitney, R. S., Burchard, W. (1980) *Makromol. Chem.* **181**, 869
Winter, H. H., Morganelli, P., Chambon, F. (1988) *Macromolecules* **21**, 535
Wu, W., Bauer, B. J., Su, W. (1989) *Polymer* **30**, 1384
Yau, W. W., Kirkland, J. J., Bly, D. D. (1979) in *Modern Size Exclusion Chromatography*, Wiley, New York
Zimm, B. H., Stockmayer, W. H. (1949) *J. Chem. Phys.* **17**, 1301

5 STRUCTURE OF GELS AS INVESTIGATED BY MEANS OF STATIC SCATTERING TECHNIQUES

J. Bastide[1] and S. J. Candau[2]
[2]Institut Charles Sadron, Strasbourg, France
[2]Laboratoire d'Ultrasons et de Dynamique des Fluides Complexes, Université Louis Pasteur, Strasbourg, France

The Physical Properties of Polymeric Gels. Edited by J. P. Cohen Addad.

5.1 Introduction and scattering basics

For many years polymeric gels have been considered simply as swollen rubbers. The first confrontation of experimental data to the rubber elasticity theory showed the extreme complexity of these materials. However, it was commonly accepted that the structural properties of gels could be described in terms of the sole mesh size (i.e. the average strand connecting first neighbour crosslinks along the network). In the 1970s, several experimental facts did suggest the possibility of effects linked with a specific network structure at length scales larger than the mesh size (Silberberg *et al* 1957, 1988, 1992). Over the years, evidence has accumulated, mainly from scattering experiments, of the existence of long-range fluctuations (Stein, 1969; Bueche, 1970; Prins 1971; Pines and Prins 1972, Wun and Prins 1974, and, more recently, many others). It is the purpose of this chapter to review a significant selection of the available scattering data and to extract from these data some general guidelines for the observed features. More precisely, the question that will be underlying the whole chapter will be: are the effects due to a true structure of the gels or, else, to overlooked thermodynamic properties? Obviously, these effects are significant only in the case of gels far above the gelation threshold.

The topics covered by this chapter concern systems with significant shear modulus and small sol fraction. Also we consider gels susceptible to undergo rather large swelling ratio, which can be achieved by using rather good solvents of the polymer network or by using electrically charged polymers. Thus we discard all the critical behaviours or volume phase-transition effects (for a review see Shibayama and Tanaka, 1993 and Onuki, 1993). Such properties are of great interest from fundamental or applied view points, but they involve a different physics.

The tools of investigation which are described in this chapter are static scattering techniques: light, X-rays and neutrons. We deal with gels in which a 'contrast' has been established between all the polymer chains constituting the network and the diluent contained in it. Other choices of labelling can be made in the case of neutron scattering: labelled meshes or labelled paths (i.e. long labelled chains linked to the network by many junctions). In such experiments, one does not investigate primarily the structure of the network but rather the deformation mechanism. Examples of data can be found in Duplessix (1975), Benoît *et al* (1976), Lutz *et al* (1976) Belkebir-Mrani *et al* (1976), Hinckley *et al* (1978), Clough, Machonnachie and Allen (1980), Beltzung, Herz and Picot (1983, 1984), Bastide (1985), Bastide, Herz and Boué (1985), Bastide, Boué and Buzier (1989), Bastide *et al* (1984), Fernandez, Sperling and Wignalt (1986), Davidson and Richards (1986), Yu *et al* (1986), Candau, Bastide and Delsanti (1982).

Gels can be studied in various states of swelling, for example the state of preparation (when the network has been formed in the presence of solvent), but also in states corresponding to a swelling or, when it is possible, to a deswelling with respect to this preparation state. In the following, we call indifferently 'maximum swelling' or 'saturation swelling' the upper limit of the expansion (for a given solvent). Recall that such a state is usually characterized by Q_{max} the ratio of the total volume of the gel to the volume of the 'dry' polymer or by the polymer volume fraction $\phi_e(=1/Q_{max})$, which is then minimum. Naturally, the better the 'quality' of the diluent, the larger the maximum swelling degree Q_{max}. Most cases that we will review in fact correspond to good solvents of the chains.

We will often compare the structure factors of the gels with those of semi-dilute solutions of the same volume fraction of polymer. This method follows naturally from the classical thermodynamic description of gels (Frenkel, 1938, 1940, Flory and Rehner 1943), where the total free energy F of a gel is written as a sum (see appendix):

$$F = F_{mix} + F_{el} \tag{1}$$

Here F_{el} is usually calculated from the classical rubber elasticity theory (Kuhn, 1936; James and Guth, 1943, 1948, 1949, Wall, 1942, 1943, Wall and Flory 1951, Flory, 1953), describing the swelling of a gel with respect to its state of preparation as a sort of stretching of a set of *ideal chains* (i.e. Gaussian) in the three dimensions of space. On the other hand, F_{mix}, is frequently assimilated with the free energy F_{sol} of a solution, of the same concentration, of linear polymer with infinite molecular weight (through, originally, the Flory–Huggins theory; Flory, 1953).

The above expression of the total free energy as a sum will be called below the *strong additivity* assumption. On formal grounds, it has been criticized by many authors (Deam and Edwards, 1976). More recently, it has also been proposed to change it into a *weak additivity assumption*, where the sum is kept but F_{max} is allowed to be different from F_{sol} (Horkay, Hecht and Geissler, 1989; see section 5.3). We accept at the present level these two possibilities as working hypotheses. The way of expressing F_{el} was also controversial, first because the chains, due to excluded volume interaction, cannot be Gaussian (de Gennes, 1979), second because the assumption made for the deformation mechanism on molecular length scales have been questioned. (see appendix). This is why, when dealing with thermodynamics associated with scattering, we will often turn to a phenomenological form of Eq. (1), where microscopic quantities are self-consistently replaced by macroscopic and measurable ones (Bastide, Candau and Leibler, 1980).

In good solvent, a solution of a chain of infinite molecular weight is structurally as well as thermodynamically very close to a more realistic system, namely a solution of chains of very large molecular weight (say $M \geqslant 5 \times 10^5$), which in a large range of polymer volume fractions fall in the category known as a 'semi-dilute' solution (de Gennes 1979; Joanny, Ch. 1, this book). Therefore, considering a gel as a semi-dilute solution perturbed by the introduction of a permanent elasticity, it is natural to compare the structure of both systems. This is why, before approaching gels, we will briefly review theoretical and experimental results relative to the scattering by semi-dilute solutions. This is done in section 5.2, that follows a short recall on the practical aspects of static scattering (in section 5.1). Throughout this chapter, we will consider separately the case of the neutral systems, whose properties are controlled by short-range van der Waals' interactions, and that of the electrically charged gels, where long-range coulombic interactions operate. In section 5.3, we discuss the theoretical predictions for idealised gels. By 'ideal', we mean gels where the whole scattering would be due to thermal fluctuations of polymer concentration. Then, we give a survey of the experimental results relative to the structure factor of gels (section 5.4). In succession, we deal with gels at maximum swelling in good solvents, we discuss the effect of crosslinking density at a given concentration and we show how the scattering curves evolve under a change of swelling degree. In section 5.5, we report experimental results of anisotropic scattering from stretched gels. Section 5.6 is devoted to the discussion of recent models and more specifically of the issue whether the observed effects result from an amplification of thermal fluctuations under strain or from quenched heterogeneities.

5.1.1 PRACTICAL ASPECTS OF STATIC SCATTERING

Gels are neither true liquids nor true solids. They exhibit on short-range scales thermal fluctuations similar to those encountered in polymer solutions, conferring on these materials their rubber–elasticity properties. However, a major difference with respect to solutions is the restriction of spatial excursions of the monomer units, due to the presence of crosslinks. Because of this, one cannot simply apply the standard procedure of scattering analysis to the gels and special care must be taken in the method of averaging and in the definition of correlation functions (Pusey and van Megen, 1989). The two relevant averages are the time average and the ensemble average.

Consider an amorphous medium which scatters a coherent radiation (in practice laser light) sufficiently weak that the first Born approximation may be applied. The instantaneous intensity $I_p(q, t)$ scattered by a set of z point-like scatterers, contained in a certain scattering volume V_p, to a point in the far field, is given by the well-known formula (see e.g. Higgins and Benoît, 1994; Maconnachie and Richards, 1978):

$$I_p(\mathbf{q}, t) = \sum_{i=1}^{z} \sum_{j=1}^{z} b_i b_j \exp\{-i\mathbf{q}[\mathbf{r}_i(t) - \mathbf{r}_j(t)]\} \tag{2}$$

where b_i, b_j, $r_i(t)$ and $r_j(t)$ represent respectively the scattering lengths and the positions of the scatterers i and j. The scattering vector amplitude q (or momentum transfer) is defined as: $q = 4\pi/\lambda_\omega \sin(\theta/2)$, where λ_ω denotes the wavelength of the radiation used in the experiment and θ the scattering angle, that is, the angle between the direction of the incident beam and the direction of observation.

Under the definition used here, $I(q)$ is in fact a scattering cross-section. Experimentally, in order to be able to compare different systems which do not necessarily have the same geometry, one usually deals with the scattering cross-section per unit scattering volume, that, for simplicity, we will also call intensity and write I(q).

Here $I_p(q, t)$ corresponds to one of the possible arrangements—that found at time t—of the z scatterers' positions in V_p. Such instantaneous positions of the scatterers represent a subset over the ensemble of all the possible arrangements (or realizations), which can be found in the whole medium, for z scatterers distributed in a volume equal to that of V_p. In a solution the polymer molecules move in Brownian motion and $I_p(q, t)$ fluctuates in time. Given enough time, the system evolves through a sequence of spatial configurations representative of the full ensemble of possible configurations in the whole solution. In such a case, a time average measurement over the particular subset of volume V_p will be equivalent to an ensemble average measurement, i.e. one averaged over a representative number of all possible spatial realizations existing in the whole medium (see Pusey and van Megen, 1989).

More explicitly, this is written

$$\langle I_p(\mathbf{q})\rangle_T = \langle I(\mathbf{q}, t)\rangle_E \tag{3}$$

where the time average reads

$$\langle I_p(\mathbf{q})\rangle_T \equiv \lim_{T\to\infty} \frac{1}{T}\int_0^T \mathrm{d}t I_p(\mathbf{q}, t) \tag{4}$$

and the ensemble average reads

$$\langle I_p(\mathbf{q}, t)\rangle_E \equiv \lim_{P\to\infty} \frac{1}{P}\sum_{p=1}^{P} I_p(\mathbf{q}, t) \tag{5}$$

The situation is entirely different if the positions of the scatterers stay completely frozen. Then, $I_p(q,t)$ will obviously become independent of time (and can therefore be written $I_p(q)$). On the other hand, $I_p(q)$ will, in general, vary a lot from one subset of scatterers to another, that is, in practice, from one scattering volume to another. Then, in order to gain some information about the structure of the whole medium, it will be necessary to perform explicitly an ensemble average, by scanning over a large number of different scattering volumes.

In order to understand this better, consider a very large number N of static and identical point-like scatterers located at random in space in a very large box of volume V (see Pusey, 1976). The scattering length is the same for all points and equal to b. These points form in a sense a frozen perfect gas. Imagine that this set of N points is divided into a large number P of subsets (with index p), each of same volume $V_p = V/P$ and containing the same number z of points ($z = N/P$). In each subset, the arrangement of the z scatterers in space will be different and so will be the intensities $I_p(q)$ scattered by these subsets.

Since all the scatterers are identical, $I_p(q)$ can also be written

$$\begin{aligned} I_{\mathrm{p}}(\mathbf{q}) &= b^2 \sum_{i=1}^{z} \sum_{j=1}^{z} \exp\{-i\mathbf{q}[\mathbf{r}_i - \mathbf{r}_j]\} \\ &= |A(\mathbf{q})|^2 = b^2 \left| \sum_{i=1}^{z} \exp(-i\mathbf{q}\mathbf{r}_i) \right|^2 \end{aligned} \tag{6}$$

where $A(\mathbf{q})$ is the amplitude of the scattered field.

The amplitude $A(q)$, a complex number, can be represented as a vector in the complex plane. Since, the positions of the scatterers are drawn at random, each value of $A(q)$ is equivalent to one particular configuration of a random walk in the complex plane; $A(q)/b$ is indeed the sum of the unit vectors with random orientations: their angles with respect to the real axis are given by the random phases $\mathbf{q}\cdot\mathbf{r}_i$. In Figure 1, we show two realizations of such a random walk.

Therefore, the probability distribution $p(A(q))$ of $A(q)$ over the ensemble of possible realizations is simply a bivariate Gaussian:

$$p(A(\mathbf{q})) = \left(\frac{1}{2\pi\langle A_r^2\rangle_E}\right)^{1/2} \exp\left\{-\frac{1}{2}\frac{A_r^2}{\langle A_r^2\rangle_E}\right\} \left(\frac{1}{2\pi\langle A_i^2\rangle_E}\right)^{1/2} \exp\left\{-\frac{1}{2}\frac{A_i^2}{\langle A_i^2\rangle}\right\} \tag{7}$$

with $A(q) = A_r(q) + iA_i(q)$ and where the brackets $\langle\,\rangle_E$, as before, denote an ensemble average over all the possible realizations.

In other words, $A(q)$ is a zero mean complex Gaussian variable when sampled over the full ensemble. Since $I(q) = A_r^2(q) + A_i^2(q)$, this Gaussian function of the field can also be written as an exponential function of the intensity. By a straightforward integration of Eq. (7), one is then led to the probability density $p(I(q))$ of detection of any specific intensity (Goodman, 1964; Pusey, 1976):

$$p(I(\mathbf{q})) = \frac{1}{\langle I(\mathbf{q})\rangle_E} \exp\{-I(\mathbf{q})/\langle I(\mathbf{q})\rangle_E\} \tag{8}$$

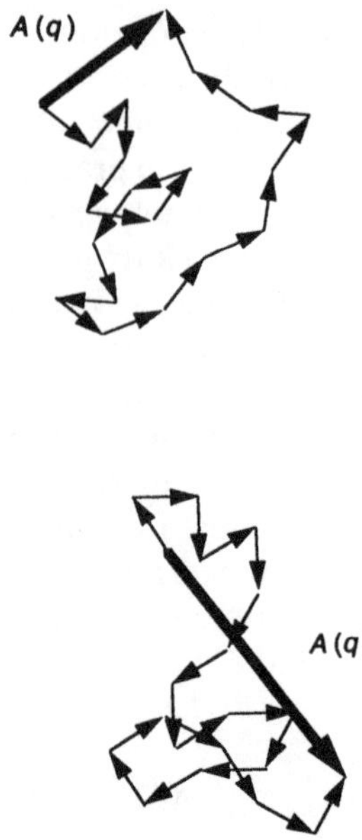

Figure 1. Schematic representations of the scattering amplitude in the complex field $A(\mathbf{q})$ for 2 configurations of a set of 18 scatterers

Thus, as stated above, if the sample is moved so that different volumes are illuminated, the detector will register intensities exhibiting very large variations, the lowest intensities being the most probable. The relevant statistical parameter is the ensemble-averaged intensity given by Eq. (5), which reduces to the well-known value characteristic of a perfect gas, which is independent of q:

$$\langle I(q)\rangle_E = zb^2 \tag{9}$$

or, when going to normalized scattered intensities,

$$\langle \mathrm{I}(q)\rangle = cb^2 \tag{10}$$

where $c = z/V_p = N/V$ is the number density of scatterers.

Recall that, up to now, we have considered only the case of a coherent radiation (in practice laser light) and a point-like detector. In practice, the scattered light is detected over a finite area. As long as this area is smaller than a coherence area (Pusey, 1976), Eqs (7) and (8) remain applicable. In the case when n coherence areas are imaged on the surface of the photomultiplier, the intensity in each coherence area is statistically independent of the intensities in the others. The total detector output is the sum of outputs produced by light impinging upon all coherence areas (Goodman, 1964). The measured intensity obeys the following first-order statistics:

$$p_n(I(\mathbf{q})) = \frac{1}{\Gamma(n)} \cdot \left[\frac{n}{\langle I(\mathbf{q}) \rangle_E} \right]^n \cdot I^{n-1} \cdot \exp\{ -nI(\mathbf{q})/\langle I(\mathbf{q}) \rangle_E \} \tag{11}$$

where Γ is the gamma function, and n is expressed by the square of signal-to-noise ratio $n = [\langle I(q) \rangle_E / \sigma]^2$, σ being the standard deviation $[\langle I(q)^2 \rangle_E - \langle I(q) \rangle_E^2]^{1/2}$; n can also be estimated through the formula: $n \approx s\Omega/\lambda_\omega^2$, where s is the illuminated area of the photocathode (determined by the pinhole diameter) and Ω is the solid angle subtended on the photocathode by a sample image at the slit (Gillies *et al*, 1987 and refs therein).

If n is very large, σ vanishes and one measures directly the ensemble-averaged scattered intensity. Another way to perform an ensemble average is the use of a non-coherent radiation, like neutrons.

Let us turn back to the case of gels. These systems are more complex since the scatterers are not entirely frozen (Pusey and van Megen, 1989). Therefore, one expects the intensity to fluctuate in time for a given scattering volume. On the other hand, depending on the number of coherence areas seen by the detector, the time averaged scattered intensity may also vary when scanning, keeping q constant, through various positions in the sample (see e.g. Sellen, 1987; Joosten, McCarthy and Pusey, 1991). Thus, a measurement over a single scattering volume may not be representative of the structure of the whole sample. If one simply rotates the detector in order to investigate different q values, keeping the sample in place, there is some risk of obtaining meaningless results: one may observe rapid intensity fluctuations versus observation angle (see Gillies *et al*, 1987). This will be discussed in more detail in the experimental sections.

In short, when dealing with the structure of gels, the main significant observable will be the *ensemble averaged* intensity $\langle I(q) \rangle_E$. The quantity which carries the information about the structure of the sample is in fact the *structure factor* of the network $S(\mathbf{q})$:

$$S(\mathbf{q}) = \left\langle \sum_{i=1}^{N_p} \sum_{j=1}^{N_p} \exp\{ -i\mathbf{q}[\mathbf{r}_i - \mathbf{r}_j] \} \right\rangle_E \tag{12}$$

There are two major differences between $S(q)$ and $\langle I(q) \rangle_E$. Firstly, the double summation extends in Eq. (12) only over the N_p monomer units (comprising those of the crosslinks) constituting the network (and not over all the scatterers, a large part of them being the solvent molecules). Secondly, $S(q)$ has been made a dimensionless quantity by removing the scattering lengths. Here, we adopted a definition where $S(q)$ is not normalized by the number of scatterers (Higgins and Benoît, 1994), but other definitions are found in the literature.

As will be recalled now, $S(q)$ is proportional to $\langle I(q) \rangle_E$, provided the system, as a whole, can be considered as incompressible binary medium. This is the case, in first approximation, for most of the studied gels, which are generally composed essentially of a homopolymer network swollen with solvent. Strictly, one should take into account a third component, namely the crosslinks. But in most cases their size in so small, their volume fraction so weak ($\approx 1\%$ of the polymer) and their structure so close to that of the polymer chains, that one can disregard them.

In order to go from $\langle I(q) \rangle_E$ to $S(q)$, the first step consists of taking advantage of the binary nature of the system (homopolymer network and solvent). For $q(v_a^{1/3}) \ll 1$ and $q(v_b^{1/3}) \ll 1$, the monomer units and the solvent molecules can be considered for the scattering as point-like scatterers, v_a and v_b denoting respectively their volumes. The scattering lengths of the monomer units or of the solvent molecules, respectively equal to b_a and b_b, are simply expressed as the sums of the scattering lengths of their constitutive atoms. Then, the gel reduces to a system composed only of two types of scatterers a and b with

$$b_a = \sum_{k \in \text{monomer unit}} b_k \quad \text{and} \quad b_b = \sum_{l \in \text{solvent mol.}} b_l \tag{13}$$

Then the scattered intensity $I(q)$ can be expressed as a sum of partial structure factors

$$\langle I(\mathbf{q}) \rangle_E = b_a^2 S_{aa}(\mathbf{q}) + b_b^2 S_{bb}(\mathbf{q}) + b_a b_b S_{ab}(\mathbf{q}) \tag{14}$$

where $S_{aa}(q)$, the polymer structure factor, is equal to $S(q)$ defined in Eq. (12).

Here $S_{bb}(q)$, the structure factor of the solvent in the presence of the polymer network, is defined in Eq. (12), the double sum extending over the positions of the solvent molecules. $S_{ab}(q)$ is a crossed term, again defined as $S(q)$, but with one summation over the monomer units and the other over the solvent molecules.

Focus for the moment on $S(q)$. Exploiting a fundamental property of the Dirac distribution, Eq. (12) can be rewritten as follows:

$$S(\mathbf{q}) = \left\langle \sum_i^{N_p} \int_V \delta(\mathbf{r} - \mathbf{r}_i) e^{-i\mathbf{qr}} \cdot \sum_j^{N_p} \int_V \delta(\mathbf{r}' - \mathbf{r}_j) e^{-i\mathbf{qr}'} \right\rangle_E \tag{15}$$

One can exchange the sums and the integrals and state that

$$\sum_i^N \delta(\mathbf{r} - \mathbf{r}_i) = n_a(\mathbf{r}) \tag{16}$$

where n_a denotes the number density of monomer units.

Then, Eq. (15) can be rewritten as follows:

$$S(\mathbf{q}) = \int_V \int_V \langle n_a(\mathbf{r}) \cdot n_a(\mathbf{r}') \rangle_E \exp[-i\mathbf{q}(\mathbf{r} - \mathbf{r}')] \mathrm{d}^3\mathbf{r}\, \mathrm{d}^3\mathbf{r}' \tag{17}$$

where $V = V'$ denotes the scattering volume.

Now, one can make use of the overall incompressibility. Consider a small volume τ centred on r, which is however, large enough to allow the use of number densities in it. Then, one obtains

$$\tau = \nu_a v_a + \nu_b v_b \tag{18}$$

where ν_a and ν_b represent respectively the number of monomer units and solvent molecules in the volume τ. Thus, dividing by τ, one obtains

$$n_a(\mathbf{r})v_a + n_b(\mathbf{r})v_b = 1 \tag{19}$$

This is the incompressibility equation which allows one to relate $S(q)$ to $S_{bb}(q)$ and to $S_{ab}(q)$. For this, one multiplies Eq. (19) by $v_a n_a(\mathbf{r}')\exp[-i\mathbf{q}(\mathbf{r}-\mathbf{r}')]$, integrates over $\mathbf{r}$ and $\mathbf{r}'$ and takes an ensemble average. The right-hand-side term is then proportional to $\delta(-q)$, that is, equal to zero for all experimentally accessible values of the scattering vector $\mathbf{q}$. Thus, one gets

$$\begin{aligned} &v_a^2 \int_V \int_V \langle n_a(\mathbf{r})\cdot n_a(\mathbf{r}')\rangle_E \exp[-i\mathbf{q}(\mathbf{r}-\mathbf{r}')]\mathrm{d}^3\mathbf{r}\mathrm{d}^3\mathbf{r}' \\ &\quad + v_a v_b \int_V \int_V \langle n_b(\mathbf{r})\cdot n_a(\mathbf{r}')\rangle_E \exp[-i\mathbf{q}(\mathbf{r}-\mathbf{r}')]\mathrm{d}^3\mathbf{r}\mathrm{d}^3\mathbf{r}' = 0 \end{aligned} \tag{20}$$

The two integrals are respectively equal to $S_{aa}(q)$—i.e. $S(q)$ and $S_{ab}(q)$, thus Eq. (20) also reads

$$v_a^2 S_{aa}(\mathbf{q}) + v_a v_b S_{ab}(\mathbf{q}) = 0 \tag{21}$$

Multiplying Eq. (19) by $v_b n_b(r')\exp[-i\mathbf{q}(\mathbf{r}-\mathbf{r}')]$, one obtains the complementary relation:

$$v_a v_b S_{ab}(\mathbf{q}) + v_b^2 S_{bb}(\mathbf{q}) = 0 \tag{22}$$

Equation (14) can then be combined with Eqs (21) and (22), which, after some rearrangement of the terms, leads to

$$I(\mathbf{q}) = b_v^2 v_a^2 S(\mathbf{q}) \tag{23}$$

where $b_v = [(b_a/v_a) - (b_b/v_b)]$ is the so-called contrast factor.

Note that, from Eqs (21) and (22), one also obtains $v_a^2 S_{aa}(q) = v_b^2 S_{bb}(q)$. In other words, the overall incompressibility of the medium means that the structure factor of the solvent in the presence of the network is proportional to the structure factor of the polymer network in the presence of solvent. Accordingly, Eq. (23) could also have been written

$$I(\mathbf{q}) = b_v^2 v_b^2 S_{bb}(\mathbf{q}) \tag{24}$$

In practice, this means that, in order to create the necessary contrast, it amounts to increasing or decreasing the scattering length of the polymer with respect to that of the solvent.

There are different ways to exploit these results. One of them consists of regrouping Eqs (17) and (23) and remarking that $v_a n_a(\mathbf{r})$ is nothing but the local value $\phi(\mathbf{r})$ of the polymer volume fraction at point $\mathbf{r}$. Then one obtains:

$$\begin{aligned} I(\mathbf{q}) &= b_v^2 \int_V \int_V \langle \phi(\mathbf{r})\cdot\phi(\mathbf{r}')\rangle_E \exp[-i\mathbf{q}(\mathbf{r}-\mathbf{r}')]\mathrm{d}^3\mathbf{r}\mathrm{d}^3\mathbf{r}' \\ &= b_v^2 \langle \phi(\mathbf{q})\cdot\phi(-\mathbf{q})\rangle_E \end{aligned} \tag{25}$$

where

$$\phi(\mathbf{q}) = \int_V \phi(\mathbf{r}) e^{-i\mathbf{qr}} d^3\mathbf{r}$$

This equation shows that a scattering experiment 'picks' up the modes $\phi(\mathbf{q})$ of wave vector $\mathbf{q}$ in the distribution of concentration fluctuations of one species with respect to the other. Moreover, $\phi(\mathbf{r})$ can be expressed as

$$\phi(\mathbf{r}) = \phi + \delta\phi(\mathbf{r}) \tag{26}$$

where ϕ is the average polymer volume fraction in the medium and $\delta\phi(r)$ is the local deviation (at point $\mathbf{r}$) with respect to this average value. Thus

$$I(\mathbf{q}) = b_\mathrm{v}^2 \langle \delta\phi(\mathbf{q}) \cdot \delta\phi(-\mathbf{q}) \rangle_E + \text{const.}\, \delta(\mathbf{q}) \tag{27}$$

The second term of the right-hand side of Eq. (27), being proportional to $\delta(\mathbf{q})$, is not accessible to the experiment (and is partly an artefact resulting from the change of discrete notation into continuous ones). Therefore, the measurable part of the scattered intensity reduces to

$$I(\mathbf{q}) = b_\mathrm{v}^2 \langle \delta\phi(\mathbf{q}) \delta\phi(-\mathbf{q}) \rangle \tag{28}$$

Let us go back to Eq. (25). Many systems, and gels in particular, exhibit some translational symmetry, which means that the correlation function $\langle \phi(r) \cdot \phi(r') \rangle$ depends only on $\mathbf{r}-\mathbf{r}'$. Then, after changing variable, integration over the variable which is left without integrand (which gives simply the scattering volume) and finally renaming as $\mathbf{r}$ the remaining dummy variable, one obtains:

$$I(\mathbf{q}) = b_\mathrm{v}^2 V \int_V \langle \phi(0) \cdot \phi(\mathbf{r}) \rangle_E e^{-i\mathbf{qr}} d^3\mathbf{r} \tag{29}$$

As stated above, in most of the experiments, one measures a normalized scattering intensity $\mathrm{I}(\mathbf{q})$ (i.e. the scattering cross-section $I(\mathbf{q})$ divided by the scattering volume V). Therefore

$$\mathrm{I}(\mathbf{q}) = b_\mathrm{v}^2 \int_V \langle \phi(0) \cdot \phi(\mathbf{r}) \rangle_E e^{-i\mathbf{qr}} d^3\mathbf{r} \tag{30}$$

or

$$\mathrm{I}(\mathbf{q}) \cong b_\mathrm{v}^2 \int_V \langle \delta\phi(0) \cdot \delta\phi(\mathbf{r}) \rangle_E e^{-i\mathbf{qr}} d^3\mathbf{r}$$

Note that the normalized intensity $I(q)$ is an area per unit volume, frequently expressed in cm^{-1}. As above, b_v is a length per volume (usually expressed in $\mathrm{cm\,cm}^{-3}$). Since $\phi(\mathbf{r})$ and $\delta\phi(\mathbf{r})$ are dimensionless quantities, the above integral has the dimension of a volume. Equation (30) is a very general formula, which may be applied, in principle, to any type of radiation. In practice, it is mostly used in this form in the field of neutron scattering.

In a light scattering experiment, the wavelength of the source λ_s is typically about 5000 Å but the wavelength $\lambda_\omega = \lambda_s/n$ (n-index of refraction) in the medium is lower (about 3500 Å). The scattering angles accessible to the experiments are in most cases ranging in between 25° and 150°, which leads to scattering vector amplitudes q typically ranging in between 0.8×10^{-3} and 3.5×10^{-3} Å^{-1}.

The radiation field is the electric field which interacts with the electronic cloud of the atoms. Thus, the scattering length b is related to the polarizability α_p of the atoms. It can be expressed in a simple way by the following equation (Cotton 1990):

$$b(\theta, \lambda, \alpha_p) = f(\theta)\alpha_p k_0^2 \tag{31}$$

where $k_0(=2\pi/\lambda_\omega)$ is the amplitude of the incident wave vector. It is the polarization of the incident and that of the scattered beam which introduce a dependence $f(\theta)$ with the scattering angle ($f(0)=1$). In the case of isotropic scatterers of concentration N/V, the polarizability α_p can be related to the refractive index n through the following equation:

$$\alpha_p = (3/4\pi)\cdot(V/N)\cdot(n^2-1)/(n^2+2) \tag{32}$$

In practice, for estimating the contrast rather than calculating differences between scattering lengths densities, one uses a 'contrast factor' $n_0^2(\partial n/\partial c)^2$, which can be shown to be proportional to b_v^2 (n_0 is the index of refraction of the solvent). Absolute values of the intensity (in cm^{-1}) are usually obtained by calibrating against toluene (whose absolute intensity in cm^{-1} is known). In this regard, it must be stressed that the assumption of overall incompressibility is never strictly valid: then a pure solvent should not scatter at all! One has actually to take into account the finite compressibility χ_T of the medium as a whole, but the density fluctuations coupled to χ_T are statistically independent from the fluctuations in composition we are interested in. Therefore they contribute to the scattered intensity through a term proportional to $k_B T\chi_T$, which has been neglected in the above derivations, and which should simply be added to $I(\mathbf{q})$ as expressed above (see des Cloizeaux and Jannink, 1987). Often, in light scattering, this contribution is not negligible, but in many cases one can get rid of it within a good approximation by subtracting the signal of the pure solvent from that of the gel.

In X-ray experiments, the wavelength λ_ω is typically ranging between 1 and 5 Å (and more frequently between 1 and 2 Å). In a current small-angle set-up, the typical q-range is 10^{-2} Å$^{-1} < 9 < 6\ 10^{-1}$ Å^{-1}. But much smaller values of q can be reached with Bonse–Hart type cameras: q can frequently be made as small as 10^{-3} Å^{-1} (Bonse and Hart, 1966; Brumberger, 1966). The secret is a high angular resolution associated with high resolution in wavelength. They are obtained by letting the incident beam pass through a double crystal ('channel cut') geometry and by analysing the scattered beam with a second double crystal.

X-rays, being an electromagnetic radiation, their scattering is achieved by the z electrons of the electronic shell. Thus the scattering length b of an atom is given by (Feigin and Svergun, 1987; Bacon, 1955; see also Cotton, 1991):

$$b = b_0 z \quad \text{with} \quad b_0 = 0.282 \times 10^{-12}\,\text{cm} \tag{33}$$

Neutrons can also be considered as waves. Their wavelength λ_ω is given by the well-known de Broglie's formula

$$\lambda_\omega = h/mv \tag{34}$$

where h represents the Planck constant and m and v denote respectively the mass and the velocity of the neutrons.

Neutrons which are extracted from the core of a nuclear reactor are fast and their average wavelength is much smaller than these values (about 1 Å). Thus, they need to be slowed down before use. This is achieved by letting them pass through a *cold source*, where they loose the most part of their kinetic energy through many interactions with a liquid at a very low temperature (e.g. liquid deuterium at 25 K). After this, the average value of λ_ω is larger, but the distribution of wavelengths is still very broad. Several types of devices can be used to reduce the width of this distribution and one has always to achieve a compromise between flux and precision in the definition of λ_ω. Currently, a mechanical selector is employed. It consists of a cylinder (or a series of discs) rotating along an axis parallel to the neutron beam. Helicoidal paths are drawn in the selector, which can be crossed only by neutrons with λ_ω (i.e. speed) in a certain interval, for a given rotation speed. For most of the experiments, a selector leading to $\Delta\lambda_\omega/\lambda_\omega = 10\%$ is chosen. Typical values of λ_ω required for experiments on gels, and on polymers in general, are 6–10 Å. Then, q can be varied typically between 0.003 and 0.3 Å^{-1}, with typical sample-detector distances of several metres. Good angular and wavelength resolutions are rarely achieved simultaneously in small angle neutron scattering because they are very expensive in terms of flux.

An almost parallel beam of neutrons (1–3 cm^2 section) is formed using a guide tube (on the walls of which neutrons are reflected) and geometrical collimation (series of irises). This beam can be considered as a plane wave 'illuminating' the sample. This wave is scattered by all the nuclei of the atoms located in the sample. The amplitude (called the scattering length amplitude) of the wave scattered by one given nucleus depends on its chemical nature, but also on which isotope it is (e.g. ^{54}Fe, ^{56}Fe, ^{57}Fe or ^{1}H, ^{2}H) and on its spin state. Thus, for a given chemical species, the scattering length amplitude usually fluctuates randomly in space around a mean value, because of statistical variations of spin state and of isotopic composition. The mean value leads to the *coherent* scattering, depending on the angle θ of observation and bringing some information on the structural properties of the sample. This is the one on which we focus our attention. The fluctuations of the scattering length amplitude which are not correlated in space give rise to an *incoherent* scattering, independent of q. In elastic small angle scattering experiments, it corresponds to a flat background which must be subtracted from the data before interpretation. The procedure for subtracting this incoherent part of the scattering is explained in Cotton (1991).

Without going into detail, it can be said that, in most cases, one can fabricate a background sample having the same incoherent scattering than the system which is studied, but without significant coherent scattering. Since the incoherent contribution arises essentially from the presence of normal hydrogen, one manages to give this background sample the same proton content as the true sample, but made with small molecules with diameter d such $q_{max}d \ll 1$ (q_{max} being the maximum scattering vector amplitude reached in the experiment). In many cases, the term corresponding to the finite compressibility of the system is approximately the same in the background sample and in the true sample. Therefore the contribution of the finite compressibility is automatically subtracted together with the incoherent part of the signal (in neutron experiments). Moreover, in the case of neutrons, its contribution is usually very small. It is not the case in light scattering, where the contrasts may be weaker.

5.2 Semi-dilute solutions

5.2.1 SEMI-DILUTE SOLUTIONS OF NEUTRAL POLYMERS IN GOOD SOLVENT

5.2.1.1 Theoretical reminder

5.2.1.1.1 'Structure'

A semi-dilute solution (see Chapter 1) is usually characterized by one single length scale, namely the correlation length ξ, defined as the characteristic size of the fluctuations of polymer concentration. More explicitly, the correlation function of the concentration (by volume) is written

$$\langle \phi(0)\phi(r) \rangle \sim 1/r \exp(-r/\xi) \quad \text{for } r \geqslant \xi \tag{35}$$

From this definition, the fluctuations of polymer concentration can be considered as essentially uncorrelated on length scales larger than ξ. In good solvent, ξ is also assumed to represent approximately the length scale beyond which the chains are interpenetrated. In other words, it means that the correlation length is roughly equal to the average distance between interchain contact points. This is no longer valid in poor solvent (Brochard and de Gennes, 1977). Finally ξ is also identified with the typical length scale above which the system can be considered as uniform.

By contrast, for $r \ll \xi$, a part of a chain behaves essentially as if it were not interacting with the other chains, i.e. its average conformation is ruled by the excluded volume statistics alone:

$$r_{ij} \sim n_{ij}^{\nu} \tag{36}$$

where n_{ij} and r_{ij} represent respectively the number of monomer units between points i and j of the chain and the average spatial distance between these points and ν denotes the so-called excluded volume (or Kuhn's) exponent ($\nu \approx 3/5$; 3/5 according to Flory, 1953 or 0.588 according to more recent renormalization group calculations; see de Gennes, 1979).

In this spatial range the correlation function of the concentration fluctuations (sometimes called the monomer density correlation function) should therefore, for $r \ll \xi$, equate that of a single chain, i.e.

$$\langle \phi(0)\phi(r) \rangle \sim r^{-4/3} \tag{37}$$

5.2.1.1.2 *Scattering*

When a 'contrast' is established between all the chains and the solvent, the neutron scattering intensity I(q) is, for $q\xi \ll 1$, proportional to the Fourier transform of Eq. (35). Thus, I(q) is expected to be given by an Ornstein–Zernicke type function

$$\mathrm{I}(q) = \frac{\mathrm{I}(0)}{(1 + q^2\xi^2)} \quad \text{for} \quad q\xi < 1 \tag{38}$$

Considering the semi-dilute solution as a true liquid (i.e. neglecting viscoelastic properties), I(0) must be proportional to $k_B T$ times the isothermal compressibility $\chi_{os} = -1/V(\partial V/\partial \Pi)_T$ (see e.g. Higgins and Benoît, 1994). In practice, this relationship is often expressed using a slightly different notation:

$$\mathrm{I}(0) = b_v^2 k_B T \phi^2 / K_{os} \tag{39}$$

where K_{os} denotes the osmotic bulk modulus, equal to $\phi\ (\partial \Pi/\partial \phi)_T$; b_v is the 'contrast factor' whose expression depends on the technique (light, X-rays or neutron), see section 5.1.

In good solvent, the osmotic pressure is well described by the des Cloizeaux (1975) law (see also de Gennes, 1979; des Cloizeaux and Jannink, 1987):

$$\Pi \sim \xi^{-3} \sim \phi^s \tag{40}$$

with $s = 3\nu/(3\nu - 1) = 9/4 = 2.25$ for $\nu = 3/5$, or 2.31 for $\nu = 0.588$. Thus K_{os} should scale with ϕ with the same exponent s as Π.

The value of exponent s is related to the dependence of the correlation length with concentration (see e.g. de Gennes, 1979):

$$\xi \sim \phi^{-\frac{\nu}{3\nu - 1}} \tag{41}$$

with $\phi^{-3/4}$ for $\nu = 3/5$ or $\phi^{-0.77}$ for $\nu = 0.588$.

Equation (41) is easily obtained from the description of the semi-dilute solution as a close packing of 'blobs' of size ξ (ϕ is identified with the overlap concentration ϕ^* of the blobs-'hyper scaling' assumption).

Combining Eqs (38)–(40), one obtains

$$I(q) \sim b_v^2 \cdot \frac{k_B T \phi^2 \xi^3}{(1 + q^2 \xi^2)} \quad \text{for } q\xi < 1 \tag{42}$$

Because of Eq. (41), it is comparable to write

$$I(q) \sim b_v^2 \frac{k_B T \phi \xi^{3-((3\nu-1)/\nu)}}{(1 + q^2 \xi^2)} \sim b_v^2 \frac{k_B T \phi \xi^{d_f}}{(1 + q^2 \xi^2)} \tag{43}$$

where $d_f = 1/\nu$ is the fractal dimension of chains with excluded volume at infinite dilution. This well-known relationship between d_f and ν follows from the inversion of the operational definition of the fractal dimension (Mandelbrot, 1977):

$$n \sim r^{d_f} \tag{44}$$

where n represents the number of elementary units of the fractal object (here a polymer chain) and r the radius of a sphere of radius r centred on the object.

Contrary to Eq. (42), Eq. (43) is directly consistent with the Guinier law (see Guinier and Fournet, 1957) of scattering by uncorrelated objects:

$$I(q) \sim \phi M(1 - q^2 R_g^2/3) \quad \text{for} \quad qR_g \ll 1 \tag{45}$$

where R_g is the radius of gyration of the uncorrelated objects and M their molecular weight.

In the present case, the uncorrelated objects are the blobs of 'size' ξ. For consistency to Eq. (45), it is enough to assume that $\xi^2 = R_g^2/3$, which makes no difficulty (see section 5.1). Here M is indeed proportional to n (Eq. (44)), which leads to the 'molecular weight of the blob' $M_\xi \sim \xi^{d_f}$. There is a difference with the conventional case, however. The absence of correlations between scattering objects in the Guinier law is not strictly valid. It is a satisfactory approximation when studying a very dilute solution (otherwise, one has to take into account inter-chain corrections depending on the second virial coefficient). Here, the absence of correlations arises from screening effects. In addition, note that Eq. 43 is expected to apply not only to polymer chains but also to other types of 'fractal objects' provided they are able to interpenetrate. An example is the case of percolation clusters considered by Daoud and Leibler (1988).

In the asymptotic regime such as $q \gg 1/\xi$ (but for $q < 1/a$, a being the statistical unit length), the scattering intensity will be proportional to the Fourier transform of Eq. (37). Thus

$$I(q) \sim \phi q^{-1/\nu} \tag{46}$$

This relationship is equivalent to the well-known law of scattering by fractal objects $- I(q) \sim \phi q^{-d_f} -$ in the range of q such as $1/R < q < 1/l_0$ (R being their size and l_0 the size of the elementary unit of the Fractal).

Between the low q regime given by Eqs (42) or (43) and the asymptotic regime defined by Eq. (46), there is a region of cross-over. Various forms have been proposed for the description of such a cross-over regime (Texeira, 1985, 1988; des Cloizeaux and Jannink, 1987; Bastide, Leibler and Prost, 1990; Shibayama, 1992; Falcao, Skov Pedersen and Mortensen, 1993), but we will not go further in the discussion of this point here.

5.2.1.2 *Experiments*

5.2.1.2.1 Standard behaviour

A typical example of a neutron scattering curve from a semi-dilute solution (polystyrene $M_W \approx 7 \times 10^5$ in deuterated toluene, $\phi = 0.048$) is given, in two representations, in Figures 2(a) and (b). In Figure 2(a), the data points corresponding to the smaller values of q probed in this experiment are plotted along a Zimm representation ($1/I(q)$ versus q^2). A straight line is obtained, which means that, in this range of q, the scattering can be described satisfactorily by the Ornstein–Zernicke function (Eq. (38)). As a result $I(q \to 0)$ can be estimated by the intercept with the y-axis, while ξ can be calculated either by the intercept with the x-axis or by the reduced slope. For this set of data, one finds $\xi = 24$ Å.

The points of Figure 2(a) appear, together with new points at larger q's, on the log-log plot of Figure 2(b), where the value $1/\xi$ obtained from the preceding plot has also been reported. On this logarithmic scale the decay is neatly visible for $q \geqslant 1/\xi$, while a levelling off is seen for $q \ll 1/\xi$. This saturation at low q is an indication that the correlations are effectively lost on length scales $r \geqslant 1/q_{\text{plateau}}$, which is essential for the meaning of a correlation distance. The presence of the quasi-plateau should therefore be checked before ξ is calculated from the 'Zimm-plot' of Figure 2(a). This is why we will often present data relative to gels in such a log-log representation.

For $q\xi > 1$, a tendency towards a slope of exponent 5/3 (or 1.7, for $\nu = 0.588$) is observed, as expected from Eq. (46). However, a true asymptotic regime $I(q) \sim q^{-1/\nu}$ is very difficult to attain since, for semi-dilute solutions of practical interest, ξ is too small to allow the existence of a real regime such as $1/\xi \ll q \ll 1/a_k$.

In a true semi-dilute regime, i.e. for $\phi \ll 1$ and nevertheless much larger than ϕ^* (which implies chains of very high molecular weight), the variation of ξ with ϕ is currently found very close to the theoretical value: one finds $\xi \sim \phi^{-z}$, with $0.7 \leqslant z \leqslant 0.8$ to be compared with $\nu_{\text{th}} = 0.77$. The experimental exponents found for the scaling law $I(q \to 0) \sim \phi^{-x}$ are slightly more scattered, ranging between 0.25 and 0.37, but they remain comparable to $x_{\text{th}} = 0.31$ (Daoud *et al*, 1975; Lapp, Picot and Strazielle, 1985; Broseta *et al*, 1986; Mendes *et al*, 1991; Geissler, Horkay and Hecht, 1991; Hecht *et al*, 1991). In a more concentrated regime, larger exponents, closer to unity are expected and found (see des Cloizeaux and Jannink, 1987) because the effective exponent ν is closer to 1/2 on short scales along the chain.

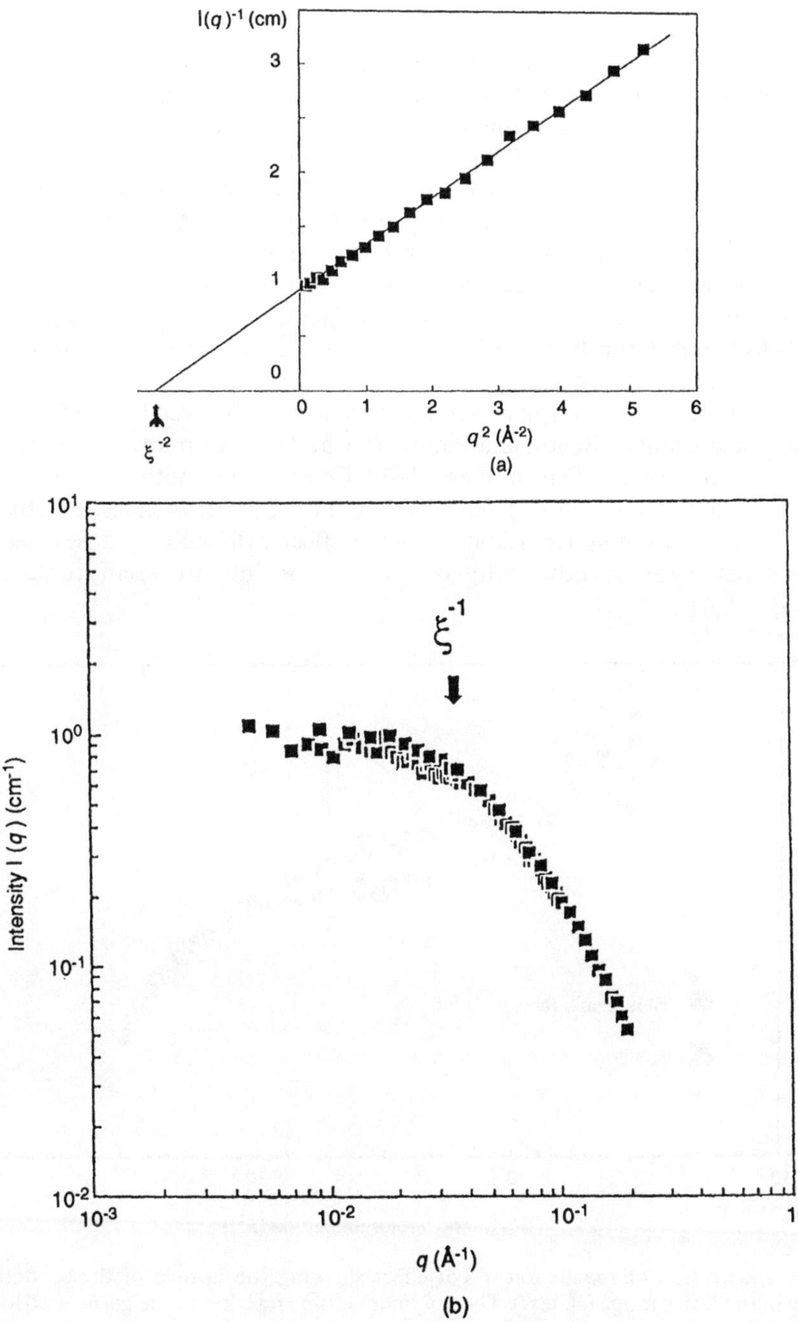

Figure 2. Scattering curves for a semi-dilute solution of polystyrene ($M_W \approx 7 \times 10^5$) in deuterated toluene at $\phi = 0.048$. (a) Zimm representation: $I^{-1}(q)$ versus q^2; (b) log–log representation; (Courtesy E. Mendes)

5.2.1.2.2 *Picot–Benoît effect*

Unfortunately, the data presented in Figure 2(b) correspond, regarding the comparison with theoretical predictions, to a favourable combination of q range and polymer concentration. If one repeats the experiment with a spectrometer allowing q values as low as $2–3 \times 10^{-3}$ Å^{-1}, one obtains, for the same polystyrene solution, an unexpected upturn at very low q. If this solution is concentrated, the same type of upturn starts to show up even for $q \approx 10^{-2}$ Å^{-1}. An example of the last situation, i.e. the case of a polystyrene solution at $\phi = 0.2$, is shown on Figure 3. On the same plot, one finds I(q) for a 'statistical gel' (see Section 5.4) of the same polymer concentration. The structure factors of gels will be discussed at length below, but it can be remarked already that comparable upturns at low q are observed in both types of systems.

This unexpected scattering at a very small angle was observed for the first time using light scattering by Benoît and Picot (1966) and later by many researchers in various systems (see e.g. Dautzenberg, 1972; Guenet, Willmoth and Ellsmore, 1983; Koberstein, Picot and Benoît, 1985; Hecht *et al*, 1991; Mendes, 1991b). Its origin is still unclear. A systematic study of this effect is difficult, because it seems to lack somehow for reproducibility. In particular, its intensity seems to depend

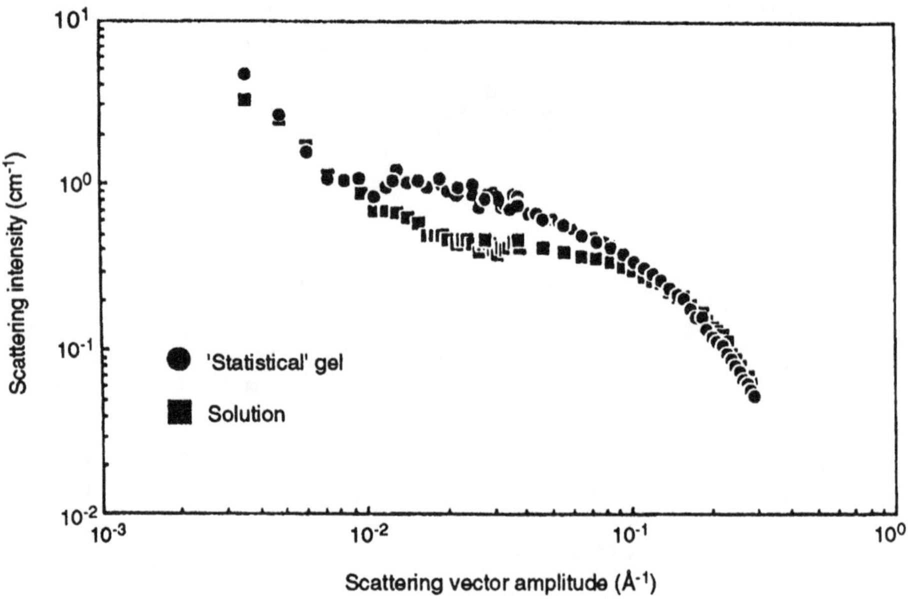

Figure 3. Illustration of the Benoît–Picot effect showing the upturn of the scattering intensity in the low q range (cf. text). The polymer volume fraction in the gel (a 'statistical gel' similar to those discussed in section 5.4 but with higher crosslinking density) and in the semi-dilute solution are equal ($\phi = 0.20$). (From Mendes (1991)

on the conditions of preparation of the solution. It may even vanish, at least temporarily, upon centrifugation and cannot be attributed to the presence of dust or impurities (Dautzenberg, 1972; Koberstein, Picot and Benoît, 1985). The phenomenon is presumably related to the existence of large-scale concentration fluctuations $\delta\phi_{ls}$. Koberstein, Picot and Benoît (1985) indeed reconstructed some neutron scattering profiles by expressing the total scattering intensity as a sum of the expected Lorentzian term for a semi-dilute solution and a so-called Debye–Bueche term (a Lorentzian squared; see Debye and Bueche, 1949; Debye, Anderson and Brumberger, 1957) associated with some random long-range non-homogeneities of correlation length a_c:

$$\mathrm{I}(q) = \frac{\mathrm{I}_{sd}(0)}{(1+q^2\xi^2)} + K' \frac{a_c^3}{(1+q^2a_c^2)^2} \tag{47}$$

The squared Lorentzian term arises from an exponential correlation function $\langle\phi(0)\phi(r)\rangle \sim \exp(-r/a_c)$ in real space, assumed to originate from random variations, with sharp boundaries, of polymer concentration on length scales bigger than a certain 'correlation distance' a_c. In this framework, K' is shown to be proportional to $\langle\delta\phi_{1s}^2\rangle$. Koberstein, Picot and Benoît (1985) found a_c to be always equal to or larger than the radius of gyration of the chains. A slightly different analysis of the problem in the framework of the Benoît–Benmouna (1984) theory is due to Medjahdi, Sarazin and François (1991). These authors base their description on the existence of an attractive potential between the chains. This was also the conclusion of Guenet, Willmoth and Ellesmore (1983) who proposed to relate this attraction to the capability, for certain atactic polymers, to form physical gels at low temperature.

The presence of the 'Picot–Benoît effect' sets an annoying problem. When present (and it be may present without being 'visible', that is, confined in an unexplored region of the spectrum), one should, in principle, conclude that a single length scale is inadequate in describing the scattering behaviour of the considered semi-dilute solution. In practice, for convenience, one usually ignores it and calculates ξ from the linear part of the $1/\mathrm{I}(q)$ versus q^2 plot, see Figure 4. Such a procedure represents a valid approximation. One may consider, that it makes sense because a reduction, or even a disappearance, of the upturn is observed after an appropriate treatment is applied to the solution (like the centrifugation method already mentioned above or by filtration, as shown in the case of polyelectrolytes by Reed *et al*, 1991). Therefore, the low angle upturn may be considered as a 'stray' effect, non-intrinsic to the semi-dilute state, superimposed on the 'pure' behaviour recalled above. This conclusion is also in agreement with the fact that the upturn strongly decreases upon dilution (possibly like ϕ), contrary to $I_{sd}(0)$ (when fitting with Eq. (47)) which always increases as the concentration is lowered. However, as pointed out by Koberstein, Picot and Benoît (1985) who calculated the correlation length before and after suppressing the upturn by centrifugation, the presence of the Picot–Benoît effect may increase

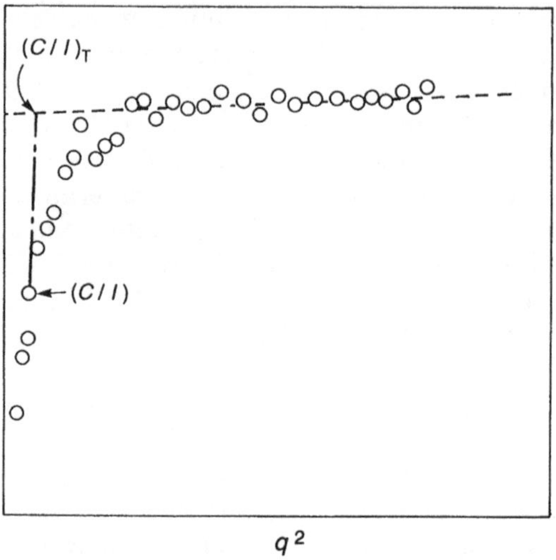

Figure 4. Graphical representation of the method used to calculate the excess low angle scattering intensity $I_{EX}(q)$ from a Zimm representation, in the case of a semi-dilute solution of polystyrene in benzene. The extrapolation of the dashed line provides an estimate of ξ (cf. Figure 2(a)). From Koberstein, Picot and Benoît (1985)

the apparent value of ξ. Let us reformulate pictorially their explanation for this enhancement. In Figure 5(a) is represented schematically a semi-dilute solution corresponding to the idealized view, with a correlation length ξ_i. On Figure 5(b) (inspired by Medjahdi, Sarazin and François, 1991), is pictured a inhomogeneous semi-dilute solution of the same ϕ. Because of the presence of regions where the polymer concentration is higher than the average, the concentration *between these regions* is lower than the average. Consequently, in the major part of the solution the local fluctuations of concentration will be larger than for the homogeneous solution.

5.2.2 SEMI-DILUTE SOLUTIONS OF POLYELECTROLYTES

Contrary to the case of neutral polymeric systems, the physical properties of polyelectrolyte solutions are poorly understood. This is due to both theoretical difficulties to describe precisely the long-range coulombic interactions and to experimental problems linked to the fact that the regime of interest where the coulombic interactions are predominant corresponds to very low polymer concentrations. In this regime the experiments are very difficult to perform; in particular the scattering from dilute polyelectrolyte solutions is very weak. In fact, most of the studies either structural or dynamical reported so far deal with semi-dilute solutions. The results obtained mainly by static and dynamic

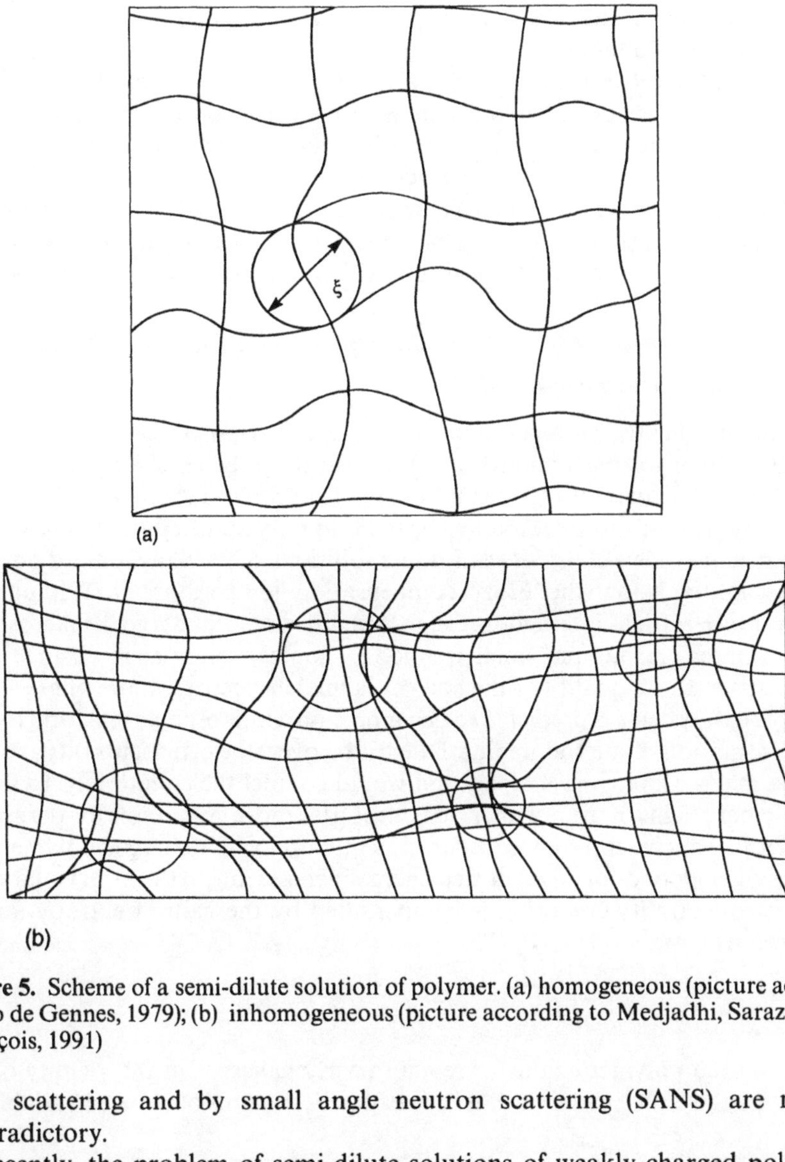

Figure 5. Scheme of a semi-dilute solution of polymer. (a) homogeneous (picture according to de Gennes, 1979); (b) inhomogeneous (picture according to Medjadhi, Sarazin and François, 1991)

light scattering and by small angle neutron scattering (SANS) are rather contradictory.

Recently, the problem of semi-dilute solutions of weakly charged polyelectrolytes in a poor solvent has received more attention. It was argued that systems, where attractive dispersion forces are nearly compensated by repulsive electrostatic interactions, should undergo a microphase separation transition. The phase separation occurs on a microscopic scale because a macroscopic phase separation would result in a too severe loss of entropy for the counterions (Borue and Erukhimovich, 1988; Joanny and Leibler, 1990).

The above models have been found relevant in the analysis of SANS and light scattering results obtained on polyacrylic acid (PAA) and polymethacrylic acid (PMA) solutions (Moussaid *et al*, 1991, 1993). Variations of the static and dynamic properties of these solutions with polymer concentration, ionization degree, temperature, and salt content were found in qualitative or even semi-quantitative agreement with the theoretical predictions.

In the following, we will recall the main results obtained by SANS and light scattering in semi-dilute solutions of weakly charged polyelectrolytes in poor solvent. This will provide a basis on which the behaviour of gels will be discussed.

5.2.2.1 Theoretical background

The detailed theoretical analysis of weakly charged polyelectrolyte systems is possible in poor solvent conditions. In this case, the polymer chain conformation remains nearly Gaussian and the random phase approximation formalism can be used safely (Borue and Erukhimovich, 1988; Joanny and Leibler, 1990).

The recent models consider semi-dilute solutions of partially ionized polymer chains slightly below the Flory compensation temperature θ. Without the presence of electrical interactions, such systems would undergo a macroscopic phase separation, but the small fraction of ionized monomers ensures their stabilization. If the quality of the solvent is further decreased, the phase separation of such systems is predicted to occur now on a microscopic scale and to lead to the formation of microdomains. Due to the overall electroneutrality requirement, a macroscopic phase separation would compel the counterions to follow the polymer chains in the polymer rich phase. It is more favourable for the system to undergo a microscopic phase separation into oppositely charged polymer-rich and polymer-poor domains. The free-energy increase due to the local violation of the electroneutrality condition is compensated by the gain of entropy for the counterions.

5.2.2.1.1 Structure factor

Assuming the polymer chains to remain nearly gaussian in the vicinity of the transition, the structure factor can be calculated. In terms of reduced variables, it reads

$$G(x) = \frac{1}{4\pi l_B r_0^2 \alpha^2} \cdot \frac{x^2 + s}{(x^2 + s)(x^2 + t) + 1} \tag{48}$$

Here l_B is the Bjerrum length, α is the fraction of ionized monomers and r_0 is a typical distance given by

$$r_0^{-2} = \left(\frac{48\pi l_B}{a_k^2}\right)^{1/2} \alpha\phi^{1/2} \tag{49}$$

where a_k is the length of the statistical unit and ϕ is the polymer concentration. The reduced wave vector amplitude x is defined as the product of the length r_0 by the usual wave vector transfer defined in a scattering experiment $q = 4\pi/\lambda_\omega \sin\theta/2$, where λ_ω is the wavelength of incident radiation in the medium and θ the scattering angle. Similarly, s is a reduced ionic strength defined as $s = \kappa^2 r_0^2$, where κ is the usual screening parameter and t is a reduced temperature given by

$$\begin{aligned} t &= -12 r_0^2 h\phi/a^2 \\ h &= -(2v\tau + 3w\phi) \end{aligned} \tag{50}$$

Here $h\phi$ is the usual virial contribution in a θ solvent, v being the excluded volume, $\tau = (T - \theta)/\theta$ the reduced distance to θ temperature and w the third virial coefficient. In a limited range of polymer concentrations ϕ, ionization degrees α and salt concentration ϕ_s, the structure factor (Eq. (48)) exhibits a maximum for a non-zero value q^* of the wave vector transfer, given by

$$q^{*2} + \kappa^2 = r_0^{-2} \quad \text{with} \quad \kappa^2 = 4\pi l_B(\alpha\phi + 2\phi_s) \tag{51}$$

This corresponds to the case where the system would undergo upon cooling a microphase separation transition with a domain periodicity given by $2\pi/q^*$. The observed peak reflects that the concentration fluctuations with wavelength $2\pi/q^*$ are favoured above the microphase separation transition. At the transition, the maximum increases without changing its position and becomes a Bragg peak, revealing the onset of an ordering in the system. Upon increasing salt concentration or decreasing polymer concentration and/or polymerization degree, Eq. (51) shows that the maximum shifts to smaller q values. When the condition $\kappa < r_0^{-2}$ is no longer fulfilled, the maximum is pinned at zero q value, and upon cooling the system undergoes a usual macroscopic phase separation.

Scattering experiments (SANS, light scattering) provide a measurement of the structure factor of semi-dilute polymer solutions. In particular, light scattering allows the determination of zero q limit that is obtained from Eq. (48):

$$S(0) = \frac{v_e^{-1}}{1 + v_e^{-1}(2v\tau + 3w\phi)} \tag{52}$$

where $v_e^{-1} = (1/4\pi l_b)\cdot(\kappa/\alpha)^2$ is the electrostatic excluded volume calculated from a screened Coulomb potential. In the limit where the electrostatic interactions are predominant one finds

$$S(0) \sim \frac{1}{4\pi l_b}(\kappa/\alpha)^2 \tag{53}$$

This is the result obtained previously by Joanny and Pincus (1980). In the absence of salt one has

$$S(0) \sim k_B T\phi/\alpha \tag{54}$$

The above expression amounts to saying that the concentration fluctuations are controlled by the osmotic pressure of the counterions, that is, equal to $k_B T$ per counterion:

$$\Pi_i \sim k_B T \phi \alpha \tag{55}$$

5.2.2.2 Experimental results

Most of the studies have been carried out on semi-dilute solutions of partially charged PAA and PMA (Moussaid *et al*, 1991, 1992). SANS experiments allow one to measure an absolute scattering intensity I(q) which is directly proportional to the structure factor of the solution:

$$\mathrm{I}(q) = KG(qr_0) \tag{56}$$

where $G(qr_0)$ is given by Eq. (48) and the proportionality constant K depends on the contrast between the polymer and the solvent relatively to the incident neutrons. This contrast can be estimated from tabulated values of the scattering length densities and the partial molar volumes of the polymerized monomers. For PAA systems, $K = 9.6$ barns.

Figure 6 shows the variations with temperature of the structure factor of a PAA solution with polymer concentration $C_p \approx 0.71\ \mathrm{M\,l^{-1}}$ and $\alpha = 0.05$. As predicted by the model, the position of the peak is, within experimental accuracy,

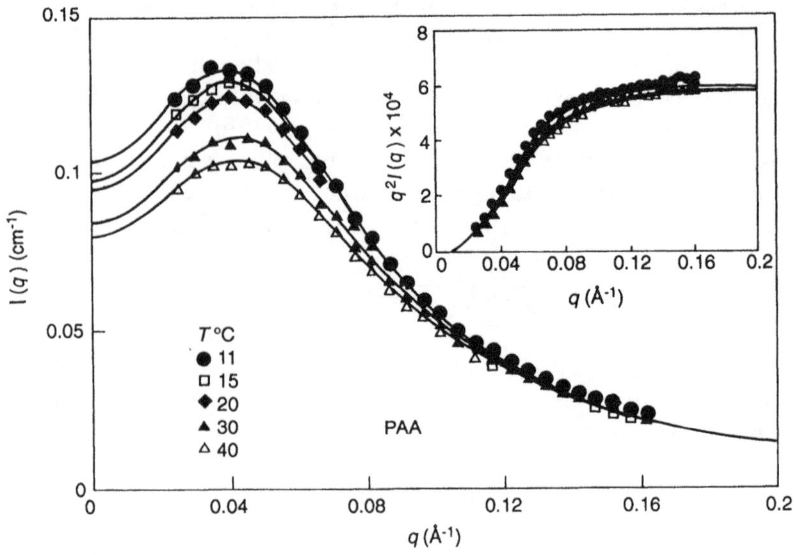

Figure 6. Temperature variation of the intensity scattered from PAA solutions with $C_p \approx 0.71$ M and $\alpha = 0.05$. The inset shows the Kratky representation of the data. Solid lines correspond to the fitted theoretical intensities

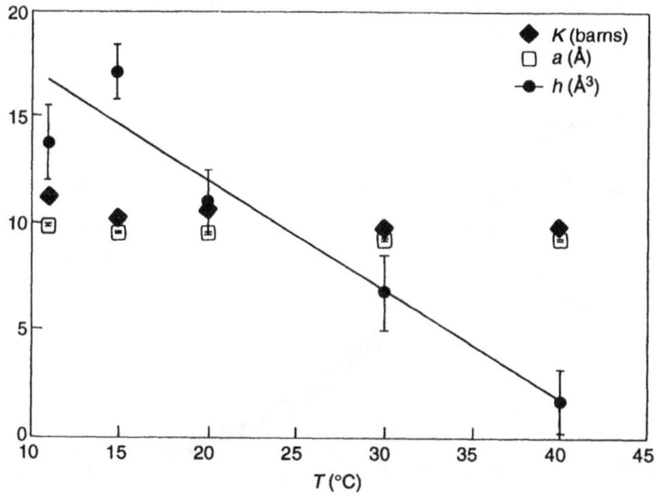

Figure 7. Temperature variation of the fitting parameters corresponding to the solid lines of Figure 6

independent of temperature. The amplitude of the peak decreases upon increasing the temperature. This is also expected since the phase diagram of the PAA–water system in acidic conditions is characterized by an upper consolute solubilization curve.

In Figure 6 are also shown the best fits to the data of Eq. (56) with three fit parameters: K, a_k and h. The temperature evolution of these parameters is given in Figure 7. It can be seen that the constant value of K is of the order of 10 barns that is a value close to that calculated, $K = 9.6$ barns. Also the parameter a_k that represents the length of the statistical unit is found constant and equal to 9 Å, that is a reasonable value. As for h, it decreases upon increasing the temperature as expected from Eq. (50). It would change its sign for a temperature between 40 and 45 °C. From the definition of h, one expects that this change occurs at a temperature close to the θ temperature. The θ temperature of PAA in protonated water containing 0.2 M HCl has been measured by Silberberg, Eliassaf and Katchlasky (1957) and Eliassaf and Silberberg (1962), and was found to be equal to 14 °C. Recent light scattering measurements (Skouri, 1994) confirm this value and show that in deuterated water, in the presence of 0.2 M DBr, the θ temperature is around 27 °C. Such a large difference in the θ temperature between protonated and deuterated water might be due to the difference in the H bonding.

It is interesting to note that light and neutron scattering experiments provide an estimate of the same quantity, i.e. the θ temperature, through measurements performed under very different conditions. In the light scattering measurement, the θ temperature is obtained by following the temperature evolution of the second virial coefficient in dilute solutions of the polymers. A strong acid is added

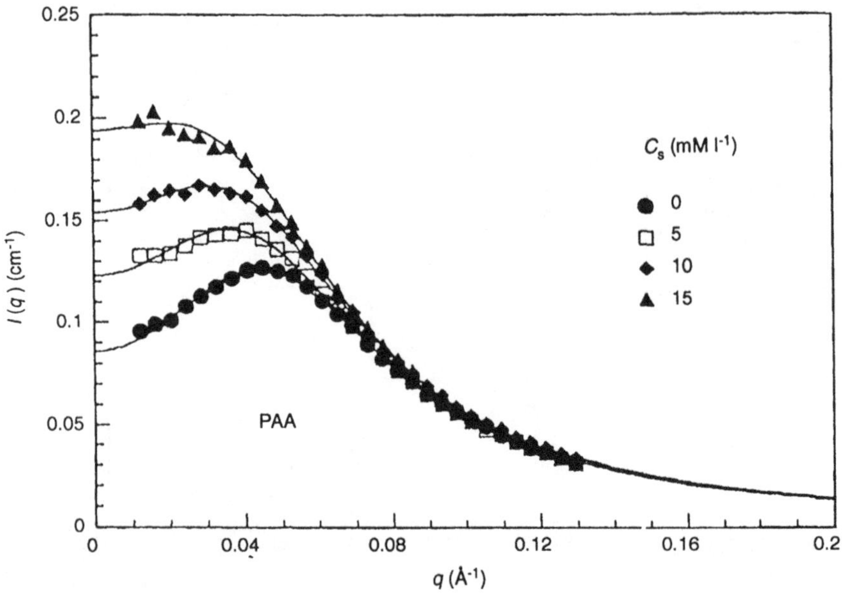

Figure 8. Salt effect on the scattering curve of PAA with $C_p \approx 0.71$ M and $\alpha = 0.05$. Solid lines are fits of the theoretical intensities to the experimental data

in order to minimize the effects of the charges. However, this acid is liable also to modify the solvent thermodynamic quality. On the other hand, the neutron scattering experiment provides a value for the overall structure factor shape as quantified by a fit to the theoretical model. The measurements are performed in rather concentrated solutions with no added acid. Thus the relatively small difference in the θ temperature obtained by the two methods appears as a rather surprising agreement that could be considered as a strong support to the theoretical model.

The same agreement between theory and experiment is observed when considering the salt effect as shown by Figure 8, in which is represented the structure factors obtained for PAA solutions with different salt contents.

The same systems have been investigated by light scattering in order to determine I(0). The variation of I(0) with characteristic parameters of the system—polymer concentration, ionization degree and salt content—are found in good agreement with the theoretical predictions of Eqs (53) and (54) (Moussaid *et al*, 1991; Skouri, 1994).

5.3 Scattering from idealized gels, according to theories

In the original theories of rubber elasticity, valid in principle for networks in the bulk as well as in the swollen state, the elastic elementary chain (that we call here the 'meshes') are assumed to interact only at the crosslinks. Their conformations are believed to be statistically independent, which is why the elastic part of the network's entropy reduces essentially to the sum of the entropies of the elementary chains. In other words, as said in the introduction, the only length scale that defines the structure is the mesh size. We will call 'idealized gels' such networks filled with solvent, in which no correlations are believed to exist beyond the scale of the mesh, and which fluctuate locally because of thermal agitation.

The question of the structure factor of such idealized gels has been addressed by several authors with a number of variant readings. We will try, at this level, to summarize a kind of 'mainstream' approach, based on the works of Tanaka, Hocker and Benedek (1973), de Gennes (1979) and Onuki (1988, 1993). However, after reviewing some experimental results and the problems set by their interpretation, we will present some modifications which have been recently introduced to this approach.

5.3.1 LOW SCATTERING VECTOR LIMIT

Contrary to the case of real liquids, the well-known relationship $\mathrm{I}(q \to 0) = b_v^2 k_\mathrm{B} T \phi^2 \chi_\mathrm{os}$ is not valid in the case of gels, because of the existence of a zero frequency shear elasticity (we return to this point in section 5.3.2). Recall that the osmotic compressibility χ_os is the inverse of the osmotic bulk modulus $K_\mathrm{os} = (1/\phi)(\partial\Pi/\partial\phi)$, where Π is the osmotic pressure (sometimes called swelling pressure in the case of gels). In the following, we write these quantities Π^gel and $K_\mathrm{os}^\mathrm{gel}$ respectively in the case of a gel, in order to distinguish them explicitly from Π^sol and $K_\mathrm{os}^\mathrm{sol}$ for a semi-dilute solution.

Describing a gel as a homogeneous solid on length scales larger than the mesh, Tanaka, Hocker and Benedek (1973) calculated, from the elasticity theory of continuous media, the time correlation function of the scattered field $\langle E(q,t)E^*(q,0)\rangle$ in the limit $q \to 0$ (which ensures that the gel can be considered as a solid). For $t \to 0$, $\langle E(q \to 0, t)E^*(q \to 0, 0)\rangle$ reduces to $\mathrm{I}(q \to 0)$, which is found to be equal to

$$\mathrm{I}(q \to 0) = \frac{b_v^2 k_\mathrm{B} T \phi^2}{(K_\mathrm{os}^\mathrm{gel} + \frac{4}{3}\mu)} \tag{57}$$

where μ denotes the shear modulus, $K_\mathrm{os}^\mathrm{gel} + \frac{4}{3}\mu$ is sometimes called the 'longitudinal' modulus M and b_v^2 is the contrast factor, see section 5.1.

As shown below, these two moduli, $K_\mathrm{os}^\mathrm{gel}$ and μ, can be calculated from the expression of the free energy of the gel. Equation (57) is therefore obtained in the framework of a *thermodynamic* theory, assuming that the only fluctuations probed by the scattering technique are *thermal*. Note that the result given in Eq. (57), which was derived in the context of light scattering, remains obviously valid in the case of neutron scattering. The field E must then be replaced by the amplitude of the scattered wave associated to the neutrons.

Experimentally, one rarely measures effectively the shear modulus of a gel whose scattering properties are studied in a parallel way. It is in general easier to determine a modulus of 'uniaxial deformation' G, defined as follows:

$$G = \sigma_0/(\lambda - \lambda^{-2}) \tag{58}$$

where σ_0 is the so-called 'nominal stress', that is, the force per unit area before deformation and where λ is the deformation ratio of the sample in the direction of the applied strain (quite often a uniaxial compression). The definition of G follows from the classical theory of rubber elasticity, which predicts a linear variation of σ_0 with $(\lambda - \lambda^{-2})$—see appendix. Such a linear dependence is frequently observed experimentally for gels in good solvents (contrary to polymer networks in the bulk, for which the dependence of σ_0 with λ has a more complicated and non-universal form).

When a gel is elongated in a reservoir of solvent, it swells. When uniaxially compressed, it deswells (see Dušek and Prins, 1969; Horkay and Zrinyi, 1988). But, the kinetics of these processes are rather slow (they may extend over several hours or even over several days). As a result, in most of the stress–strain experiments, the volume of the sample can be considered as conserved. The Poisson ratio is then equal to 1/2 and G must strictly equate the shear modulus μ (see Landau and Lifshitz, 1973).

The osmotic bulk modulus K_{os}^{gel} relates to the osmotic pressure of the gel (sometimes called the swelling pressure—see Horkay, Hecht and Geissler, 1989) $\Pi^{gel}(\phi)$ the same way the osmotic bulk modulus K_{os} of a solution relates to its osmotic pressure Π. But, contrary to a solution, a gel reaches a state of swelling equilibrium, for which the dilution tendency of the polymer is exactly balanced by the tri-dimensional elastic retraction of the network. The swelling pressure Π^{gel} is therefore coupled to the elasticity. Making two important simplifying assumptions, namely, (i) the 'strong' additivity hypothesis for the free energy density (Eq. (1) and (ii)) the expression of the elastic part as a scaling law ($F_{el} = B\phi^l$; Bastide, Candau and Leibler, 1980), one gets (see Appendix):

$$\Pi^{gel}(\phi) = \Pi^{sol}(\phi) - gG(\phi) \quad \text{with } g = \frac{(1-l)}{2/3} \tag{59}$$

For one given gel at different swelling degrees in the same solvent (i.e., different ϕ), $G(\phi)$ is expected to vary, in the framework of this simple model, like ϕ^l. According to the classical picture $l = 1/3$, because of a combination of the density of elastic chains, which varies like ϕ, with the effect of their assumed extension, which leads to a $\phi^{-2/3}$ correction. Experimentally, the elastic modulus G of many gels varies effectively like $\phi^{1/3}$ (see e.g. Horkay, Hecht and Geissler, 1989). However, different values of the apparent exponent are sometimes reported (ranging typically between 1/3 and 2/3). In another connection, the question of the chain extension upon swelling is a delicate and controversial one (see Bastide,

1985a; Bastide, Picot and Candau, 1981; Bastide *et al*, 1984; Candau, Bastide and Delsanti, 1982). Altogether, this is why we prefer to start with the more general form of Eq. (59). Thus one expects, in principle

$$K_{os}^{gel} = \phi(\partial\Pi_{gel}/\partial\phi) = K_{os}^{sol} - \frac{(1-l)l}{2/3} G(\phi) \tag{60}$$

The state of maximum swelling of a gel corresponds to zero swelling pressure Π_{gel}. As a result, for a series of various gels at maximum swelling in good solvent (reaching different equilibrium values ϕ_e of the concentration), one should obtain

$$G(\phi_e) = \left[\frac{2/3}{1-l}\right] \cdot \Pi_{sol}(\phi_e) \sim \phi_e^{2.31} \tag{61}$$

The second part of this relation, $G(\phi_e) \sim \phi_e^{2.31}$ should be satisfied under the condition that $\Pi_{sol}(\phi)$ follows the des Cloizeaux law (according to des Cloizeaux, 1975, the exponent 2.31 should be written $\nu d/(\nu d - 1)$, where ν is the excluded volume exponent and $d = 3$, the dimensionality of current space).

The experimental values of the modulus for a series of gels with different crosslink densities generally follow an apparent scaling law $G(\phi_e) \sim \phi_e^n$ with an exponent n effectively very close to 2.31 (Munch *et al*, 1976b; Zrinyi and Horkay, 1980; Horkay, Hecht and Geissler, 1989; Horkay *et al*, 1991). But one obtains generally measured values of $G(\phi_e)$ smaller than $\Pi_{sol}(\phi_e)$,typically by 30%, even for gels whose modulus G varies like $\phi^{1/3}$ under deswelling, which, according to Eq. 61, should lead to $G(\phi_e) = \Pi_{sol}(\phi_e)$.

Such types of comparisons were carried on systematically by Horkay, Hecht and Geissler, 1989; Horkay *et al*, 1991), Zrinyi and Horkay (1980), Geissler *et al* (1988), Geissler, Horkay and Hecht (1991) and Hecht *et al* (1992), who performed combined osmotic deswelling and stress–strain experiments on various sorts of gels. In the interpretation of their results, they adopted a 'weak' additivity assumption instead of the 'strong' one made in Eq. (1): they allowed F_{mix} in a gel to be different from F_{mix} for a semi-dilute solution. Accordingly, they generalized Eq. (59) (with $l = 1/3$, a value which is consistent with most of their measurements of G) by writing

$$\Pi_{gel}(\phi) = \Pi_{mix}(\phi) - G(\phi) \tag{62}$$

and they compared the value of $\Pi_{mix}(\phi)$ and calculated from the sum of the measured values $\Pi_{gel}(\phi)$ and $G(\phi)$ with the values of $\Pi_{sol}(\phi)$ from the literature. They found $\Pi_{mix}(\phi)$ systematically smaller than $\Pi_{sol}(\phi)$, by 30–40%, in the case of polyvinylacetate and polydimethylsiloxane gels, but always varying the same way with polymer concentration (i.e. when scaling laws are observed the exponents have been found to be the same). In section 5.6, we return to the discussion of the physical origin of this difference of prefactor. At the present level, we only integrate it phenomenologically in the estimate of $I(q \to 0)$ by writing:

$\Pi_{mix}(\phi) = r'.\Pi_{sol}(\phi)$, where the 'reduction parameter' r' is equal to unity in the framework of the 'strong' additivity assumption and $r' < 1$, in the framework of the 'weak' one (we call this the GHH correction). Since the 'mixing pressure' $\Pi_{mix}(\phi)$ apparently varies with ϕ in the same way as $\Pi_{sol}(\phi)$ (Hecht *et al*, 1992) and since $\Pi_{sol}(\phi)$ can, in most cases, be described by a scaling law, K_{os}^{sol} should be replaced in Eq. (60) by $K_{mix} = r' \cdot K_{os}^{sol}$. Combining this modified Eq. (60) with Eq. (57), we find

$$\mathrm{I}(q \to 0) = \frac{b_v^2 \phi^2}{K_{os}^{gel} + \mu} = \frac{b_v^2 k_B T \phi^2}{K_{mix} + \left[4/3 - \dfrac{l(1-l)}{2/3}\right] G} \tag{63}$$

Note that if l has the classical value 1/3, Eq. (63) takes a very simple form

$$\mathrm{I}(q \to 0) = \frac{b_v^2 k_B T \phi^2}{r' \cdot K_{os}^{sol} + G} \tag{64}$$

5.3.1.1 $\mathrm{I_{gel}}$*(0) for idealized gels at maximum swelling versus* $\mathrm{I_{sol}}$*(0) for solutions*

Still assuming that $\Pi_{sol}(\phi)$ can be represented by a scaling law, one obtains $K_{os}^{sol} = s\,\Pi_{sol}(\phi)$ from $K_{os}^{sol} = \phi[\partial \Pi_{sol}(\phi)/\partial \phi]_T$, with $s = 2.31$ in the framework of the des Cloizeaux law. Since $\Pi_{gel} = 0$ at swelling equilibrium, Eq. (62) implies $G = \Pi_{mix}(\phi_e) = r' \cdot \Pi_{sol}(\phi_e)$. We can therefore express $\mathrm{I}(q \to 0)$ as a function of $\Pi_{sol}(\phi_e)$; one gets from Eq. (63)

$$\mathrm{I_{gel}}(q \to 0) \sim \frac{b_v^2 k_B T \phi^2}{r' g(\phi_e) K_{sol}(\phi_e)} \sim \frac{I_{sol}(q)}{r' g(\phi_e)} \tag{65}$$

with

$$g(\phi_e) = \left[1 + \frac{1}{s}\left(\frac{8/9}{1-l} - l\right)\right]$$

$$g(\phi_e) = (1 + l/s) \approx 1.4 \text{ for } l = 1/3; \approx 1.55 \text{ for } l = 1/2; \approx 1.7 \text{ for } l = 0.6$$

Thus we get the ratio $\mathrm{I_{gel}}(0)/\mathrm{I_{sol}}(0) = 1/r'\, g(\phi_e)$, for gels at maximum swelling. The lower limit, corresponding to the classical 'strong additivity' assumption for the free energy ($r' = 1$) leads to values of the ratio $\mathrm{I_{gel}}(0)/\mathrm{I_{sol}}(0)$ ranging from 0.6 to 0.7, for l decreasing from 0.6 to 1/3. Therefore, due to the elastic contribution, the classical picture predicts that the gels should scatter less than solutions. Taking into account the Geissler–Hecht–Horkay argument leading to $r' \approx 0.65$, $\mathrm{I_{gel}}(0)/\mathrm{I_{sol}}(0)$ should become very close to unity (0.9 for $l = 0.6$, 1 for $l = 1/2$, 1.1 for $l = 1/3$). Even with this correction, the "thermal scattering" of gels should never become significantly larger than that of equivalent solutions.

5.3.1.2 $I_{gel}(0)$ *for idealized gels at incomplete swelling versus* $I_{sol}(0)$ *for solutions*

The above estimates can easily be generalized to the case of gels in incomplete swelling conditions. According to Eq. (61) coupled with the GHH correction, one expects

$$G(\phi) = \frac{2/3}{1-l}\cdot \Pi_{mix}(\phi_e)\cdot(\phi/\phi_e)^{1/3} = \frac{2/3}{1-l}\cdot r'\cdot \Pi_{sol}\cdot(\phi_e)\cdot(\phi/\phi_e)^{1/3}$$

$$= \frac{2/3}{1-l}\cdot r'\cdot \Pi_{sol}\cdot(\phi)\cdot(\phi/\phi_e)^{1/3-s} \qquad (66)$$

Combining Eqs. (66) and (67), one gets a form similar to Eq. (65)

$$I_{gel}(0) = \frac{I_{sol}(0)}{r' g(\phi)} \qquad (67)$$

with, now,

$$g(\phi) = \left[1 + \left(\frac{1}{s}\right)\left(\frac{8/9}{1-l} - l\right)(\phi/\phi_e)^{1/3-s}\right]$$

Under deswelling of the gel, since $1/3 - s \approx -2$, the term

$$\left(\frac{1}{s}\right)\left(\frac{8/9}{1-l} - l\right)(\phi/\phi_e)^{1/3-s}$$

becomes rapidly negligible with respect to unity as ϕ/ϕ_e is increased. Thus, under the *strong additivity* assumption ($r' = 1$), the scattering of the gel, initially below that of the solution, should get progressively closer to that of the equivalent semi-dilute solution: typically the ratio $I_{gel}(0)/I_{sol}(0)$ should increase from 0.7 to 1. Taking into account the GHH correction, in most situations, the ratio $I_{gel}(0)/I_{sol}(0)$ should typically increase from roughly unity at maximum swelling (as above) to 1.4 at maximum in deswollen conditions.

With respect to the experimental data which will be discussed in the following, it is also relevant to consider the case of a series of gels at a fixed concentration (e.g. that of the reaction bath), but with increasing crosslinking densities. Since, in our approach, we eliminated the molecular quantities, this change must be expressed in Eq. (67) by a change of ϕ_e, keeping ϕ fixed: the larger the crosslinking density, the lower ϕ_e and therefore the larger $g(\phi)$. In other words, whatever the presence or the absence of the GHH correction, and whatever the value of l, an increase of the crosslinking density should lead, keeping the concentration constant, to some decrease of scattering intensity (typically up to 40%).

In general, whatever the concentration, the extrapolated intensity $I(q \to 0)$ of an idealized gel should be smaller than that of the comparable semi-dilute solution, if the free energy satisfies the strong additivity assumption. Taking into account the GHH correction which is required by the analysis of some thermodynamical data, one might imagine $I_{gel}(q \to 0)/I_{sol}(q \to 0)$ to become sometimes larger than unity. However, in any case this ratio should remain less than 1.4. In other words, according to the classical picture, and consistently to more recent thermodynamic data, a gel in which all the fluctuations are of 'thermal' origin should not scatter at very low q in a very different way than a semi-dilute solution of same concentration.

5.3.2 STATIC STRUCTURE FACTOR OF ISOTROPIC GELS

It has been recalled in section 5.1 that, in the case of a binary system, a scattering experiment 'picks' up the modes of wave vector $\mathbf{q}$ in the distribution of concentration of one species with respect to the other (see Eq. (25)):

$$I(q) = b_v^2 \langle \phi(\mathbf{q})\phi(-\mathbf{q}) \rangle \tag{68a}$$

$$= b_v^2 \langle \delta\phi(\mathbf{q})\delta\phi(-\mathbf{q}) \rangle \quad \text{for } q \neq 0 \tag{68b}$$

By analogy with the situation occurring at $q = 0$, one may guess that $\langle \delta\phi(\mathbf{q})\delta\phi(-\mathbf{q}) \rangle$, often written $\langle \delta\phi_q \delta\phi_{-q} \rangle$ is proportional to some kind of generalized compressibility which should be related to thermodynamics.

The free energy we have used up to now (see Eq. (1) and appendix) applies to systems for which the concentration is assumed to be uniform throughout the gel. This is sufficient to estimate most of the thermodynamical properties at zero frequency when the systems are far from critical points. But in order to deal with fluctuations of finite wavelength $2\pi/q$, it is necessary to include, in the free energy, the effect of the inhomogeneity they represent (recall that here we deal only with *annealed* heterogeneities corresponding to thermal fluctuations). This is done by transposing to gels the Ginzburg–Landau theory of critical phenomena (de Gennes, 1979; Onuki, 1988, 1993). We follow here the derivation by Onuki. See also the presentations made by Ramzi *et al* (1995) and Boué (1995). Equation (1) then becomes

$$F = F_{\text{mix}} + F_{\text{el}} + F_{\text{inh}} \tag{69}$$

where

$$F_{\text{inh}} = \frac{1}{2} \int_V C |\nabla\phi|^2 d^3r \tag{70}$$

and C is a constant.

On the basis of the earlier work by Tanaka, Hocker and Benedek (1973), Onuki expressed the elastic part of the free energy F_{el} in the framework of the theory of elasticity of continuous media. F_{el} was taken in the Flory-Rehner form (implying strong additivity, affine deformation and logarithmic term in the free energy). Since we stay at a qualitative level when dealing with the variation with q, we will stick to the Flory–Rehner presentation, only dropping, as before, the logarithmic term. For generality, we start with the case when the network is submitted to any sort of deformation, not necessarily isotropic, with respect to the *reference state* (usually identified with the state of preparation). Let $\mathbf{R}$ (R_1, R_2, R_3) be the instantaneous position of a point of the deformed network. The average position of the same point would have been $\mathbf{r}^0$ (r_1^0, r_2^0, r_3^0) in the reference state. Thus one can define the local and instantaneous deformation tensor $\Gamma_{ij} = \partial R_i/\partial r_j^0$ (with respect to this reference state). After rewriting the Flory-Rehner elasticity in these notations, one can express F_{el} as a simple function of the first invariant

$$I_1 = \sum_i \sum_j \left(\frac{\partial R_i}{\partial r_j^0}\right)$$

$$F_{\text{el}} = \frac{1}{2}\nu_0 k_B T \int_{V_0} \sum_i \sum_j \left(\frac{\partial R_i}{\partial r_j^0}\right)^2 d^3r^0 \tag{71}$$

where ν_0 is the number of chemical meshes (or elastic elementary chains) per unit volume in the reference state.

The instantaneous position R of a point can be written as the following sum:

$$\mathbf{R} = \mathbf{r} + \mathbf{u} \tag{72}$$

where $\boldsymbol{r}$ is the average position in the deformed state and $\boldsymbol{u}$ the instantaneous deviation. Therefore the first invariant I_1 can be expressed as

$$I_1 = \sum_i \sum_j \mathbf{A}_{ij} + 2\sum_i \sum_j \mathbf{E}_{ij}\left(\frac{\partial u_i}{\partial r_j}\right) + \sum_i \sum_j \sum_k \mathbf{E}_{ij}\left(\frac{\partial u_k}{\partial r_i}\right)\left(\frac{\partial u_k}{\partial r_j}\right) \tag{73}$$

where A_{ij} is the average deformation tensor (such as $r_i = \sum_j \mathbf{A}_{ij} r_j^0$) based on the assumption of affine deformation and where $\mathbf{E}_{ij} = \sum_m \mathbf{A}_{im}\mathbf{A}_{jm}$.

Expressing A_{ij} with respect to its principal axes, one obtain

$$\{\mathbf{A}_{ij}\} = \begin{pmatrix} \Lambda_1 & 0 & 0 \\ 0 & \Lambda_2 & 0 \\ 0 & 0 & \Lambda_3 \end{pmatrix} \tag{74}$$

where $\Lambda_1, \Lambda_2, \Lambda_3$ denote the deformation ratios, with respect to the reference state, along the three principal axes. Thus

$$\{\mathbf{E}_{ij}\} = \begin{pmatrix} \Lambda_1^2 & 0 & 0 \\ 0 & \Lambda_2^2 & 0 \\ 0 & 0 & \Lambda_3^2 \end{pmatrix}$$

Using these simplifications, the first invariant

$$\sum_i \sum_j \left(\frac{\partial R_i}{\partial r_j^0}\right)^2$$

can be rewritten in a more simple form:

$$\sum_i\sum_j\left(\frac{\partial R_i}{\partial r_j^0}\right)^2=\sum_j\Lambda_j^2+2\sum_j\Lambda_j^2\left(\frac{\partial u_j}{\partial r_j}\right)+\sum_i\sum_j\Lambda_j^2\left(\frac{\partial u_i}{\partial r_j}\right)^2 \tag{75}$$

Let us now turn to the case when the deformation of the network is an isotropic swelling with respect to the reference state: all the Λ_j terms are identical in Eq. (75) ($\Lambda_1=\Lambda_2=\Lambda_3=\Lambda; \Lambda^3=Q/Q_0=\phi_0\phi$). Then, using some classical definitions of elasticity theory (Landau and Lifshitz, 1973) and remembering that according to the Flory–Rehner theory

$$\mu=v_0k_BT(\phi/\phi_0)\Lambda^2=v_0k_BT(\phi/\phi_0)^{1/3}$$

one can rewrite the deviation of F, up to bilinear order, as

$$\delta F=\int_V\left\{\frac{1}{2}K_{os}^{gel}(\nabla\cdot\mathbf{u})^2+\frac{1}{4}\mu\sum_i\sum_j\left(\frac{\partial u_i}{\partial r_j}+\frac{\partial u_j}{\partial r_i}-\frac{2}{3}\delta_{ij}\cdot\nabla\cdot\mathbf{u}\right)^2+\frac{1}{2}C|\nabla\delta\phi|^2\right\}d^3r \tag{76}$$

where K_{os}^{gel} and μ are the mean osmotic bulk modulus and shear modulus, i.e. the same quantities as those used up to now.

Then, δF may be written in the Fourier space as

$$\delta F=\frac{1}{2}(2\pi)^{-3}\int\left\{\left(K_{os}^{gel}+\frac{\mu}{3}+C\phi^2q^2\right)|\mathbf{q}\cdot\mathbf{u}_q|^2+\mu q^2|\mathbf{u}_q|^2\right\}d^3q \tag{77}$$

where $\mathbf{u}_q$ denotes the Fourier transform of $\mathbf{u}(\mathbf{r})$.

Since $\delta\phi(r)/\phi=-\nabla\cdot\mathbf{u}$, where ϕ is the average polymer concentration in the gel, $\delta\phi_q$ is related to $\boldsymbol{u}_q$ through the equation $\delta\phi_q/\phi=-i\boldsymbol{q}\cdot\boldsymbol{u}_q$.

Since each mode of wave vector q is assumed to have an energy $(1/2)k_BT$ (equipartition theorem), one can finally derive the structure factor

$$S(\mathbf{q})\sim\langle\delta\phi_q\delta\phi_{-q}\rangle_E\sim-k_BT\left[\frac{\delta^2F}{\delta\phi_q\delta\phi_{-q}}\right]^{-1} \tag{78}$$

which leads to

$$S_{gel}(q)\sim k_BT\frac{\phi^2}{K_{osm}^{gel}+\frac{4}{3}\mu+C\phi^2q^2} \tag{79}$$

It can be readily seen that, for $q\to0$, Eq. (79) reduces to Eq. (57), i.e. to $S(q\to0)\sim\phi^2/[K_{os}^{gel}+4/3\mu]$. Moreover, identifying Eq. (79) with the Ornstein–Zernicke function—a Lorentzian of form $S(0)/(1+q^2\xi_g^2)$—one obtains

$$\xi_g=\sqrt{\frac{C\phi^2}{K_{osm}^{gel}+\frac{4}{3}\mu}} \tag{80}$$

Unfortunately, this relation does not allow the estimation of ξ_g precisely because very little is known about the coefficient C. Moreover, there is a built-in limitation: this approach is based on a so-called mean-field approximation which is well known in several fields of statistical physics (and in particular for semi-dilute solutions), not to lead to the experimentally observed scaling laws. At present there is no better way to handle mathematically the fluctuations in gels, but, taking advantage of what is known for semi-dilute solutions, it is possible to make yet further predictions.

5.3.2.1 c theorem*

This was carried out by de Gennes (1979), who focused on a simple situation. He considered the case of gels made by end-linking of precursor chains (with N_c monomer units) at a concentration c close to their overlap concentration $c^*(N_c)$. He argued first that such a gel should not swell significantly and that the final volume fractions ϕ_e (at swelling equilibrium), for a series of gels of different chemical meshes, should be proportional to the respective $\phi^*(N_c)$. This reads

$$\phi_e = k(f_c) N_c^{-4/5} v^{-3/5} a_k^{-6/5} \tag{81}$$

where v denotes the excluded volume parameter, a_k the statistical unit length and $k(f_c)$ is a prefactor depending on the functionality f_c.

At the same time, the average size (say, the radius of gyration) of the chemical meshes of the network was expected to stay approximately equal to R_F, the value for free chains in a good solvent. Finally de Gennes concluded that, since R_F is the only available characteristic length scale, ξ_g should be identified with (through, at least, a prefactor $1/\sqrt{3}$) with R_F. Another way to express the same thing is to state that this description of a gel at maximum swelling is very similar to that of a semi-dilute solution. Thus ξ_g should be approximately equal to ξ_{sd}, the semi-dilute solution of the same polymer concentration.

5.3.2.1.1 Generalized c theorem*

Along the above line of arguments, the relation $\xi_g \approx \xi_{sd}$ is expected to be valid only for gels prepared at a concentration close to ϕ^* of the mesh and at swelling equilibrium. It is, however, tempting to extend the validity domain of this approximate equality $\xi_g \approx \xi_{sd}$ to any gel in any state of swelling. Although this can scarcely be found written in an explicit form, it is often assumed in the course of oral discussions by a part of the scientific community dealing with gels. This does not come without posing a problem summarized below in the form of geometrical remarks.

5.3.2.1.2 Geometrical arguments

The point we want to emphasize is that assuming ξ_g roughly equal to ξ_{sd} at various states of swelling is somewhat contradictory with the current assumption of affine deformation for at least the mean positions of the junctions.

Imagine that the same amount of polymer is put in several cubic boxes of variable sizes L^3 (L being the length of an edge). The boxes are assumed to be filled completely with a good solvent of the chains. If the polymer is made of linear chains in the semi-dilute regime, $\xi_{sd} \sim \phi^{-3/4}$ means $\xi_{sd} \sim L^{9/4}$ ($L^{2.31}$ if $v = 0.588$). If now the polymer is a network in several states of incomplete swelling, the analogy proposed in the preceding paragraph leads to the same variation, i.e. $\xi_g \sim \xi_{sd} \sim L^{9/4}$. The point is that such a variation law represents a strongly non-affine change.

In order to better understand the problem, assume, temporarily, slightly integrated viewpoint with regard to affineness. Imagine a gel made by end-linking at a concentration $\phi_c \gg \phi^*$ (i.e. the meshes are interpenetrated: $R_{mesh} > \xi_g$, where R_{mesh} denotes the average end-to-end distance of the meshes) and suppose that the swelling of this gel is associated with a strong stretching, with no unfolding, of the chains. In the preparation conditions one could have, however, $\xi_g \approx \xi_{sd}$. But postulating a deformation affine up to the scale of the correlation length, one would get $\xi_g \sim L \sim \phi^{-1/3}$ (i.e. far less than for the semi-dilute solution with same polymer concentration). As a result, ξ_g would become rapidly smaller than ξ_{sd} as the degree of swelling increases $\xi_g \approx \xi_{sd}$ could not hold for any concentration.

To be sure, the current assumption of affineness $R_{mesh} \sim \phi^{-1/3}$ should lead to a dependence weaker than $\xi_g \sim \phi^{-1/3}$: some rearrangement might occur between R_{mesh} and ξ_g. Nevertheless, affineness up to the end-to-end vector of the mesh should essentially prevent some sliding of the meshes with respect to each other. Thus, even if ξ_g does not vary in a strictly affine manner (because of some loss of affineness between the scale of the mesh and the scale of ξ_g) it seems unusual that it could vary as $L^{9/4}$ instead of L. In summary, we believe that the generalized c^* theorem implies a non-affine unfolding of the network.

5.3.2.2 Structure factor of an idealized gel under uniaxial deformation

We turn now to the case when the network is first swollen (or deswollen) by a factor Λ^3 with respect to the reference state and then uniaxilly deformed at constant volume with respect to the new isotropic state. More explicitly:

$$\{\mathbf{A}_{ij}\} = \begin{pmatrix} \Lambda_1 & 0 & 0 \\ 0 & \Lambda_2 & 0 \\ 0 & 0 & \Lambda_3 \end{pmatrix} = \begin{pmatrix} \lambda & 0 & 0 \\ 0 & \lambda^{-1/2} & 0 \\ 0 & 0 & \lambda^{-1/2} \end{pmatrix} \cdot \begin{pmatrix} \Lambda & 0 & 0 \\ 0 & \Lambda & 0 \\ 0 & 0 & \Lambda \end{pmatrix} \tag{82}$$

Then Eq. (72) reads

$$R_i = r_i + u_i = \Lambda_i r_i^0 + u_i = \lambda_i \Lambda r_i^0 + u_i \tag{83}$$

with $i = x, y, z$.

We follow the same lines as before, but without the simplification arising from the fact that the Λ_j terms are identical. The fact that $\Lambda_1 \neq \Lambda_2 = \Lambda_3$ acts on the

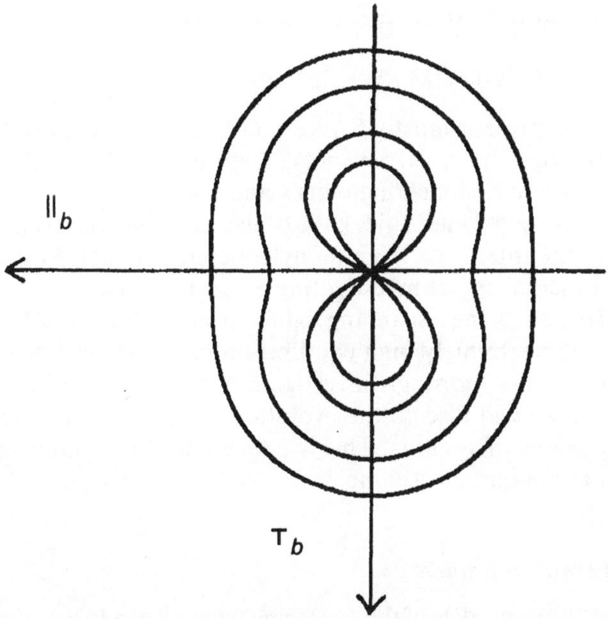

Figure 9. 'Normal' butterfly scattering patterns from uniaxially stretched gels, as calculated by Onuki (1988) for an idealized gel. The stretching direction is horizontal

shear term and introduces a dependence of the scattering intensity with the direction of **q**. The result can be written in a simple form:

$$S_{\text{gel}}(q) \sim k_{\text{B}} T \frac{\phi^2}{K_{\text{osm}}^{\text{gel}} + \frac{1}{3}\mu + \mu J(\beta) + C\phi^2 q^2} \tag{84}$$

where

$$\begin{aligned} J(\beta) &= (\lambda^2 - 1/\lambda)\cos^2\beta + 1/\lambda \\ &= \lambda^2 \cos^2\beta + 1/\lambda \sin^2\beta \end{aligned} \tag{85}$$

where β denotes the angle between q and the stretching axis.

According to Eq. (84), $I_{\text{gel}}(q \to 0)$ is expected to be reduced for q parallel to the stretching direction and enhanced for q perpendicular. Note that the physical origin of this effect is an increase of rigidity of the network in the direction of stretching, which is expected to suppress the thermal fluctuations. Moreover, the iso-intensity lines should exhibit a double-winged shape (often called 'butterfly patterns' (Oeser, Picot and Herz, 1988)), with here a large axis perpendicular to the stretching direction (see Figure 9). Some comparisons of this clear qualitative prediction with experimental results are presented in section 5.5.

5.4 Experimental structure factors of gels

5.4.1 GELS AT MAXIMUM SWELLING

In this section, we present mostly SANS and SAXS results about neutral gels swollen at saturation by good solvents. Regarding light scattering, we will summarize briefly some pioneering works and discuss a more recent study, in which ensemble-averaged intensities have been measured directly (i.e., by adding, on the detector, the intensities from many coherence areas). At present several groups of experimentalists are investigating the statistics of fluctuations (thermal and quenched) in gels, using a scanning light scattering device. As for the charged gels, because of the extremely high equilibrium swelling ratio, the scattering is very weak, so that no thorough investigation has been performed as yet at saturation. Later we will discuss the evolution of the structure factor upon an increase of the degree of swelling. This will give a hint of what can be expected experimentally at maximum swelling.

5.4.1.1 Light scattering studies

Even before the above models of the scattering from idealized gels were completely developed, several series of light scattering experiments were performed on gels swollen at saturation, by Bueche (1970), Pines and Prins (1972) and Wun and Prins (1974). The chemical gels studied by these authors (Pines and Prins considered also 'physical' gels like gelatine) were prepared only from monomers and not from polymers). Their light scattering spectra all exhibited a strong angular dependence, which was not caused by dust. Moreover, the absolute intensity was, in most cases, much larger than expected from the thermodynamical estimates made at that time. It was therefore concluded that the strong scattering by gels originated from 'frozen' heterogeneities of polymer concentration. The origin of the latter was attributed either to the *non-randomness* of the crosslinking process (Wun and Prins, 1974) or to a non-uniform swelling resulting from the *random* but *non-homogeneous* character of the crosslinking (Bueche, 1970). As we will see in the following, these reasons are not equivalent. In addition, in a theoretical paper, R. Stein (1969) predicted that the scattering by a gel should reflect the non-uniformity of the crosslinking and that the *excess scattering* with respect to a comparable solution should permit the characterization of the network inhomogeneity. Thus, in these papers, the main questions about gels were posed from comparison of the scattering with some thermodynamical predictions: the effect of the non-randomness and even the difference between *random* and *uniform*, regarding the distribution of crosslinks in space. Apparently, this distinction, introduced by Bueche, was not fully appreciated before being rediscovered (see section 5.6). Some of the conclusions were also rather close to those which are currently reached. However, we will not present and discuss these pioneering works in detail: it would force us to return to ways of thinking and representations of the data which subsequently evolved.

We will show in more detail the results of only one recent light scattering study (Soni and Stein, 1990). This work deals with poly(dimethylsiloxane) networks prepared in the bulk by end-linking of chains with tetrafunctional molecules tetrakis(dimethylsiloxane) in the presence of a platinum catalyst. In addition to unimodal networks, prepared with chains of molecular weight equal to $M_n = 22\,500$ and 770 ($M_w/M_n \approx 1.8$ in both cases) bimodal networks were also studied. In the latter, the proportions of short chains varied between 0 and 100%. The samples were subsequently allowed to swell at saturation in toluene or in benzene and their equilibrium swelling degrees were found to decrease from 5 to 1.9 approximately upon increasing the proportion of short chains. For the light scattering measurements, large scattering volumes were chosen in order to ensure small speckles (or, in other words, small coherence areas) and therefore allow many of them to be seen at the same time by the detector. In most cases, the experimental scattering intensities were found to be larger than calculated for an idealized network, having the same degree of swelling and the same crosslinking density. It is worth noting that the theory used for the calculation did not consider the effect of the shear modulus, as in Eq. (63). Thus the excess scattering with respect to idealized gels seems to be slightly underestimated by this approach. Examples of values obtained under such assumptions for the excess scattering intensities, for short chains content (SC) varying between 0 and 100%, are shown on Figure 10. Note that the excess scattering is maximum for 50% short chains and that the 100% short chains network is the least swollen.

As in the earlier studies, this excess scattering was attributed to the presence of crosslinking heterogeneities producing nearly frozen-in spatial fluctuations of polymer concentration. The data were analysed within the framework of the Debye–Bueche theory, which, indeed, was developed for describing heterogeneous media (Debye and Bueche, 1949; Debye, Anderson and Brumberger, 1957). In this approach, already encountered in section 5.2, one postulates an exponential form for the spatial correlation function of the number density of scatterers, which leads to a Lorentzian squared for the scattered intensity (see above Eq. (47)). In practice, with this scheme, one tries to plot $(\mathrm{I}(q))^{-1/2}$ as function of q^2 and, if a straight line is obtained, one obtains from its slope an estimate of the Debye–Bueche correlation length a_c. For these data, approximate straight lines were effectively found (in the rather narrow range of q allowed by light scattering technique), and values of a_c ranging between 300 and 400 Å were obtained for nearly all the gels of the series, whatever the short chains content. Such values are rather large when compared with the average size of the chemical mesh of the network, which agrees well with an interpretation in terms of crosslinking heterogeneities: the correlation length a_c corresponds to several mesh sizes. From the same approach, one may also estimate $\langle \delta\phi^2 \rangle$ from $\mathrm{I}(q \to 0)$ (one expects $\mathrm{I}(q \to 0) \sim \langle \delta\phi^2 \rangle a_c^3$; see Eq. (47). In the framework of the Frenkel–Flory–Rehner

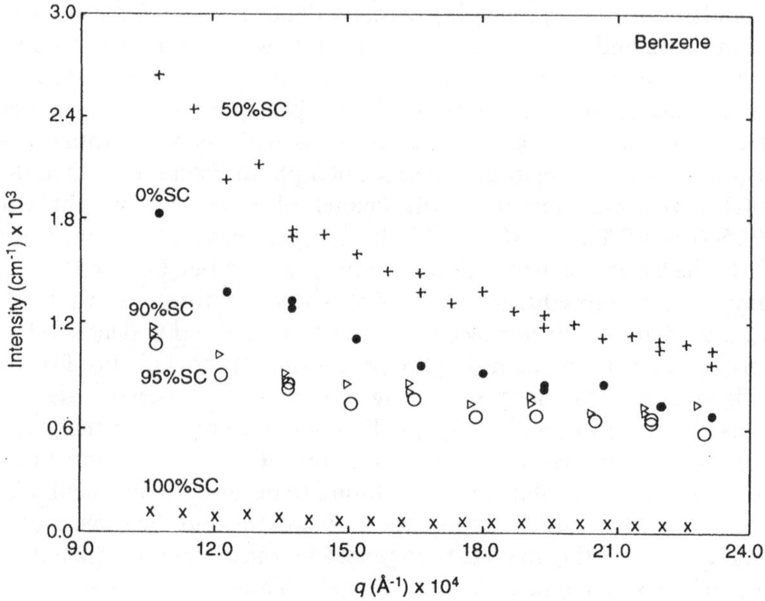

Figure 10. Absolute light scattering intensities versus scattering vector amplitude for bimodal PDMS swollen in benzene; % SC stands for the content of small precursor chains in the network (see text). From Soni and Stein (1990)

theory (see appendix), local fluctuations of swelling degree were previously related to local fluctuations of an effective and local mesh size M_c (Stein, 1969). With this scheme, one can therefore associate with $I(q \to 0)$ an heterogeneity parameter

$$H = [\langle \Delta M_c^2 \rangle]^{1/2} / \langle M_c \rangle$$

through $\langle \delta\phi^2 \rangle$. But, in the case of bimodal networks, the values obtained for H were found to be smaller than that calculated from the known proportions of long and short meshes, describing the gel as a close packing of them. This discrepancy was attributed to the interspersion of the short and large chains. In summary, it was concluded that the non-uniformities of polymer concentration in the bimodal gels were weaker than expected, but spanned surprisingly large length scales (with respect to the larger precursor chains).

Comments After confronting classical thermodynamical theories, the attribution of the excess scattering to the presence of crosslinking heterogeneities is the more natural interpretation. However, from such *static* and *ensemble-averaged* scattering experiments, one cannot discard another explanation proposed by Rabin and Bruinsma (1992), and by Bruinsma and Rabin (1994). These authors indeed developed a theory which predicts the amplification upon swelling of the *thermal fluctuations* of polymer concentration (see section 5.6). In gels significantly swollen with respect to their preparation state (it is the case here), such a mechanism should also lead to a large excess of scattering intensity with respect to solutions which would have nothing to do with structural heterogeneities. But this excess intensity would then fluctuate in time instead of being quenched, as it should be in the case of overscattering associated with structural non-uniformities.

Other types of experiments are therefore needed to distinguish between the two possibilities.

A controversial point in this study is the use of the Debye–Bueche plot for extracting a_c and $I(q \to 0)$. It means indeed that the whole structure factor could be fitted by a squared Lorentzian. In the limited q range of light scattering experiments, it is not possible to test this assumption: here one remains typically in the regime $qa_c < 1$.

5.4.1.2 *SANS and X-ray studies—typical examples of gel scattering spectra*

We show in Figures 11(a)–(d), typical examples of scattering spectra (plotted on a log-log scale) of neutral gels at maximum swelling. The measurements have been performed essentially by small angle neutron scattering in a domain of q ranging approximately in between 0.006 and 0.01 Å^{-1}, but in one case some light-scattering data are added. In order to a produce a contrast, the networks, made of protonated polymer, were swollen in a *deuterated* good solvent. It can be seen at once that the scattering intensity from the gels is in all cases much larger for the smaller q values than that from the semi-dilute solution of the same (or of comparable) polymer concentration. Conversely, on the larger q values, the signals of the gels and those of the comparable semi-dilute solution are approximately superimposed. The slight differences which can be seen in these examples are of the same order of magnitude as the uncertainties involved in the experiment. The latter are those which affect the estimate of the absolute intensity (through the measurement of the transmission and the sample thickness: $\pm 10\%$) and that (typically $\pm 7\%$) arising from the difficulty in measuring precisely the maximum swelling degree of a small sample. Note, however, that the systematic superposition, at high q's, of the spectra of gels at maximum swelling and those of comparable solutions is controversed, since in some cases, larger differences are apparently found (Geissler, Horkay and Hecht, 1991). For the discussion of this point, see sections 5.4.3 and 5.6.

Beyond these general characteristics of the gel structure factors, there are some noticeable differences in detail that will now be reviewed.

Figure 11(a) (from Mendes *et al*, 1991; see also Bastide, Boué and Buzier, 1989; Bastide *et al*, 1990) shows the scattering spectrum of a 'statistical gel' prepared by crosslinking of a semi-dilute solution (in 1,2-dichloroethane) of polystyrene ($M_W \approx 7 \times 10^5$) at a polymer volume fraction $\phi_c = 0.1$. Molecules of 1,4-bis(chloromethyl)benzene at a concentration $R_c = 0.8\%$ (in moles per mole of monomer units) were employed as crosslinkers: they established, at random, tie-points between the chains. The reaction of the Friedel and Crafts type and involving a co-reagent (SnC14, 2% mole mole^{-1}), underwent upon heating at 60 °C. For more detail about such syntheses, see Antonietti and Sillescu (1985) or Collette *et al* (1988). The obtained gel was washed in tetrahydrofuran (THF), dried and then swollen at saturation in deuterated toluene ($\phi_e = 0.042$, or $Q_{max} = 1/\phi_e = 24$, which means an 'overswelling' of 2.4 by volume with respect to the preparation conditions). Despite the larger intensity at low q, the shape of the scattering spectrum of this gel looks rather similar to that of a semi-dilute solution. On the log-log plot presented, one can indeed observe a tendency towards a saturation at low q, followed, at larger q, by a crossover towards an apparent (and rather short) 'fractal regime': I(q) seems to be tending towards an

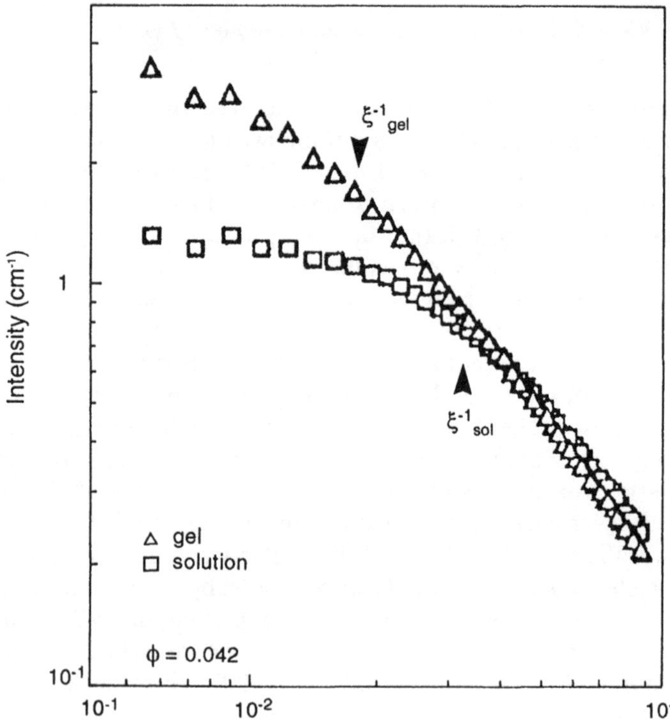

(a)

Figure 11. (*Contd*)

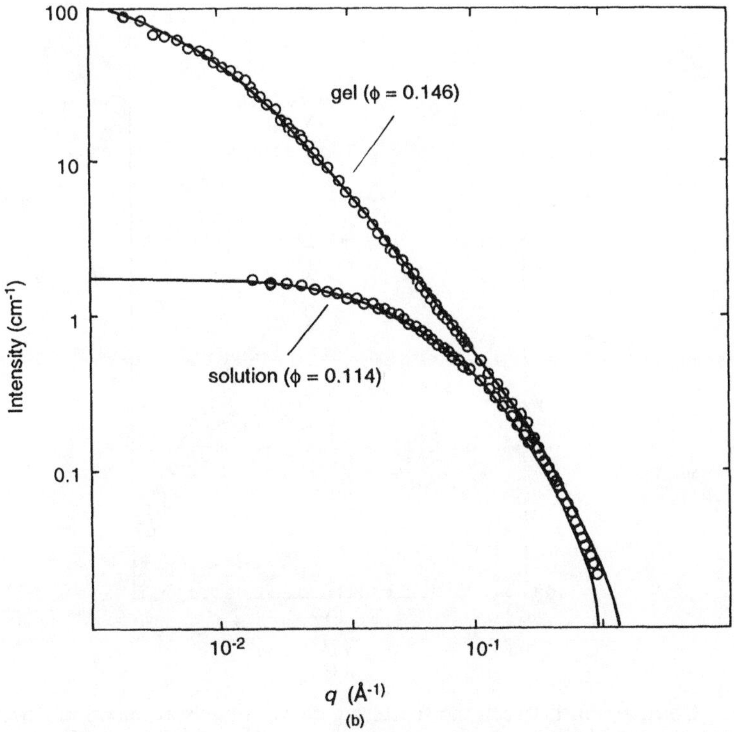

(b)

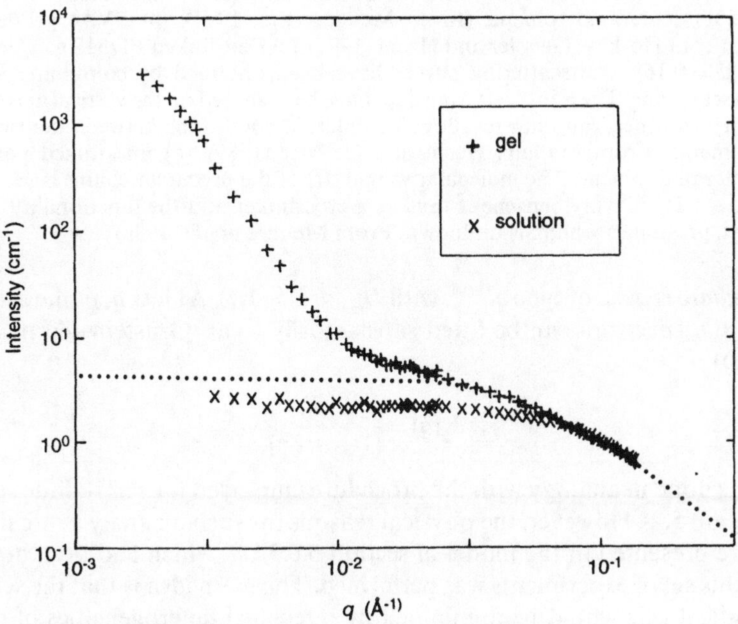

(c)

Figure 11. *(Contd)*

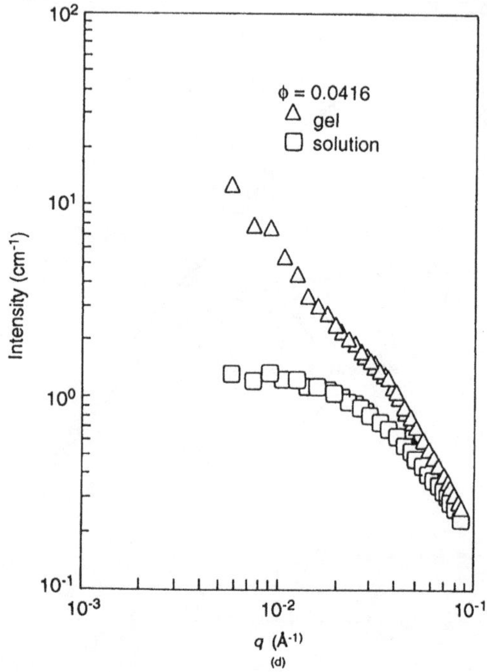

Figure 11. Comparison between the scattering curves of gels at maximum swelling and those of solutions at about the same polymer concentration. (a) 'Statistical' polystyrene gel ($R_c = 0.8\%$)/deuterated toluene. (From Mendes *et al*, 1991). (b) PVAc gel/deuterated toluene. From Horkay, Geissler and Hecht (1992). (c) End-linked PDMS gels/deuterated octane ($\phi = 0.16$). The scattering curves have been obtained by combining light and neutron scattering. The light scattering data have been shifted on the vertical axis to obtain continuity with neutron data to allow for different conditions between the two sets of measurements. From Mallam, Hecht and Geissler (1989). (d) End-linked polystyrene gels/deuterated toluene. The molecular weight M_c of the precursor chains is $M_c = 22\,000$ ($M_w/M_n \leqslant 1.15$). Divinylbenzene is used as a crosslinker and the functionality f_x of the junctions, presumably high, is unknown. From Mendes *et al* (1993a)

intermediate regime of type $q^{-D_{\rm eff}}$, with $D_{\rm eff} \approx 1.5$–1.7). At low q, in the saturation region, the spectrum can be fitted satisfactorily to an Ornstein–Zernicke form (Eq. (38))

$$\mathrm{I_g}(q) = \frac{\mathrm{I_g}(0)}{(1 + q^2\xi_g^2)} \tag{86}$$

This was done in analogy with the procedure employed for semi-dilute solutions (see section 5.2). However, the physical reasons for such an analysis are different. They are presented in the model of section 5.6.3.1.2, which had been developed before this set of experiments was performed. The main idea is that the scattering of statistical gels would be dominated by quenched heterogeneities of polymer concentration, coupled to cluster-like heterogeneities of crosslinking density. The former as well as the latter would result from the random character of the gelation process, but the former would be largely enhanced by the overswelling with respect to the state of preparation. With this procedure, one extracts from the data an extrapolated intensity $I_g(q \to 0)$ and a 'correlation length' ξ_g. Here ξ_g is assumed to represent a sort of inter-cluster correlation length and not the size of the heterogeneities. Obviously, because of the differences in the two levels of intensity in the two systems, one finds rather different values for the gel and for the comparable solution: $\xi_g = 62$ Å and $\xi_{sd} = 26$ Å; $I_{gel}(0) = 3.7\,cm^{-1}$ and $I_{sd}(0) = 1.35\,cm^{-1}$.

As mentioned above in section 5.2.1.2.2, polystyrene semi-dilute solutions often exhibit an unexpected upturn at very low q. This effect is not visible on Figure 11(a) because, at the concentration $\phi = 0.042$ of both the solution and the gel, the upturn superimposed on the low q levelling off appears only for $q < 0.005$ Å^{-1}, a range of q which was not reached in this experiment. It is also present in the case of the same polystyrene gels (Mendes, 1991), and the magnitudes of the effect are comparable in gels and solutions (see above Figure 3). As stated above, we believe that this effect has little to do with the gel 'structure' and will, in practice, be ignored in the following, especially when discussing the extrapolated intensity $I(q \to 0)$ and the correlation length ξ_g. We recall, however, that this effect (possibly arising from interchain associations) might perturb the estimation of $I(q \to 0)$ and ξ_g.

Comments In short, these statistical polystyrene gels exhibit scattering curves having a general aspect resembling those of semi-dilute solutions. The complete signal of the gel cannot be fitted satisfactorily by a single squared Lorentzian as proposed by Stein (see above), when studying light scattering by PDMS gels.

However, as in Stein's experiments, the signal from the gel is much stronger at low q than that of the comparable solution and exceeds largely what is expected from classical thermodynamic estimates: recall that in section 5.3, it was shown that $I_g(q \to 0)$ for idealized gels should be essentially smaller than or equal to $I_{sol}(q \to 0)$. But, as before, one cannot be sure that the overscattering arises from structural heterogeneities: the Rabin–Bruinsma explanation (see section 5.6) could also qualitatively account for the observations. This is valid for all data presented in subsection 5.4.1.

As comparison is performed at the same polymer concentration, an approximate superposition at high q is found. Note that no effect of local chain stretching can be seen on the scattering curves. In fact, a significant stretching would lead to a softened decay of $I(q)$ with q, for $q\xi_{sd} > 1$. Although such an isotropic elongation associated with swelling has never been detected by scattering experiments, one cannot exclude a masking of the effect by the overscattering at lower q's.

We will refer below to this category as that of type 1 gels. Swollen networks which seem to fall in this class are as follows:

1. Gels made from semi-dilute solutions of weakly chloro-methylated polystyrene (fixed at random on the chains) crosslinked through a Friedel and Crafts reaction (Zielinski *et al*, 1992). We will refer later to them as chloromethylated polystyrene gels.
2. Polystrene gels prepared by end-linking from chains (with a narrow distribution of molecular weights) terminated by hydroxyl groups with tri-isocyanates (Hakiki, 1991; Mendes *et al*, 1993). We will refer to them later as urethane-polystyrene gels.
3. Some of the polydimethylsiloxane gels (those with the higher crosslinking density) studied (Falcao, Skov Pedersen and Mortensen, 1993) and prepared by random crosslinking (electron irradiation) in the bulk of PDMS chains ($M_w = 77\,500$, $M_n = 16\,000$). These networks were swollen at saturation in deuterated xylene. The data are analysed in the framework of a model assuming the existence of cluster-like and fractal heterogeneities, similar, to some extent to the frozen-blob model (see section 5.6). Apparently, the variation of $I(q)$ was not suitable to be fitted with a squared Lorentzian (see p. 183).
4. Poly-methylmethacrylate gels made by a radical copolymerization of a monomer and a small proportion of a crosslinker also seem to fall in this category (Lal *et al*, 1993), but the experiments have not been performed at maximum swelling.

Figure 11(b) (from Geissler *et al*, 1992; see also Horkay *et al*, 1991a, 1993a, b; Hecht *et al*, 1991; Geissler, Horkay and Hecht, 1993, 1994) shows the scattering curve of a gel of poly(vinyl acetate) prepared by acetylation of a chemically crosslinked poly(vinyl alcohol) gel. This precursor gel was made by random crosslinking of an aqueous solution of polymer concentration 6% w/w. The concentration of crosslinks was $R_c = 1\%$ which, assuming complete reaction, leads to a ratio of monomer units to crosslinker of 50. Note that the concentration $\phi_e = 0.146$ at maximum swelling is larger than in the preceding case and rather close to the concentration of preparation. Figure 11(b) also represents the structure factor of a semi-dilute solution of concentration $\phi = 0.114$, i.e. not equal to that of the gel, but not radically different. At high q, there is a downwards curvature in both cases, which was not visible in Figure 10. It may arise from the fact that the crossover regime around $1/\xi_{sd}$ is shifted in the present case to larger q values: ϕ_e being larger, ξ_{sd} is smaller. According to Horkay *et al* (1991a), this curvature arises, in addition, from the finite thickness of the chain, which may lead to a steeper decay for $q > 0.1$ Å^{-1} (see Rawiso, Duplessix and Picot, 1987). This question is of secondary importance to our main concern which is the gel structure and will not be discussed further.

On low q's ($q < 0.01$ Å^{-1}), the gel structure factor starts to level off, which means that the larger correlation distance in the system must be of a few hundreds of angstroms. In this respect, the form of I(q) for this gel is comparable with that of Figure 10, while correlations seem, however, to extend here over larger length scales. But there is, on the other hand, a difference which is visible at intermediate q values: the structure factor has—very slightly—the aspect of a double shoulder. We will see below that this character is enhanced as the same gel is deswollen. This double shoulder aspect has been interpreted by Mallam *et al* (1989), Horkay *et al* (1991a), Hecht *et al* (1992) and Geissler, Horkay and Hecht (1993) through the introduction of two more or less independent scattering contributions (a Lorentzian, I_F, comparable with that of semi-dilute solutions and a stretched exponential $-I_S$). They proposed to write the total intensity $I(q)$ scattered by the gel as follows:

$$I(q) = I_S(0)\exp[-(q\Xi)^p] + \frac{I_F(0)}{(1 + q^2\xi_F^2)} \tag{87}$$

where $I_S(q)$ and $I_F(0)$ represent respectively the 'solid-like' contribution (i.e. arising from *static* fluctuations related to heterogeneities of crosslinking) and the fluctuating one (i.e. that due to random fluctuations resulting from thermal motion). Here $I_F(0)$ is identified with the scattering of an idealized gel (see section 5.3 (Eq. (57)), that is

$$I(0) = \frac{b_v^2 k_B T \phi^2}{(K_{os}^{gel} + 4/3\mu)}$$

We discuss in more detail this approach in section 5.6. At this level, it involves two separate correlations distances Ξ and ξ_F. The former, Ξ, is frequently associated by Geissler *et al* with the size of the heterogeneities.

Comments In short, such samples again show a small angle scattering which is much larger than that expected from an idealised gel. The difference is even bigger than for the gel of Figure 10 and the shape of the signal is slightly different.

Because of these differences, we call 1′ the class of gels whose structure factor, like that of Figure 11b, 'saturates' at low q but cannot, however, be described by a single Lorentzian, nor by a squared Lorentzian. The feature we associate typically with this category 1′ is the presence of a kind of inflexion point in the middle of the 'intermediate regime' (or 'double shoulder'; see more typical examples in Figure 16d). Equation (87) can be used to describe such behaviour, but is not the only one. For instance, the scattering envelope proposed by Falcao, Skov Pedersen and Mortensen (1993) is also able to fit satisfactorily scattering curves with double shoulders.

In this class 1′ fall apparently the following systems:

1. Polyacrylamide gels prepared by radical copolymerization with small amounts of methylene bisacrylamide (Hecht, Duplessix and Geissler, 1985; Mallam, Hecht and Geissler, 1989). But the thermodynamical behaviour of these polyacrylamide gels (variation of the elastic and osmotic bulk modulus with the equilibrium concentration) seems to deviate frequently from those of

other gels. This may be related to a non-random distribution of the bisacrylamide in the gel (see section 5.4.2).

2. Some polydimethylsiloxane (PDMS) gels prepared in the bulk either by statistical crosslinking (γ-irradiation, electron irradiation) (Mendes *et al*, 1993b; Falcao, Skov Pedersen and Mortensen, 1993; Rouf (1995)) or end-linking of short chains (prepared, e.g. by reaction of α, ω, diSilvinyl + cyclic tetramers carrying 4SiH groups (Rouf 1995)). It is worth noting that, when increasing the crosslinking density in the same PDMS system, Falcao *et al.* observed a transition from the double-shoulder (1′) behaviour to the single-shoulder behaviour (1; see above).
3. Possibly polystyrene networks prepared by γ-irradiation in the bulk and swollen afterwards (Davidson, Richards and Machonnachie, 1986). In fact, Davidson *et al.* do not display scattering curves in a log-log representation. They show, for $q > 3 \times 10^{-2}$ Å^{-1}, Ornstein–Zernicke plots (or Zimm plots) that is $1/I(q)$ versus q^2, from which they extract values of a correlation length ξ. They report the existence of a strong increase of scattered intensity at lower q's, but they discard this signal for the determination of ξ. The feasibility of such an analysis suggests the presence of a second 'shoulder' at rather large q's, similar to those first reported by Geissler *et al.* Therefore, the values of ξ they find are very likely comparable with the small correlation length of Geissler *et al.* or of Falcao, Skov Pederson and Mortensen (1993). See section 5.6.

On Figure 11(c) (taken from Mallam, Hecht and Geissler, 1989) is represented a scattering spectrum which looks somewhat different from the preceding ones: it is composed of two distinct parts, a very strong signal at low q and, at higher q's, a smoother one. It is that of a network of polydimethylsiloxane (PDMS) prepared, in bulk, by reaction of the hydroxylated end groups of PDMS chains with 4% w/w of ethyl triacetoxy silane (ETAS). From viscosity measurements the average molecular weight of the precursor chains was estimated to be $M_c = 40\,000$. The network is swollen at saturation in deuterated octane. In its smoother part, for $q > 10^{-2}$ Å^{-1}, the scattering intensity of this gel resembles that of gels of classes 1 or 1′. The high scattering for $q \leqslant 10^{-2}$ Å^{-1} looks quite similar to the Picot–Benoît effect mentioned above. It is not connected to the gel structure itself, but, its origin is different. It was indeed shown that it can be attributed to the presence of poly(ethylsiloxane) particles (PES) embedded in the gel matrix. These particles were formed by excess ETAS molecules which reacted with each other. In order to demonstrate this, Mallam *et al.* utilized the method of contrast variation which is summarized below.

When performing a neutron scattering experiment on a binary system (in principle, like a network swollen by a diluent), one measures a scattering intensity I(q) proportional to the structure factor $S(q)$ via the contrast factor $b_v^2 = (b_p/v_p - b_s/v_s)^2$, where b_p and v_p denote respectively the scattering length and the volume of the scattering unit of the polymer, b_s and v_s represent the corresponding quantities in the case of the solvent (see section 5.1). In the case of the polymer, the scattering unit is, in general, identified with the repeat unit (or 'monomer unit') of the chain. In the case of a pure solvent, it is natural to chose the solvent molecule. In order to obtain a high contrast, one of these two components, say the solvent, is in general deuterated. But studies can also be performed using a mixture of solvents, for instance a mixture of deuterated and non-deuterated solvents (with scattering length, per molecule, respectively equal to b_s^h and b_s^d). In this situation, since the volume of the solvent molecule is the same in both cases, one defines an 'effective' scattering unit for the solvent whose scattering length density b_s is given by

$$b_s = [n^d b_s^d + (1 - n^d) b_s^h]/v_s \tag{88}$$

where n^d represents here the number fraction of deuterated solvent molecules in the solvent mixture.

If the system studied is effectively binary, by changing n^d one simply varies the contrast and, eventually, finds a composition for which the coherent signal vanishes: a *contrast-match* is then achieved. It is useful to determine experimentally this point when one desires to introduce a third component in the system in order to study its structure or conformation. The data are then simpler to analyse since, owing to the contrast-match of the two first components, the ternary system is transformed, as far as scattering is concerned, into a binary one.

Another use of the contrast variation method is to check whether a system is really a binary one. Suppose that it is not the case and that some dusts or some 'foreign' residues are present in the gel. These objects may give rise to a strong 'stray' scattering even if their number density is low, provided their size is large. For simplicity, we assume that there is no important 'cross-term' between the contributions to the scattering of the gel itself and that from these defects. Thus, we write the total scattering intensity $I_T(q)$ which is measured as a sum of two terms

$$I_T(q) = b_v^2 S_{gel}(q) + b_v'^2 S_{imp}(q) \tag{89}$$

where $S_{gel}(q)$ and $S_{imp}(q)$ denote respectively the structure factors of the gel alone and of the 'impurities'; b_v^2 and $b_v'^2$ are the corresponding contrast factors.

Incidentally, note that if the impurities are sufficiently dilute, S_{imp} can be expressed approximately as

$$S_{imp}(q) = \phi_i V_w P(qR_g) \tag{90}$$

where $P(qR_g)$ denote the 'form factor' of the impurities, V_w and ϕ_i are respectively the (weight average) molecular volume (in cm^3) of the impurities and their volume fraction.

If the scattering length density $b_i/v_i = \mathrm{b}_i$ of the impurities is different from that of the network chains, when varying the contrast the two terms in the right-hand side of Eq. (89) will not change on the same way. It might be possible to cancel each of them for different proportions of deuterated solvent. This is what Mallam *et al.* succeeded to do: they showed that the highest q region of the spectrum is cancelled when the scattering length density of the solvent is equal to that of the pure PDMS, whereas the lowest q region disappears for a different composition. In the last case, the extinction conditions were consistent with a signal originating from poly(ethyl siloxane) particles (whose scattering length density was calculated). The low q signal was, moreover, shown to vary as q^{-4} (Porod scattering) in the range of q's probed by the neutron scattering which is in agreement with an interpretation in terms of particles having rather well-defined boundaries.

As can be seen on Figure 11(c), Mallam, *et al* (1989) also performed light scattering experiments on the same samples, from which they determined an average radius of gyration R_g (PES) of the PES particles. Here R_g (PES) was found equal to 810 Å. They finally confirmed their interpretation by performing a contrast variation experiment by light scattering: they studied the same gel (at the same swelling degree $Q = 4$) in five different solvents of increasing refraction index. The contrast factor $b_v'^2$ is then proportional to $(n_{\mathrm{imp}}^2 - n_g^2)^2$, where n_{imp} and n_g denote respectively the refraction index of the impurities and that of the gel (polymer + solvent). The signal was found to be minimum for a refraction index of the gel equal to that of pure PES.

The contrast variation is therefore a valuable tool for the study of gels. It allows us to check whether the measured signal arises from the polymer network structure only or is combined with that of trapped impurities. This method was applied by Horkay *et al* (1991) to the case of poly(vinyl acetate) gels (similar to that of Figure 11) and permitted to show that, in such systems, all the scattered intensity must be related to the gel structure only.

Comments In summary, the PDMS gel whose scattering intensity in given on Figure 11(c) does not seem to belong to a class of networks significantly different from those of Figures 11(a) and (b). We have seen that the upturn at low q does not originate from the gel structure itself but from trapped impurities. If it was not present, one would observe a saturation at low q, with, however a scattered intensity significantly larger than that of the solution of same polymer concentration. It is not easy to decide whether the true structure factor (i.e. without the stray signal from the impurities) of this gel could be fitted like gels of class 1 by a single Lorentzian (plus, possibly, a crossover to a 'fractal' regime) or requires the use of two correlation distances as suggested for gels of class 1′. According to Mallam *et al* (1989a) the second option is more satisfactory, therefore a subcategory $1' + I$ (I for impurities) may be created for ranking this gel.

In a comparable category, can be put some 'amino-methylated' polystyrene gels prepared in two steps: (1) amino-methyl groups are fixed at random to the phenyl rings of the PS chains; (2), crosslinking is achieved by reaction of these groups with benzyl-dialdehyde (see Zielinski, 1991; Zielinski *et al*, 1992). Such gels exhibit a strong upturn in a range of q ($5 \times 10^{-3}\,\text{Å}^{-1} < q < 2 \times 10^{-2}\,\text{Å}^{-1}$) where a tendency towards a saturation is observed for two other types of polystyrene gels prepared by random crosslinking ('statistical' and chloromethylated gels) and also for the urethane polystyrene gels prepared by end-linking (the maximum swelling degrees and elastic moduli of these various types of PS gels are comparable). The scattering intensity of the 'amino-methylated' gels could be fitted by the sum of two contributions:

$$\mathrm{I}(q) = \mathrm{I}_1(q) + \mathrm{I}_2(q) = \frac{A_0}{(1 + q^2\xi_L^2)^2} + \frac{I_g(0)}{(1 + q^2\xi_g^2)} \tag{91}$$

where $I_2(q)$ is comparable with the structure factor of type 1 gels. On the other hand, $I_1(q)$, meant to describe the low q upturn, looks similar to the one employed by Koberstein, Picot and Benoît (1985) for the description of the Picot–Benoît effect (it has the same Debye–Bueche form, see above Eq. (47)). But remember that the unexplained overscattering at low q in semi-dilute solutions (and possibly in gels of type A) occurs at lower q values. Anticipating a bit with respect to section 5.4.2, it can be stated that whereas $\mathrm{I}_g(0)$ and ξ_g both *increase* upon swelling, ξ_L and A_0 both *decrease* upon swelling. Therefore the physical origins behind ξ_L and ξ_g, A_0 and $I_g(0)$ seem to be very different; $I_1(q)$ might again be a stray signal and was attributed to the presence of non-reacted polar amine groups which, by association in a non-polar solvent, might provoke the formation of heterogeneities having little to do with the network topology. This is why we will not insist on these kinds of gels in the following.

On Figure 11(d) is presented the scattering spectrum from another kind of gel prepared by end-linking (Bastide, Boué and Buzier, 1989; Mendes *et al*, 1993). Such networks are formed by anionic block copolymerization of styrene and divinylbenzene (DVB). In the present case, three DVB molecules per chain end are added after the polymerization of the chains is completed. The weight average molecular weight of the chains is 22 000 and the size distribution is narrow ($M_w/M_n \approx 1.2$). The DVB molecules can react with the living extremities of the chains and between themselves, thus they are believed to form some 'nodules' acting as crosslinks. The dependence with q of the scattering intensity from this gel is rather different from that of those preceding. Going from the larger values of q to the smaller ones, one observes first a rather fast increase from an approximate superposition with the solution. Note that there is here a small upward concavity which did not appear on the spectra of Figures 11(a)–(c). Then, one observes a neat change of slope (a more abrupt 'shoulder' that for the preceding gels) followed by a second regime of fast increase, which starts again with an upward concavity. At very low q the apparent variation of the intensity is close to q^{-3}. It was not possible to fit satisfactorily such a spectrum with either the sum of two

Lorentzians, the sum of a Lorentzian and a stretched exponential (Geissler, Horkay and Hecht, 1993) or the sum of a Lorentzian and a Lorentzian squared (as in Eq. 36). The abrupt shoulder around $q \approx 4 \times 10^{-2}$ Å^{-1} is, in particular, a source of difficulty. It was proposed to consist of a smoothened local maximum. Comparable maxima, i.e. at approximately the same q values, are indeed observed for solutions of star-shaped molecules (Duplessix, 1975; Richter *et al*, 1992) and are also predicted theoretically (Witten, Pincus and Cates, 1986). Following this idea, gels have been prepared either by statistical crosslinking or by end-linking of stars and their scattering intensity revealed, as expected, either a local maximum or a shoulder comparable to that of Figure 11(d), for approximately the same value of q (Mendes, 1991; Mendes *et al*, 1995a). Some excess intensity at low q was also observed in the case of gels of stars with respect to the solutions of stars of the same concentration. As a result, it was proposed to describe such DVB end-linked PS gels as networks whose 'basic element' is not a linear mesh, but a star of rather high functionality f ($f \geqslant 10$). Moreover, these stars are believed to have a large distribution of functionality, which might account for the large signal at low q.

Comment Since such gels cannot be ranked in the classes 1 or 1′, we open provisionally a category 2. In the following subsections we test the necessity or otherwise of a separate category for these gels.

5.4.2 EFFECTS OF THE CROSSLINKING DENSITY FOR GELS AT GIVEN ϕ

We will now review the main trends which have been observed in the scattering by gels as the crosslinking density is varied at constant polymer concentration. For simplicity, we choose the concentration at which the gels were prepared. This means that the experiment is performed either effectively in the reaction bath or after swelling, washing, exchange of solvent and deswelling back to the concentration of preparation.

5.4.2.1 Neutron and X-ray scattering

Our first example (see Figure 12(a)) concerns some gels which have already been mentioned above and which were ranked in category 1. They are made from semi-dilute solutions of weakly chloro-methylated polystyrene (fixed at random on the chains) crosslinked through a Friedel and Crafts reaction (Zielinski *et al*, 1992). As before, the solvent (toluene) is deuterated, and one investigates here the effect of the crosslinking on the correlations in the reaction bath itself at constant polymer concentration (equal to $\phi_{\text{prep}} = 0.085$, or $Q_{\text{prep}} = 11.75$). Also on Figure 12(b) is plotted (filled circles) the signal from the precursor semi-dilute solution. As the crosslinking density is increased (which is measured here by the decrease of the swelling degree at saturation Q_{max} indicated in the figure caption), there is an increase of the scattering intensity at low q which is not very strong but still clear. This means that the fluctuations in space of polymer concentration are amplified by the introduction of crosslinks. The shape of the signal is not strongly affected: it can be fitted by a Lorentzian like that of Eq. (86). On doing so, one finds that $I(q \to 0) = 1$, 1.08 and 1.73 cm^{-1} and $\xi_g = 16.8$, 17.9 and 24 Å for the three densities of crosslinking: the correlation length ξ_g associated with the spatial fluctuations also appears to be enhanced.

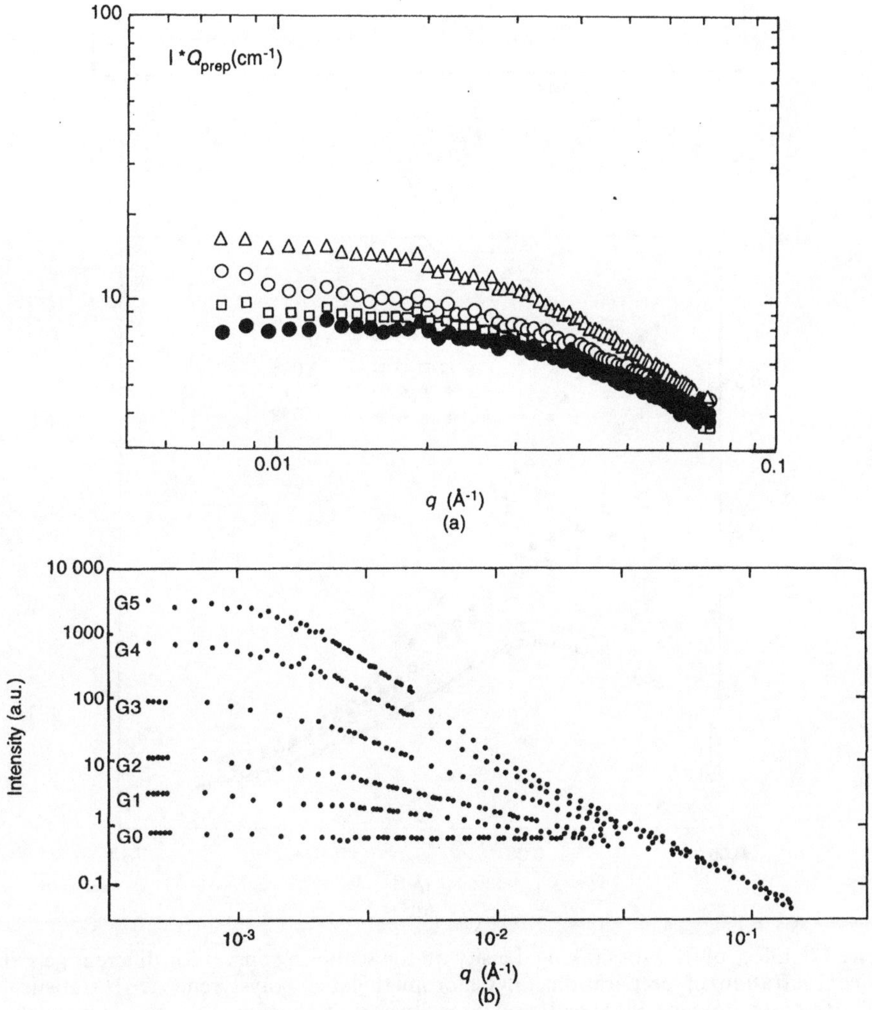

Figure 12. (*Contd*)

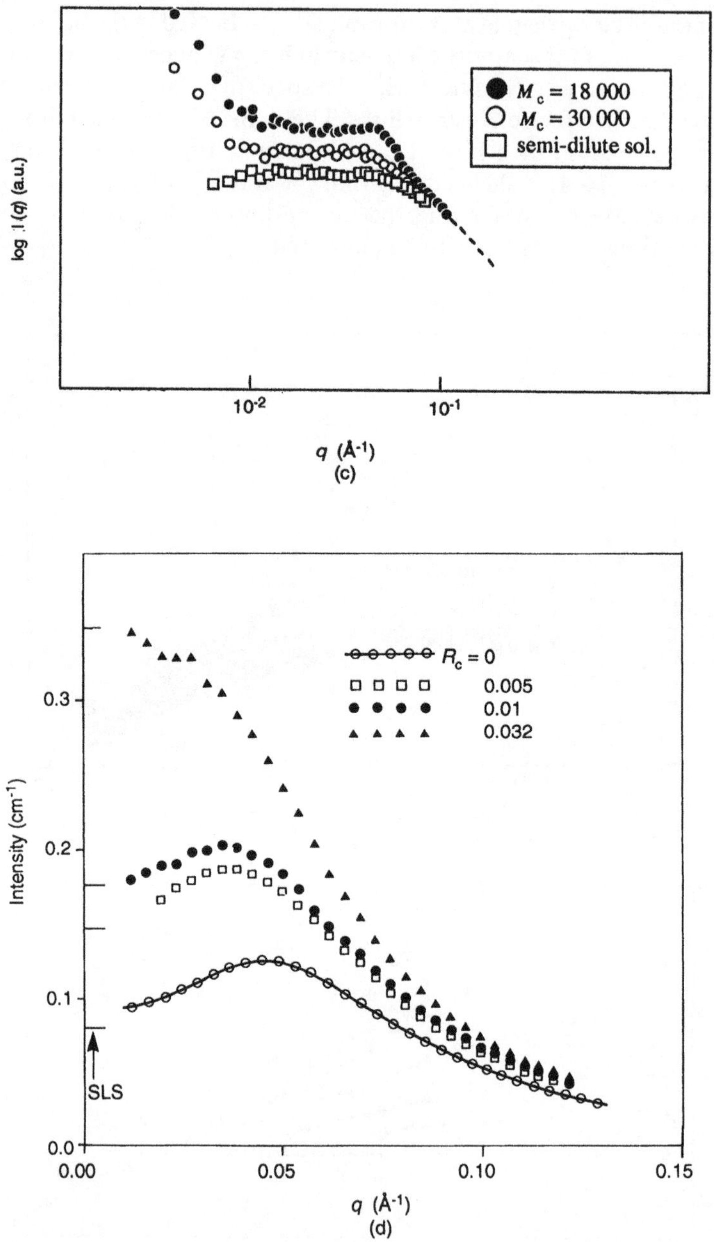

Figure 12. Effect of the crosslinking density on the scattering curves for different gels at the concentration of preparation. (a) Chloromethylated polystyrene gels ('statistical gels')/deuterated toluene. Here scattered intensity is multiplied by Q_{prep}, or, equivalently, normalized by ϕ_p the polymer volume fraction in the reaction bath. From Zielinski *et al*

(Contd.)

Comments As in those of the preceding section, this behaviour does not correspond to that of 'idealized gels'. According to Eqs (66) and (67), when increasing the crosslinking density, one should, in principle, act only on the shear modulus and observe a decrease of $I(q \to 0)$. Since the evolution is opposite, the first option is to assume that a raise of the crosslinking density amplifies the 'thermal' fluctuations of polymer concentration because of a mechanism overlooked in the derivation of Eq. (66). In this regard, one may first imagine a decrease of the overall solubility of the polymer (or an increase of χ) when the network is formed. Such a process has been proposed by a number of authors (e.g. Yen and Eichinger, 1978; Brotzmann and Eichinger, 1981, 1982, 1983; Shibayama *et al*, 1992; Cohen *et al*, 1992). Still with the option of enhanced dynamical fluctuations, one may invoke the coupling mechanism (between elasticity and fluctuations of concentration) proposed by Rabin and Bruinsma (1992, Rabin, Perkarski, Bruinsma 1993, Bruinsma and Rabin 1994), which is different in terms of physical origin but comparable in its practical manifestation. We summarize this model in section 5.6.1. But note that, in order to be effective, such explanations of the scattering enhancement should involve a decrease of K_{mix} under crosslinking larger than that measured by Geissler–Hecht–Horkay–Zrinyi (which is, in general, limited to roughly 35%).

The other option is, again, that *quenched fluctuations* of polymer concentration are formed and enhanced under the introduction of crosslinks in the system. In this case the question remains why the correlation distance seems to be enhanced in a rather continuous manner, from a starting-point which is the correlation length of the semi-dilute solution. The formation of more or less isolated 'lumps' (that is 'convex islands' of higher crosslinking density) might produce a growth of a signal in a completely different range of q (as observed in the case of crosslinks aggregates in PDMS gels, see preceding section).

What makes such a discussion specially meaningful is the fact that other gels, ranked above in category 1, seem to behave in a very comparable manner. In the case of the 'statistical' polystyrene gels, the behaviour is almost exactly the same, except that, for the smaller crosslinking densities (which correspond, however, to

---

(1992). The gels are identified by their swelling degree at saturation in toluene Q_{max} (the larger Q_{max}, the smaller the crosslinking density). (●) semi-dilute solution (□) $Q_{max} = 40$, (○) $Q_{max} = 37$, (△) $Q_{max} = 17$. (b) Polyacrylamide gels/D_2O in the reaction bath. G_0: $R_c = 0$ (semi-dilute solution); (G_1): $R_c = 0.53\%$, (G_2): $R_c = 1.06\%$, (G_3): $R_c = 1.58\%$, (G_4): $R_c = 2.11\%$, (G_5): $R_c = 2.64\%$ in mole of crosslinker/mole of monomer. From Hecht, Duplessix and Geissler (1985). (c) End-linked 'anionic' polystyrene gels/deuterated toluene. The gels have been prepared by anionic block copolymerization of styrene and divinylbenzene. Molecular weights M_c of the precursor chains are as indicated in the figure. The larger M_c the smaller the crosslinking density. (From Bastide-Boué (1988), unpublished results.) (d) Polyacrylic acid gels/D_2O. $C_p = 0.707$ M and $\alpha = 5.1 \times 10^{-2}$. The values of R_c are as indicated. The small bars at very small q represent the light scattering results. From Moussaid, Candau and Joosten (1994).

'good' gels with essentially no sol fraction), the difference with respect to the solution can be weaker (Mendes 1991). On the contrary, in the case of poly-(methylmethacrylate) gels (Lal *et al*, 1993) the difference between the gel and the solution is bigger, although a Lorentzian fit still seems satisfactory. As we have noticed, the fact that such gels are directly synthesized from monomers—and not from pre-existing chains—might be of importance.

We turn now to our second example of behaviour in the reaction bath, namely the scattering by polyacrylamide gels, prepared from random copolymerization of acrylamide (AA) and methylene bisacrylamide (BAA). (See Hecht, Duplessix and Geissler (1985), Mallam *et al* (1989)). In Figure 12(b) is presented the evolution of the scattering spectrum as the amount of bisacrylamide (1, 2, 3, 4, 5 $\mathrm{mg\,ml^{-1}}$ or $C = 1.25\%$, 2.5%, 3.75%, 5%, 6.25% weight fraction BAA/AA or, $R_c = 0.53\%$, 1.06%, 1.58%, 2.11%, 2.64% in $\mathrm{mol\,mol^{-1}}$) in the gel is enhanced. The lower curve corresponds to an uncrosslinked solution. The polymer concentration in the reaction bath $c_p = 0.08\ \mathrm{g\,cm^{-3}}$ (or $\phi_{prep} \approx 0.056$) is the same for all samples. The spectra were obtained by a combination of light, X-ray and neutron scattering experiments. It can be seen that the variation with the degree of crosslinking is drastic, since the excess with respect to the solution, for the largest amount of bisacrylamide, passes three decades. Therefore the spatial fluctuations of polymer concentration are here strongly amplified as a result of the introduction of crosslinks. Note, in addition, that the gel spectra exhibit a change of slope in the middle of the intermediate regime, for $q \approx 10^6\ \mathrm{cm^{-1}}$ (or $q \approx 10^{-2}\ \mathrm{Å^{-1}}$), which allows us to classify such a system in category 1′. There is, however, a difference with other systems, like polyvinylacetate gels, which exhibit a comparable aspect. In the present case, contrast variation experiments (using SANS) did not lead to a linear dependence of the square root of the intensity with the deuteron content in the solvent ($H_2O + D_2O$) (see above). This anomaly was interpreted as a non-homogeneous distribution of the crosslinker, whose scattering length density is different from that of the polymer. It is therefore clear that the excess scattering in these gels at low q (with respect to the signal of the semi-dilute solutions) cannot be explained solely by an amplification of thermal fluctuations.

Comments Such gels exhibit a dependence of the scattering spectrum with the crosslinking density which is even further at variance with the predictions of the 'idealized gel' model. Owing to the contrast variation, there is here no doubt that a part, at least, of the excess scattering with respect to a semi-dilute solution arises from the building up, with crosslinking, of quasi-static fluctuations in space of the polymer concentration. Other clues are given in the section 5.4.2.2 devoted more specifically to light scattering experiments. Another point of interest is that, again, there are wide differences between the scattering of a solution and that of gels which were prepared from monomers and not from chains. Here, there is some direct evidence that quenched fluctuations are involved in this excess.

Therefore, one may wonder whether the synthesis from monomers leads more especially to the formation of quasi-static heterogeneities.

Our third example will deal with end-linked polystyrene–DVB gels, which have been ranked above in category 2. Two scattering spectra are presented in Figure 12(c), together with that, as usual, of the semi-dilute solution of the same polymer concentration. The latter ($\phi = 0.1$) is very close to preparation concentration, but this is not an experiment actually performed in the reaction bath: the gels have been washed, dried and reswelled with the appropriate amount of deuterated toluene. The synthesis has been achieved with three DVB molecules per living end in both cases. But the average molecular weight of the precursor chains is 30 000 in one case (lower spectrum) and in the other 18 000 (upper spectrum). Therefore the crosslinking density is increased from the first situation to the second. As a result, similar to preceding examples, the scattering intensity is clearly enhanced at low q.

Comments An important point to notice in the present behaviour is that the evolution is qualitatively comparable to those preceding, although the shape of the signal is entirely different: one recognizes the abrupt 'shoulder' mentioned, round 7×10^{-2} Å^{-1}, which is here more pronounced than for the gel at maximum swelling shown in section 5.4.1), and the upturn at very low q. In both these regimes the intensity is increased as the average meshes are shortened. It is difficult to image how such a complicated shape of signal could be attributed solely to dynamical fluctuations. Therefore, in this system also, the enhancement of the scattered intensity is very likely almost entirely attributable to some amplification of 'quenched fluctuations' of polymer concentration.

Our last example of neutron scattering spectra in this section is devoted to polyelectrolyte gels, namely partially charged poly(acrylic acid) gels (Ilmain, 1991; Moussaid *et al*, 1993). See Fig. 12(d). Such systems have been prepared by radical copolymerization of acrylic acid and methylene bis-acrylamide in heavy water (D_2O). The experiment shown here is actually performed in the reaction bath (whose polymer concentration is $\phi_{\text{prep}} \approx 0.05$, or $c_p = 0.7$ M) and the ionization degree is $\alpha = 0.051$ for all samples (it has been obtained by addition of NaOH). Three spectra of gels with crosslinking densities R_c (in mole of crosslinker per mole of monomer) respectively equal to 0.005, 0.01 and 0.32 are presented together with that of the semi-dilute solution of the same ϕ. The latter exhibits, at a finite value of q, a maximum which has been discussed above. The gels with $R_c = 0.005$ and $R_c = 0.01$ also exhibit maxima whose positions are slightly shifted towards lower q values, and whose intensity is enhanced. In the case of the third gel, with the higher crosslinking density, the increase of intensity is even stronger and the maximum is no longer visible in the q range explored here. As previously, and although the shape of the scattering spectra is significantly different from that of neutral gels, the scattering intensity appears to be neatly enhanced at low q under an increase of the crosslinking density.

Comments As in the case of neutral networks, the behaviour which is observed here does not correspond to the one expected for 'idealized gels'. Equations (66) and (67) should indeed apply as well to polyelectrolyte systems and one should observe a decrease of $I(q=0)$ as the crosslinking density is enhanced, because of the resulting rise of the shear modulus. Note, however, that the effect should be less important than in neutral gels: the osmotic bulk modulus is here much bigger, due to the contribution of counterions (Schosseler Ilmain, Candau, 1991). Since $I(q \to 0)$ clearly increases, one faces the same question as before: does it come from the building up of quenched fluctuations or from an unexpected amplification of the thermal fluctuations? In this respect, an interesting observation was made recently by Schosseler *et al* (1994), namely that, by subtracting the scattering function of the solution from those of the gels, one finds in the high q regime a common asymptote in q^{-x}, with $x \approx 4$. This suggests the existence in the gels of domains limited by sharp interfaces supporting the interpretation in terms of heterogeneities. The next paragraph will provide some more clues in this respect.

5.4.2.2 *Light scattering results*

Some light scattering results have been presented already in this section and we will not come back to them in this paragraph. Our aim is to focus here on the results of experiments in which the sample is scanned in the laser beam (see section 5.1.1). This method, which has been applied as yet only to a limited number of systems, is well suited to determine which part of the scattered intensity must be attributed to dynamical fluctuations and which part to quasi-static ones. It must be recalled that a slightly different method (scanning the detector) has been developed earlier by Sellen (1987). For homogeneity and conciseness of presentation we do not discuss these earlier experiments, whose conclusions are comparable to the ones we give below.

In the first example of this section, we return to polyacrylamide gels prepared according to the standard method (see above in section 5.4.2.1) directly in light scattering cells, with $c_p = 2.5\%$ ($\phi \approx 0.018$), $R_c = 0.67\%$. An analysis along the lines given in section 5.1 revealed strong variations of the time-averaged scattered intensity $\langle I(q) \rangle_T$ with the position of the scattering volume, as shown in Figure 13 (from Joosten, McCarthy and Pusey, 1991). Figure 13 also shows a histogram representing an estimate of the probability distribution function $P(\langle I(q) \rangle_T)$ of the scattered intensities. The curve shows a cut-off at low intensities, which means that the detected intensity is never smaller than a certain threshold value. Since the detection was almost exactly performed on one single coherence area, it would have been possible, for entirely frozen scatterers, to obtain

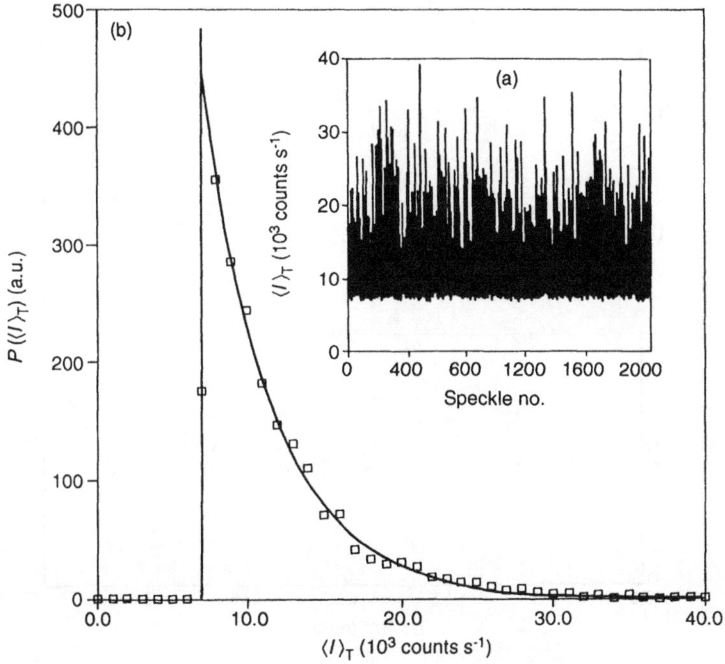

Figure 13. Inset (a) $\langle I(q)\rangle_T$ for various speckles as obtained by scanning through various positions in the sample. Data are taken from a polyacrylamide gel with $c_p = 0.025$ ($\phi \approx 0.018$) and $R_c = 0.68\%$ at a scattering angle $\phi = 150°$. (b) Probability distribution function, $P(\langle I(q)\rangle_T)$, of the scattered intensities shown in (a). Solid line: least squares fit to Eq. (92). From Joosten, McCarthy and Pusey (1991)

completely dark speckles on the detector (i.e. zero intensities) in some positions of the sample. Since this is not observed, the cut-off value can be attributed to the contribution of the thermal fluctuations to the scattered intensity $I_F(q)$, which therefore allows us to measure this quantity. In addition to this cut-off, one observes a negative exponential part, very similar to that observed for a frozen perfect gas or a glass (see section 5). In general, one may satisfactorily fit the probability distribution with the following expression:

$$p(\langle I(q)\rangle_T \sim H[(\langle I(q)\rangle_T) - I_F(q)]\exp\left\{\frac{\langle I(q)\rangle_T - I_F(q)}{\langle I(q)\rangle_E - I_F(q)}\right\} \qquad (92)$$

where $H(x)$ is the Heaviside step function.

Focus first on the ensemble-average scattered intensity $\langle I(q)\rangle_E$. As explained in section 5, $\langle I(q)\rangle_E$ can be determined from such scanning experiments, either by a direct averaging or from the fit of the data to Eq. (92). The two estimates are generally found to coincide very satisfactorily. In Figure 14(a) (from Joosten McCarthy and Pusey, 1991) is displayed the variation with q of the $\langle I(q)\rangle_E$,

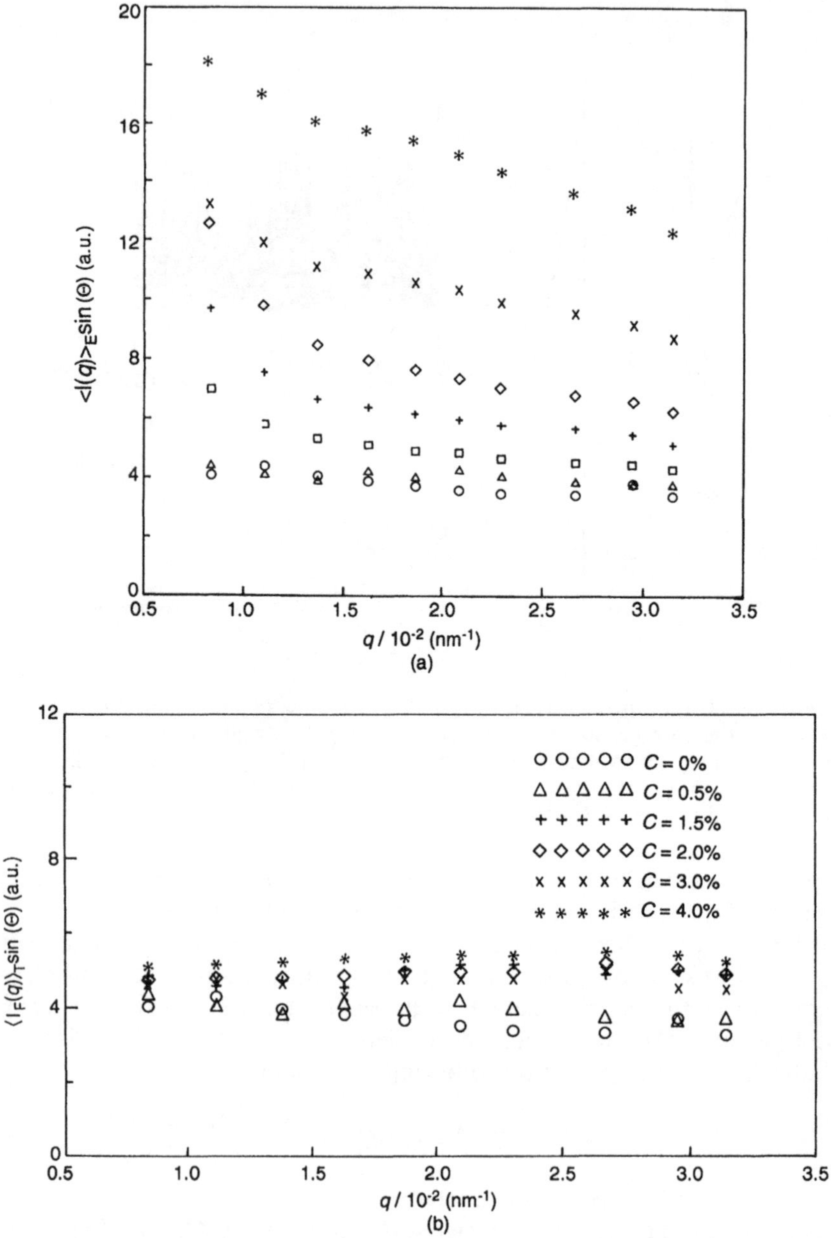

Figure 14. Ensemble averaged scattered intensity $\langle I(q)\rangle_E$ (a) and fluctuating component $\langle I_F(q)\rangle$; (b) versus scattering vector amplitude for polyacrylamide gels with constant polymer concentration and varying crosslink content. From Joosten, McCarthy and Pusey (1991)

obtained for gels of various crosslinking densities (weight fraction BAA/AA: $C = 0.5\%$, 1%, 1.5%, 2%, 3%, 4%, that is $R_c = 0.21\%$, 0.42%, 0.63%, 0.85%, 1.26%, 1.68%) and a semi-dilute solution. The polymer concentration is the same for all the samples ($c_p = 2.5\%$ in g g^{-1} or $\phi_{prep} \approx 0.018$). As for all the situations already encountered, $I(q \to 0)$ increases significantly with the crosslinking density. If one refers back to the SANS data reported in the preceding paragraph, one observes that the effect here is larger than for the various polystyrene gels prepared by crosslinking of pre-existing polymer chains. On the other hand, it is comparable to that observed for PMMA gels prepared from monomers. More surprisingly, it seems, at first sight, to be smaller than for the PAM gels already discussed above (Figure 12(b)). This may be due to the fact that (1) the weight fractions C (or R_c) of bisacrylamide crosslinks per acrylamide monomer unit are smaller for most of the samples, or (2) the overall content of bisacrylamide in the reaction bath is smaller here than for the situation investigated by Hecht Duplessix and Geissler (1985). The total polymer concentration ϕ is indeed lowered by a factor of roughly 3 in the present gels. This difference may have a non-negligible effect on the network structure since the crosslinker is possibly less soluble than acrylamide.

More precisely, it can be seen that the signal for the gel of lower crosslinking density ($C = 0.5\%$, $R_c = 0.21\%$) and that of the polymer solution almost exactly coincide. Therefore, we recover here the behaviour observed for several types of polystyrene gels. Moreover, $\langle I(q) \rangle_E$ appears to be almost independent of q, which is natural since the larger correlation distance is expected to be much smaller than $1/q_{max}$ (q_{max} being the maximum q values in this series of experiments, roughly 3×10^{-2} Å^{-1}). As the crosslinking density is raised, one observes, in a parallel way with the enhancement of the intensity, an increasing dependence of $\langle I(q) \rangle_E$ with q. This probably means that the greater distance of correlation present in the system increases and becomes comparable with $1/q_{max}$ (here $q_{max} \approx 3 \times 10^{-3}$ Å^{-1}).

Comment Again, we speculate whether this enhancement is governed by an increase of the thermal fluctuations or results from the formation of 'quenched fluctuations' of polymer concentration.

Regarding this alternative, we dispose of very important additional information, with this technique namely an estimate of the *fluctuating part* of the scattering intensity $\langle I_F(q) \rangle$, that can be calculated from the probability distribution of scattered intensities (Eq. (92)). It can also be obtained from the asymptotic value of the intermediate scattering function obtained by dynamic light scattering (Joosten McCarthy and Pusey, 1991; Pusey and van Megen, 1989).

The values of $\langle I(q) \rangle_F$ obtained using this procedure for the different gels whose total intensity $\langle I(q) \rangle_E$ was shown on Fig. 14(a) are presented on Figure 14(b). Here $\langle I(q) \rangle_F$ is now independent of q, which is expected from the calculations of section 5.3.2 provided the fluctuating part is identified with the intensity from idealised gels for which the whole scattered intensity is supposed to arise from dynamical fluctuations. But, in a surprising manner if one follows the same idea, $\langle I(q) \rangle_F$ is slightly larger than for the semi-dilute solution with same concentration and seems to increase slightly with the crosslinking density (up to 30%). Remember that for idealized gels one should observe a decrease (according to Eqs (66) and (67)).

Comment Therefore these results allow us to conclude that, for the present system at least, most of the increase of the scattering intensity at low q arises from an enhancement of 'quenched fluctuating'. Since this phenomenon is not accompanied by a reduction of the fluctuations part, as would be expected from a progressive 'freezing' of the degrees of liberty of the gel (one of the possible mechanisms considered by Pusey and van Megen, 1989), formation of 'quenched fluctuations' probably means the building up of quasi-static concentration heterogeneities. But, at the same time, the small but detectable amplification of thermal fluctuations discards the description of the phenomenon as the exact superposition of a signal from heterogeneities superimposed on that of idealized gels of the same polymer concentration. Something more complicated happens, which is consistent with the reduction of osmotic modulus observed, in other systems, by Geissler–Hecht–Horkay *et al.* It remains to determine whether it comes from an increase of the Flory–Huggins parameter χ, or from the coupling mechanism proposed by Rabin and Bruinsma (1992) or, finally, from a feedback effect of the growing heterogeneity of the system. We will discuss this question further in section 5.6.

Polyacrylic acid gels with different ionization degrees and crosslink densities have also been investigated by means of speckle scanning. The general features of the results are the same as for the polyacrylamide gels: the scattered intensity usually varies with the position of the sample and the distribution of the intensities has the shape of a truncated exponential (as Eq. (92)). A feature common to all the results is that the ensemble-averaged scattering intensity of these gels is q independent, as illustrated by Figures 15(a) and (b) (Moussaid, Candau and Joosten, 1994). In detail, one observes that the intensity is a decreasing function of the ionization degree for a given crosslink density (cf. Figure 15(a)). Furthermore, it increases with the crosslink density in agreement with the neutron scattering data of (cf. Figure 15(b)). As far as the fluctuating component is concerned, it is found to be equal to, or very slightly larger than that of the solution of the same concentration.

Comments Again, most overscattering of the gel with respect to the semi-dilute solution appears to originate in some quenched fluctuations of polymer concentration. Thus, this behaviour is apparently common to any kind of gel, neutral or charged.

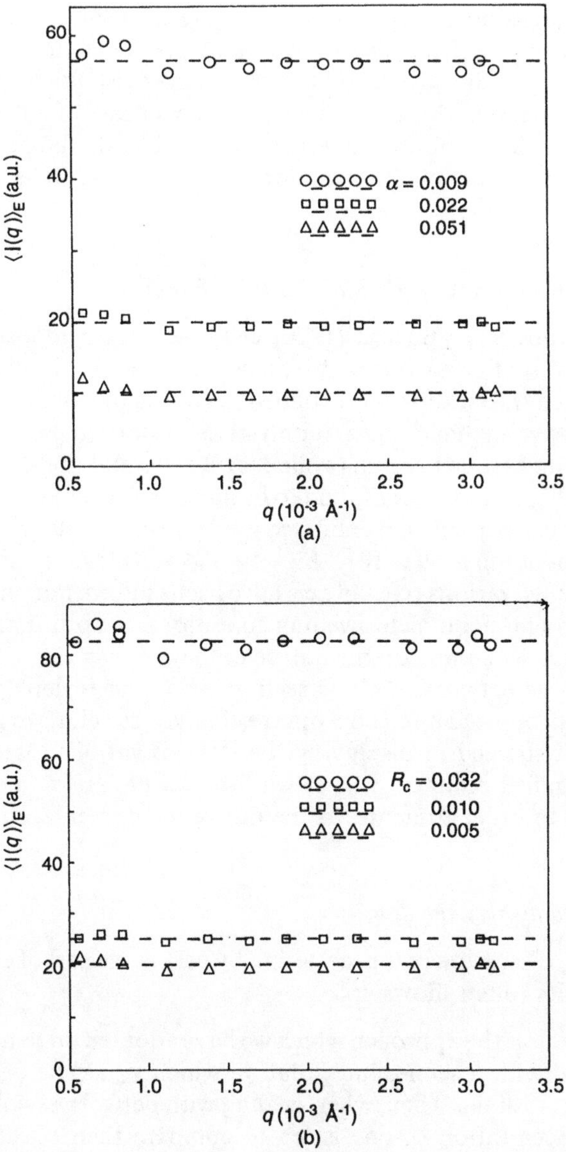

Figure 15. Ensemble averaged scattered intensity versus scattering vector amplitude for polyacrylic acid gels. (a) $R_c = 0.005$ and α as indicated in the figure; (b) $\alpha = 0.022$ and R_c as indicated in the figure. From Moussaid, Candau and Joosten (1994)

Nevertheless, it is noteworthy that these quenched fluctuations are reduced upon increasing the ionization degree of the network. Thus, one may imagine that the coupling between the quenched fluctuations of crosslinking density (or, more generally, of 'local rigidity') and the quenched fluctuations of polymer concentration depends on the osmotic properties. Recall that the latter are drastically changed upon a variation of the ionization degree, because of the large entropy of the counterions.

5.4.3 INFLUENCE OF THE SWELLING RATIO

As far as we known, F. Bueche (1970) was the first to report a spectacular enhancement of scattering intensity associated with the swelling of polymer networks. In fact, this author, who studied networks prepared in the bulk from monomers (methyl methacrylate and polystyrene), noticed essentially the difference of scattering between the unswollen (dry) state and the state of maximum swelling: the light scattering was found to be hundreds of times larger than that of the polymer alone. Moreover, it exhibited a very strong angular dependence (i.e., a large variation of $I(q)$ for $9 \times 10^{-4}\,Å^{-1} \leqslant q \leqslant 4 \times 10^{-3}\,Å^{-1}$). Since these pioneering experiments, various types of gels have been studied, not only at maximum swelling but also in incomplete swelling conditions. Several typical behaviours have been found, depending on the nature of the polymer and the conditions of preparation of the networks. In this section, we try to review the main trends which have been reported up to now upon a change of swelling degree. But to this end we did not succeed in displaying the data of various research groups in a completely unified manner. This is why the results are somewhat split into several subsets in order to adapt to the different representations found in the literature.

5.4.3.1 How to display the data?

Basically, in the discussion of the scattering by gels, some sort of conflict between two points of view often shows up:

1. First—and this is the approach which we have adopted up to now—a gel may be compared with a semi-dilute solution which would be perturbed by the presence of crosslinks. Then, when dealing with networks swollen at different polymer concentration ϕ, one is led to compare their scattering functions normalized by ϕ, as currently done for solutions. In the latter case indeed, one writes, for large enough q values, $I(q) \sim \phi q^{-1/\nu}$ (Eq. 46): in a given regime of q, one expects to probe essentially the correlation on length scales $\approx 1/q$, therefore, for $q\xi \gg 1$, one should detect the statistics of a single excluded volume chain. In gels, a perturbation of chain statistics due to network effects is classically expected, but it should have vanished on length scales comparable with the size of the Kuhn step length. For most polymers the latter is

ranging between 5 and 10 Å; therefore, with this scheme, the normalized scattering curves $I(q)/\phi$ of both solutions and gels should meet for $q \geqslant 1 - 2 \times 10^{-1}\,\text{Å}^{-1}$.

2. There is, however, a rather different point of view which may lead, ultimately, to the abandonment of the comparison to semi-dilute solutions of the same concentration and even the normalization by ϕ. At this level we only summarize the main ideas; more details are given in section 5.6. The key ingredient of this approach is the second moment Q^* of the scattering intensity, which is defined as follows (see Porod, 1982; Mallam *et al*, 1989a and b; Geissler, Horkay and Hecht, 1991, 1993; Horkay *et al*, 1991a):

$$Q^* = \int q^2 I(q) dq \tag{93}$$

Using the properties of Fourier inversion, Q^* should be equal to

$$Q^* = b_v^2 2\pi^2 \langle \delta\phi^2 \rangle^{\text{loc}} \tag{94}$$

where

$$\langle \delta\phi^2 \rangle^{\text{loc}} = \lim_{r \to 0} \langle \delta\phi(0)\delta\phi(r) \rangle$$

We add the index^{loc} (for local) because $\langle \delta\phi^2 \rangle$ defined through $\lim_{r\to 0}\langle \delta\phi(0)\delta\cdot\phi(r)\rangle$ is obviously a very local quantity: it represents the mean square amplitude of the concentration fluctuations in 'cells' of vanishingly small sizes. In the case of scattering objects formed of a homogeneous medium, having well-defined boundaries and representing a total volume fraction equal to ϕ, $\langle \delta\phi^2 \rangle^{\text{loc}} = \phi(1-\phi)$. Therefore, Q^* should then represent a 'scattering invariant':

$$Q^* = \int q^2 I(q) dq = b_v^2 2\pi^2 \phi(1-\phi) \tag{95}$$

This relation has an important consequence: if valid, the scattering at large q's is never independent of the scattering at very small q's. In other words, if different systems, with the same overall concentration of scatterers, have very different scattering at very low q (because of differences in the long-range spatial fluctuations), they should not exhibit exactly the same scattering at large q: strictly, the scattering curves should cross somewhere. This argument has been used by Hecht *et al* (1991) and Geissler, Horkay and Hecht (1991) to explain the crossing of scattering curves from gels and semi-dilute solutions with the same polymer concentration. Furthermore, the superposition of scattering curves after normalization by the concentration should break down, *stricto sensu*, in the case of systems for which the signal at very low q is strongly non-linear with concentration (as semi-dilute solutions). From the available experimental results, it seems at present difficult to draw unambiguous conclusions: as will be shown, concentration normalized scattering curves often merge nicely at large q's. This may be due to the fact that, even if a 'scattering invariant' applies to polymers, the q^2 term in Eq. (95) often weakens the effects.

A second issue concerns the use of the 'second moment' of the scattering intensity to evaluate the mean square amplitude of the static fluctuations of concentration $\langle \delta\phi^2 \rangle_s$. Geissler, Horkay and Hecht (1993) assume that $I(q) = I_F(q) + I_s(q)$, where $I_s(q)$ is related to static spatial fluctuations of polymer concentration (Eq. 87). Despite the fact that the overall second moment Q^* should be independent of the structure (since it is expressed only as a function of the concentration ϕ and the contrast b_v^2), these authors proposed that the second moment of the static part $I_s(q)$ of the intensity can be used to calculate $\langle \delta\phi^2 \rangle_s$:

$$\langle \delta\phi^2 \rangle_s = \frac{1}{b_v^2 2\pi^2} \int q^2 I_s(q) dq \tag{96}$$

where $I_s(q)$ is the static part evaluated by a fit to the form proposed in Eq. (87).

We discuss this approach further in section 5.6. At this level, we will present representations, with and without normalization, for the spectra of the gels which have been obtained.

5.4.3.2 Effect of the swelling degree on the scattering functions obtained from SANS and X-ray

As stated above, gels are often synthesized in the presence of solvent. The polymer concentration can therefore be changed either by deswelling or by swelling with respect to the preparation state. In this section we first review typical scattering behaviours of gels which were swollen with respect to their preparation state.

5.4.3.2.1 Examples of non-normalized curves (neutron scattering)

Neutral gels We consider first some direct superpositions, given on Figures 16(a)–(d) of scattering curves from gels at different swelling degrees, together with those of semi-dilute solutions studied in a comparable range of concentrations. It can be seen first (see Figure 16(a)) that the scattering intensities of the solutions at different concentrations cross each other in the middle of the q range investigated here. Such a feature arises from the change of behaviour upon dilution at large and small q values: in the 'intermediate regime' $1/\xi < q < 1/b$, the scattering is expected to scale like $\phi q^{1/\nu}$; this means that in this range of q, the larger the concentration (by volume) ϕ, the larger the scattering, which is observed. Conversely, in the limit of q tending towards zero, one expects $I(q \to 0) \sim \phi^{-0.31}$, which means that the larger the concentration, the smaller the scattering, which is also observed (in practice, for these data, one finds $I(q \to 0) \sim \phi^{-0.37}$). Recall that the rise of the scattering at low q's together with a decrease of the concentration is attributed to the progressive 'unscreening', upon dilution, of the correlations carried by the chains.

In Figure 16(a) are displayed the scattering intensities from polystyrene 'statistical' gels (swollen in deuterated toluene; see section 5.4.1) corresponding to a crosslinking density of $R_c = 0.8\%$ (mol. mol.$^{-1}$). As the curves of the solutions, they cross each other clearly. An immediately apparent difference with respect to the behaviour of the solutions is that the enhancement of the intensity at low q, as the swelling degree is increased, is much stronger. In fact, for these data, an apparent scaling law with a rather large exponent is found for the zero q scattered intensity: $I(q \to 0) \sim \phi^{-t}$, with $t = 1.5$. Such a variation law is found after fitting the data points at very low q to an Ornstein–Zernicke function (Eq. (86)). For another series of gels prepared according to the same method, but with a smaller crosslinking density ($R_c = 0.5\%$), the apparent exponent m is reduced (e.g. $t = 0.92$, see Bastide *et al* (1990)), though remaining much larger than that of the solutions. Using the same Ornstein–Zernicke fit at small q's, it is also possible to determine a value ξ_g meant to represent the larger correlation distance present in the system. For $R_c = 0.8\%$ and 0.5%, one finds $\xi_g \sim \phi^{-m}$ with m respectively equal to 1.68 and 1.2. Such a variation is very strong and much larger than that found for the solutions. Note, for instance, that the variation $\xi_g \sim \phi^{-1.68}$ corresponds approximately to a dependence with the fifth power of the lateral size L of the gel (since $L \sim \phi^{-1/3}$, $\xi_g \sim L^{\approx 5}$).

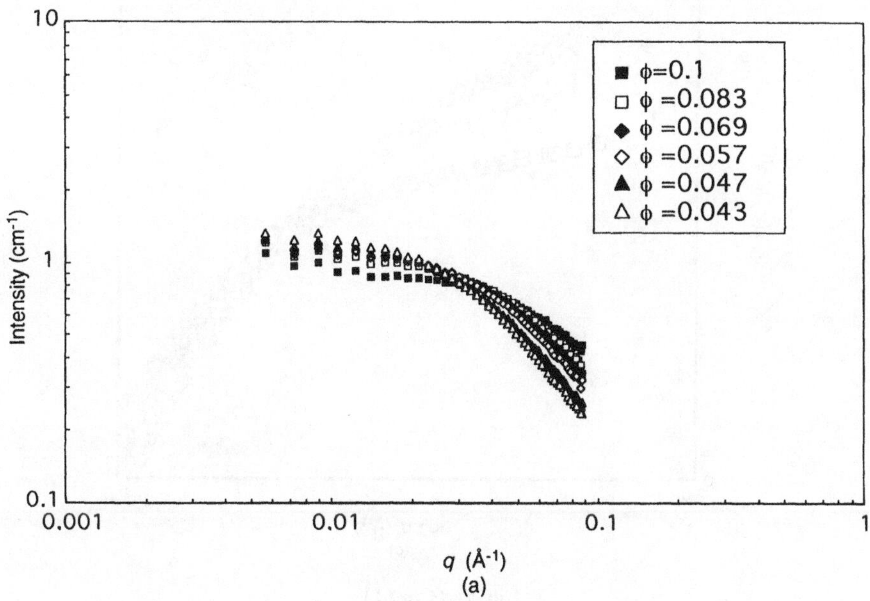

(a)

Figure 16. (*Contd*)

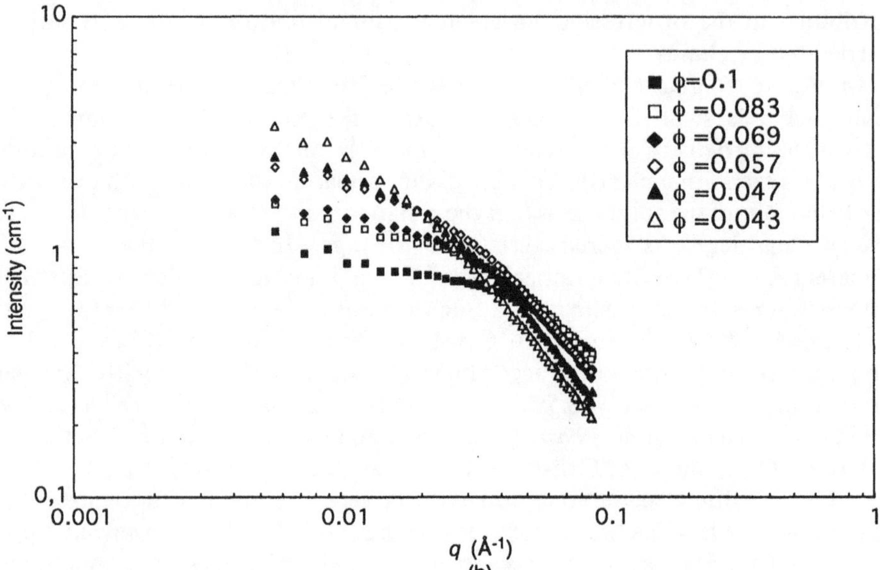

(b)

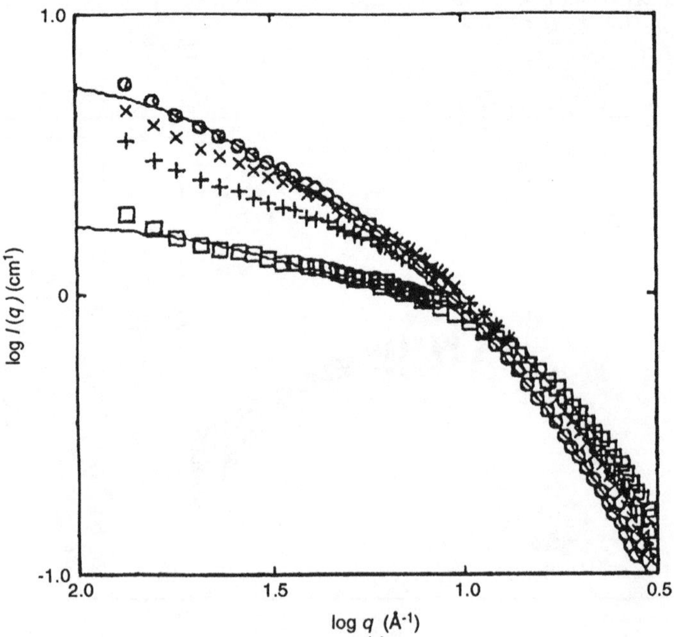

(c)

Fig. 16.(*Contd.*)

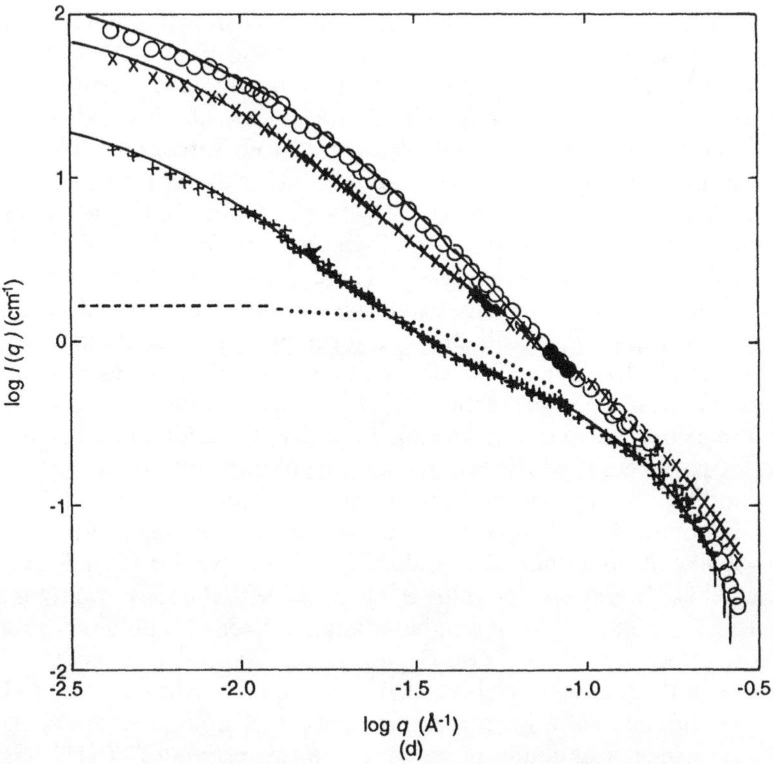

Figure 16. Non-normalized scattering curves of systems with different polymer concentrations. (a) Semi-dilute solutions of polystyrene ($M \approx 7 \times 10^5$)/deuterated toluene. Courtesy of E. Mendes. (b) 'Statistical' polystyrene gels deuterated toluene. ($R_c = 0.8\%$; as in section 5.4.1). Courtesy of E. Mendes. (c) PDMS gels/deuterated octane (as in section 5.4.1). (□) $\phi = 0.353$; (+) $\phi = 0.254$; (×) $\phi = 0.223$; (○) $\phi = 0.166$ (maximum swelling). From Geissler, Horkay and Hecht (1994). (d) PVAc gels/deuterated toluene. (+) $\phi = 0.399$; 1 (×) $\phi = 0.208$; (○) $\phi = 0.146$. From Geissler, Horkay and Hecht (1993)

Other types of gels exhibit very similar behaviour. For example, the previously mentioned gels prepared by end-linking of polystyrene chains of narrow molecular distribution with tri-functional crosslinkers (see section 5.4.1) exhibit scattering curves with a comparable shape. When analysing the data in the same way, apparent scaling laws are found for ξ_g and $I(q \to 0)$, with nearly the same exponents ($m \approx 1.5$ and $t \approx 1.5$) for three different values of the molecular weights of the chemical mesh ($M_c \approx 5000$, 10 000 and 15 000) (Hakiki, 1991; Mendes *et al*, 1993b). In other words, when the swelling degrees of the three gels are equal, the values found for ξ_g and $I(q \to 0)$ are approximately the same, but naturally the larger the chemical mesh the larger the maximum swelling degree.

Although analysed in a different manner, the data shown in Figure 16(c) also exhibit rather similar features (Geissler, Horkay and Hecht, 1994). They have been obtained with PDMS gels-swollen in deuterated toluene-identical to those already discussed in section 5.2. Recall that the curves exhibited a strong upturn at very small q's, which were proved to originate from a stray scattering (clusters of crosslinking molecules). Here, the range of q where this upturn is visible is not displayed: the curves start at $q \approx 10^{-2}$ Å^{-1}. As in Figure 16(b), the curves cross each other and the intensity at low q's increases significantly with the swelling degree of the gel (i.e., as ϕ decreases). These scattering intensities have been described as being a sum of two contributions—a static one $I_s(q)$ and a dynamical one $I_F(q)$ (see Eq. (87)). Recall that the form proposed for the static part is $I_s(0)\exp[-(q\Xi)^p$, the 'dynamical' part being supposed to correspond with the scattering by an idealized gel (section 5.3). Therefore, in order to describe the data in this framework, one has to determine $I_s(0)$, Ξ, p, $I_F(0)$, ξ, that is, a rather large number of parameters, which can be a source of difficulty. However, Geissler, Horkay and Hecht proposed to check the validity of the fit they made by relating the values obtained for $I_F(0)$ with some thermodynamic information. Indeed, $I_F(0)$ must give the longitudinal modulus M of the gel (see Eq. (57)). From direct measurements of the elastic modulus G (identified with the shear modulus μ), the difference $M - 4/3\mu = K_{os}^{gel}$ can first be estimated. From the obtained values for K_{os}^{gel}, those of the osmotic pressure Π_{os}^{gel} of the gel can then be computed (in the framework of the scaling model recalled in section 5.2). Such values of Π_{os} can finally be compared with direct measurements of osmotic pressure of the gels. A good agreement was found in between the two estimates of Π_{os}^{gel} (Geissler, Horkay and Hecht, 1994).

When describing the 'static' part $I_s(q) = I_g(q) - I_F(q)$ with a stretched exponential, the best fit is obtained for $p = 2$, that is, for a Gaussian form. As the polymer concentration in the gel is changed from 0.353 to 0.166, $I_s(0)$ fitted from this form varies from 0.35 to 1.77 cm^{-1}. Therefore, these values confirm that the relative enhancement of 'static' intensity at low q is rather important. However, the absolute values of $I_s(0)$ are smaller (by a factor 2 to 3) than the values found for $I_F(0)$: according to this analysis, the quenched fluctuations do not seem to be very important in this system. Moreover, the values found for the larger correlation length Ξ vary only from 56 to 59 Å (the variation law was fitted to $\Xi \sim \phi^{-0.12}$).

As stated above, another analysis of the data can be carried out by evaluating the integral (Eq. (95)):

$$\frac{1}{b_v^2 2\pi^2} \int q^2 \{I_g(q) - I_F(q)\} \, dq = \langle \Delta\phi^2 \rangle_s^{loc}$$

Conclusions consistent with the preceding ones are found: using this procedure, the quantity $\langle \Delta\phi^2 \rangle_s^{loc}/\phi^2$ is found rather small in all cases, but varies significantly from 0.06×10^{-3} to 0.91×10^{-3} as the degree of swelling is increased.

Polyvinylacetate gels at different swelling ratios is deuterated toluene have been studied by the same group (Horkay *et al*, 1991a; Hecht *et al*, 1991; Geissler,

Horkay and Hecht, 1993, 1994) and three examples of the obtained scattering curves are shown in Figure 16(d). The samples are the same as the ones presented in section 5.4 (Figure 11(b)). As in the preceding examples of the present section, a strong enhancement of scattering intensity associated with the swelling of the gel can be seen at low q. However, at variance with the preceding cases, the curves do not cross each other distinctly at high q. As before, a decomposition into a static part $I_s(q)$ and a fluctuating part $I_F(q)$ was attempted, with a stretched exponential form for the static part; a 'stretching' exponent $p = 0.7$ was then found. Again, a satisfactory agreement was obtained between the estimates of the swelling pressure Π_{os}^{gel} calculated from the best fit for $I_F(0)$ and the directly measured values. Contrary to the PDMS case, the fitted values for $I_s(0)$ is now much larger than the fluctuating part: for instance for $\phi = 0.146$, $I_s(0) = 223\ \text{cm}^{-1}$, whereas $I_F(0) = 1.58\ \text{cm}^{-1}$. The values found for the dynamical correlation length ξ_F remain practically constant as the swelling degree is changed by a factor of 2.7. At the same time the values found for the 'static' correlation length Ξ in the framework of this description increase slightly as the gel is deswollen. Note that the absolute values of Ξ are much larger than in the case of the PDMS networks (from 215 to 271 Å). Again, $\langle \delta\phi^2 \rangle_s^{loc}/\phi^2$ was evaluated by integration of $q^2\{I_g(q) - I_F(q)\}$. The values obtained for $\langle \Delta\phi^2 \rangle_s^{loc}/\phi^2$ are also much larger than in the PDMS case but, contrary to Ξ, are increased strongly with the degree of swelling (from 0.5×10^{-3} to 35×10^{-3}, for ϕ decreasing from 0.399 to 0.146).

Comments (1) As shown immediately above, the curves of Figures 12(b)–(d) exhibit a certain degree of similarity: the scattering intensity at low q's is clearly enhanced as the swelling degree is increased. An agreement is found regarding the prime cause of such behaviour: it is viewed mostly as the result of an inhomogeneous swelling process (or, the building up of 'concentration polydispersity', Geissler, Horkay and Hecht, 1993). The scanning light scattering experiments reported in section 5.4.2 and in section 5.4.3.3 lead independently to the same interpretation. However, in the detail, the quenched heterogeneities of concentration associated with the non-homogeneity of the swelling are not described in the same way. On the one hand, the analysis is focused (for simplicity) on a correlation distance (written ξ_g) which is assumed to be the larger one in the system. It is found to increase strongly with the swelling degree of the networks. The underlying physical picture is the Bastide, Leibler and Prost model (1990): some 'more crosslinked regions', kinds of branched 'clusters' formed by chance, which may be interconnected and build a 'superstructure', which is progressively revealed, as the gel swells, through two mechanisms in parallel: (a) an enhancement of the difference of polymer concentration between the more and the less crosslinked regions; (b) a progressive 'unscreening' of spatial correlations carried, on rather large scales, by the 'superstructure' (for more details, see section 5.6). Thus, ξ_g is not believed to represent their size but a distance characteristic of the balance between intra-cluster and inter-cluster correlations. In the other approach of the problem, by Mallam *et al*, 1989a and b, Geissler *et al*,

1990, the data are, from the beginning, decomposed into two parts. A secondary assumption is that the part associated with the static fluctuations is described as a more or less stretched exponential (Horkay *et al*, 1991a). In these conditions, the corresponding correlation length Ξ (which is then identified with a size) of non-uniformities does not change much with the swelling degree, it can even decrease as the gel is swollen. Therefore, in such a description, there is no unscreening of correlations (or very little), the increase of $I_s(q \to 0)$ being almost entirely contained in the enhancement of $\langle \delta\phi^2 \rangle_s^{loc}$ (evaluated through the modified Porod invariant method) through the relation $I_s(q \to 0) \sim \langle \delta\phi^2 \rangle_s^{loc} \cdot \Xi^3$. At the present stage, it is very difficult to discriminate between these two approaches.

(2) Note that in Figure 16(d) the smooth double-shoulder aspect of the curves is clearly reinforced as the polymer concentration is increased. Therefore, it is difficult to describe them through a single characteristic length scale. Thus, at least two characteristic length scales become necessary to describe these systems. Nevertheless, it is not a priori sure that the these two length scales correspond to the 'static' and the 'dynamical' scattering. In the analysis performed by Geissler, Horkay and Hecht, such a decomposition is cross-checked by thermodynamic measurements: the estimates of the gel swelling pressure are consistent with the values extracted by the fit for $I_F(0)$. Shibayama, Tanaka and Han (1992a and b), using a comparable decomposition for poly(*N*- isopropylacrylamide) gels, were led to the conclusion that the fluctuating part of the intensity is much stronger in gels than in solutions. They interpreted this finding in arguing that the Flory–Huggins parameter χ was significantly raised by the presence of crosslinks. Note that this conclusion is based on an unverified assumption, namely that the Lorentzian part of the scattering, after decomposition, is entirely associated with dynamical fluctuations.

(3) The scattering curves of Figure 16(c) are also described as the sum of two well-separated contributions, but the part attributed to the static scattering is smaller than for other systems. In particular, it seems to be smaller than the one which could be extracted using the same analysis for other types of PDMS gels (Falcao, Skov Petersen and Mortensen, 1993; Rouf (1995). It must be recalled that gels of Figure 16(c) exhibit another type of upturn at very low q (not shown here), arising not from the network but from clusters of excess crosslinking molecules (see section 5.4.1). It is difficult to be sure that a part of the scattering from the network itself is not masked by this stray scattering. Another possibility (and a more interesting one!) would be that the excess crosslinking molecules reduce the structural heterogeneity.

Turning to the charged gels, the studies are rather scarce, but they reveal a behaviour quite different from that of the neutral ones. This is illustrated by the results of a study by Shibayama, Tanaka and Han (1992b) on poly(*N*-isopropylacrylamide-*co*-acrylic acid) gels (NIPA/AAc). The sample investigated is a statistical copolymer having 668 mM of neutral NIPA and 32 mM of AAc that is supposed to be fully ionized, leading to an ionisation degree of 0.0457. The samples were studied under both swelling and deswelling conditions with respect to the state of preparation. The scattering curves obtained at two different temperatures are given on Figures 17(a) and (b). At low temperature (25 °C), the behaviour looks at first sight similar to that observed for some of the neutral gels (see above, Figure 16(b)): in particular one observes a crossing of the curves

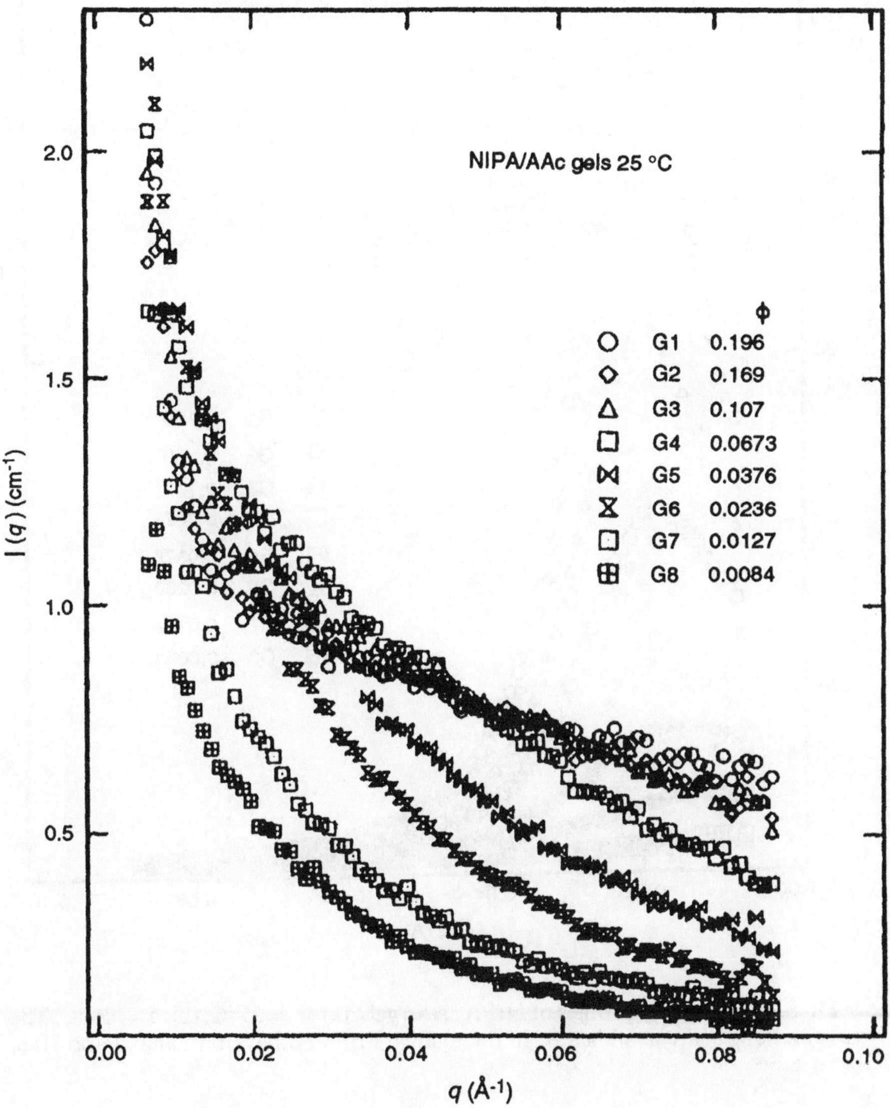

Figure 17. (*Contd*)

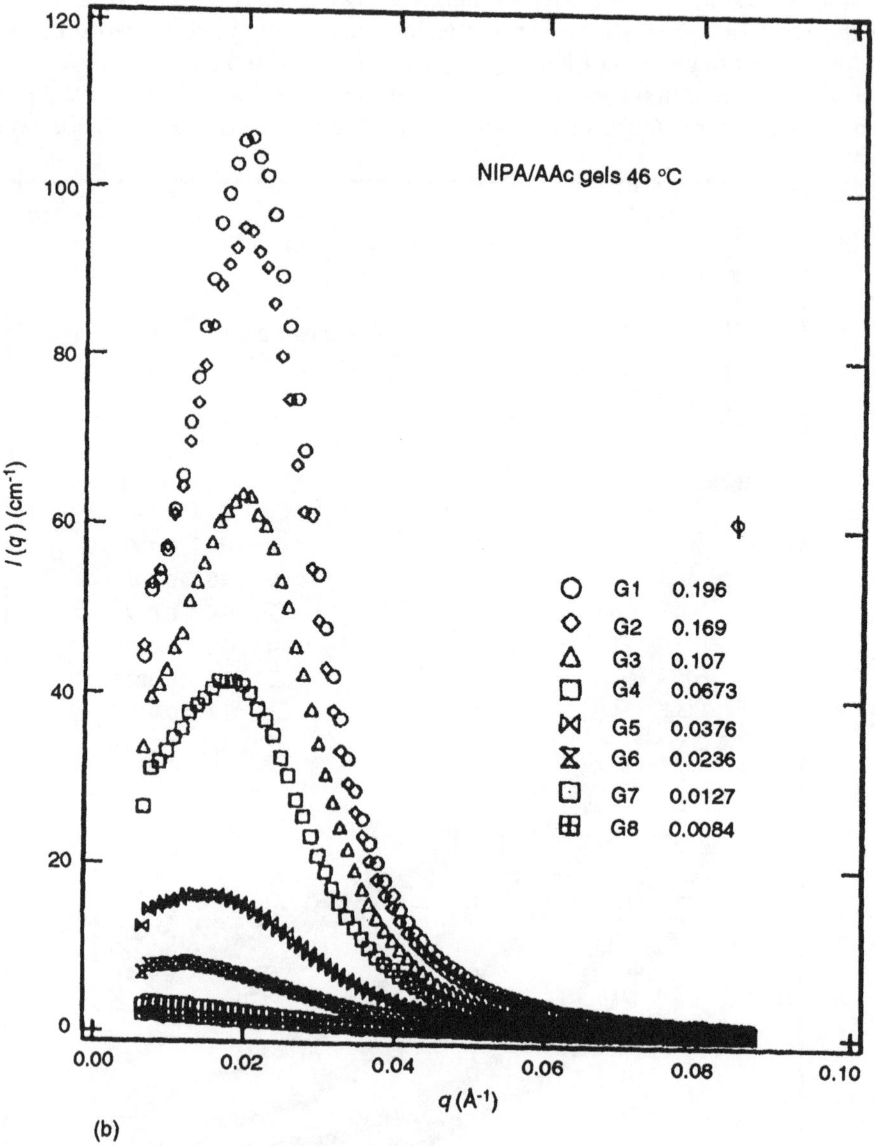

Figure 17. SANS intensity profiles of NIPA–AAc gels. (a) $T = 25\,°\mathrm{C}$; (b) $T = 46\,°\mathrm{C}$. The polymer fractions ϕ are as indicated in the figures. From Shibayama Tanaka and Han, 1992b

associated with a change of swelling degree. A closer inspection of Figure 17(a) reveals that this crossing does not concern the samples which are swollen with respect to the state of preparation ($\phi_c \approx 0.065$), contrary to what happens for neutral gels. On the other hand, under deswelling, there is a decrease of scattering intensity at low q which is then comparable to the behaviour of neutral gels. By comparison with the results shown above for polyelectrolyte solutions and gels, one might be surprised not to observe a peak in the scattering function. This is likely to be due to the predominant effect of the static feature of the gel structure, to be discussed later. Again we face the problem of the relation between the local architecture and the submicroscopic structure (here the copolymerization leads to charged sequences quenched at given places along the chains, contrary to the case of partially neutralized polyacrylic acid where the charges are delocalized).

As the temperature is raised, a spectacular change occurs, characterised by an enhancement of the scattering and the appearance of a peak in the SANS profile (cf. Figure 17(b)). This effect has been interpreted as due to the approach of a micro-phase separation such as that described for polyelectrolyte solutions (section 5.2). In fact, if one further increases the temperature (at $T = 50.8\,°C$), these gels undergo a discrete volume phase transition from swollen to shrunken states. Concerning the effect of the swelling degree, the scattering increases with polymer concentration. This seems to be a feature common to all charged gels exhibiting a peak, as discussed in section 5.4.3.2.4 .

5.4.3.2.2 Comparison with solutions

A systematic comparison between gels swollen to different extents and the corresponding solutions has also been made for a limited number of systems and we have selected three representative examples. These examples refer to systems for which the scattering curves normalized by the concentration superimpose in the high q limit. This means that in the non-normalized representation, the curves cross each other.

Statistical gels These gels are the same polystyrene networks swollen in deuterated toluene, as presented or quoted in section 5.4.1, 5.4.2 and above in section 5.4.3.2. (Mendes, 1991; Mendes *et al*, 1991; Bastide *et al*, 1990). On Figure 18 are displayed the scattering curves obtained for a gel with crosslinking density equal to 0.8% (mol mol^{-1}), at different degrees of swelling, ranging between that of preparation and that of maximum swelling. At the concentration of preparation ($Q = 10$, $\phi = 0.1$, not actually in the reaction bath but after washing, drying and then reswelling the network in the deuterated solvent), the spectrum of the gel is almost exactly superimposed for the whole range of q's, on that of the semi-dilute solution of the same concentration. The latter was prepared with the same polymer chains as those used for making the network,

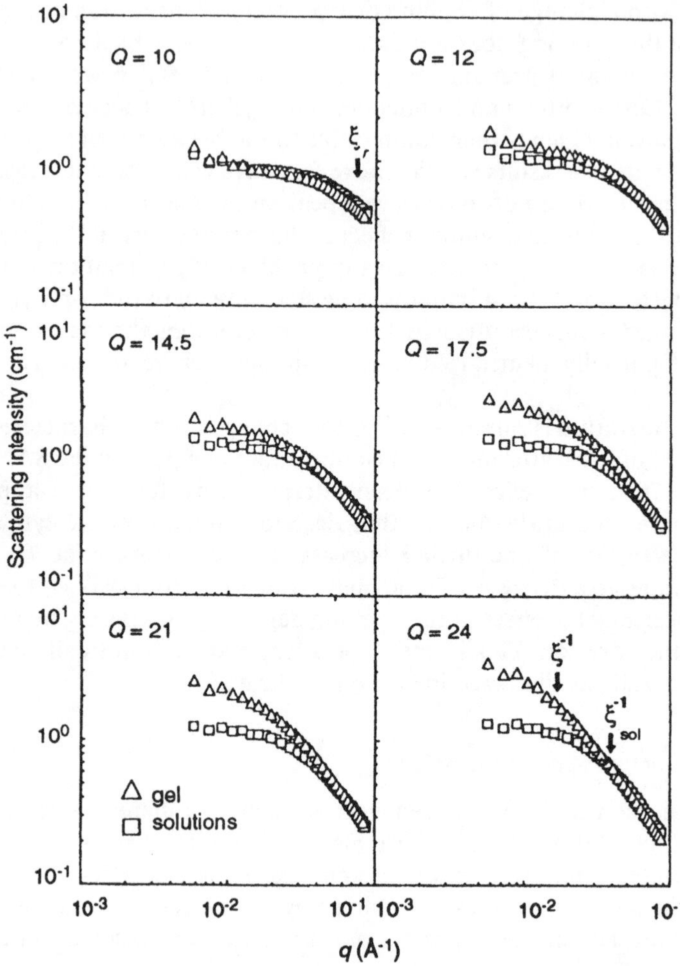

Figure 18. Comparison between intensity scattered by a polystyrene 'statistical' gel (as in section 5.4.1; $R_c = 0.8\%$) at various swelling degrees and those of semi-dilute solutions of the same polymer concentration. On the top left is indicated the inverse of the correlation length ξ_r^{-1} determined at the concentration of preparation ($\phi_{prep} = 0.10$, that is $Q_{prep} = 10$) from an Ornstein–Zernicke fit; ξ_r is found to be approximately the same for both gel and semi-dilute solution. Bottom right: at maximum swelling ξ^{-1} is found larger for the solution than for the gel. Adapted from Mendes *et al* (1991) and Mendes (1991)

namely polystyrene with $M_W \approx 7 \times 10^5$, $M_W/M_n \approx 1.15$. However, upon some swelling with respect to the state of preparation (sometimes called 'overswelling'), one observes the building up of a clear difference, at low q's, between the gel and the semi-dilute spectra. As stated above, the gel curves can be fitted at low angle with an Ornstein–Zernicke function, which allows us to determine a correlation length ξ, varying as $\phi^{-1.7}$ and an extrapolated intensity $I(q \to 0)$ varying as $\phi^{-1.5}$. The values found for these exponents (quoted above), being larger than those found for the solutions are obviously implying the separation of curves at low q's which is shown on Figure 18. On the other hand, the spectra of gels and solutions coincide in the high q limit, within experimental accuracy.

Our second example, shown on Figure 19 deals with a statistical gel prepared according to the same method, but with a higher crosslinking density (2% mol mol^{-1}), and submitted to a deswelling with respect to the state of preparation (Mendes *et al*, 1991). At the concentration of preparation ($\phi = 0.1$), one observes, contrary to the preceding example, some excess scattering at low q with respect to the solution of the same concentration. However, this observation is consistent with the behaviours reported in section 5.4.2: the enhancement of crosslinking density seems to induce, above some threshold and whatever the type of gel, an increase of scattering intensity. However, upon deswelling, the difference between the gel and the solution is progressively reduced and eventually vanishes for $Q < 5$. One can also observe that the appearance under deswelling of some 'Picot–Benoît effect is nearly identical for the semi-dilute solution and for the gel.

Comments The overscattering in the preparation conditions, when present, can therefore be suppressed by deswelling of the gel, and no further increase has been observed as yet upon further deswelling, beyond the state where the scattering from the gel and the solution merge. This point will be discussed in section 5.6.3.1.

Note that, in the low q regime, one detects the occurrence of the Picot–Benoît effect (the scattering upturn at low q; see section 5.2, Figure 3), the latter being enhanced as the polymer concentration is increased. As mentioned above, this upturn is not specific of gel materials as it also shows up in solutions.

5.4.3.2.3 "Anionic" gels

The neutron scattering of gels prepared by anionic block copolymerization of styrene and DVB has also been compared systematically to that of semi-dilute solutions of same concentration (Mendes *et al*, 1993a). Some spectra relative to a gel prepared by end-linking of chains with molecular weight $M_W = 22\,000$ ($M_W/M_n \leqslant 1.2$) and studied at several degrees of 'overswelling' with respect to the concentration of preparation are shown in Figure 20. At first sight it can be seen that the difference between the scattering curves of the gel and that of the equivalent solution is important but does not change very much with swelling degree, contrary to what observed in the case of statistical gels. In more detail, it appears that the distinct shoulder exhibited by the gel spectra is more visible in incomplete swelling conditions than at maximum swelling. In fact, depending on

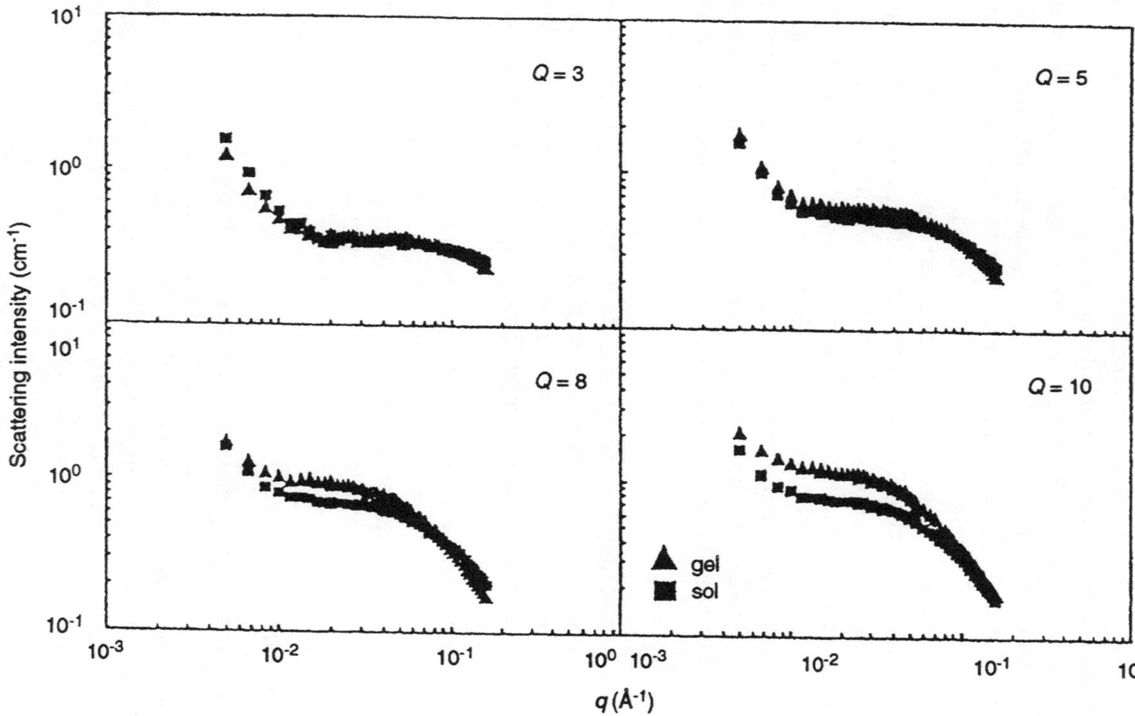

Figure 19. Comparison between intensity scattered by a 'statistical' gel (same as in section 5.4.1, but $R_c = 2\%$) at various swelling degrees $Q < Q_{prep}$ and those of semi-dilute solutions of the same polymer concentration (deswelling with respect to the state of preparation of the gel). The Picot–Benoît upturn is visible at low q. Mendes–Boué–Bastide, unpublished data

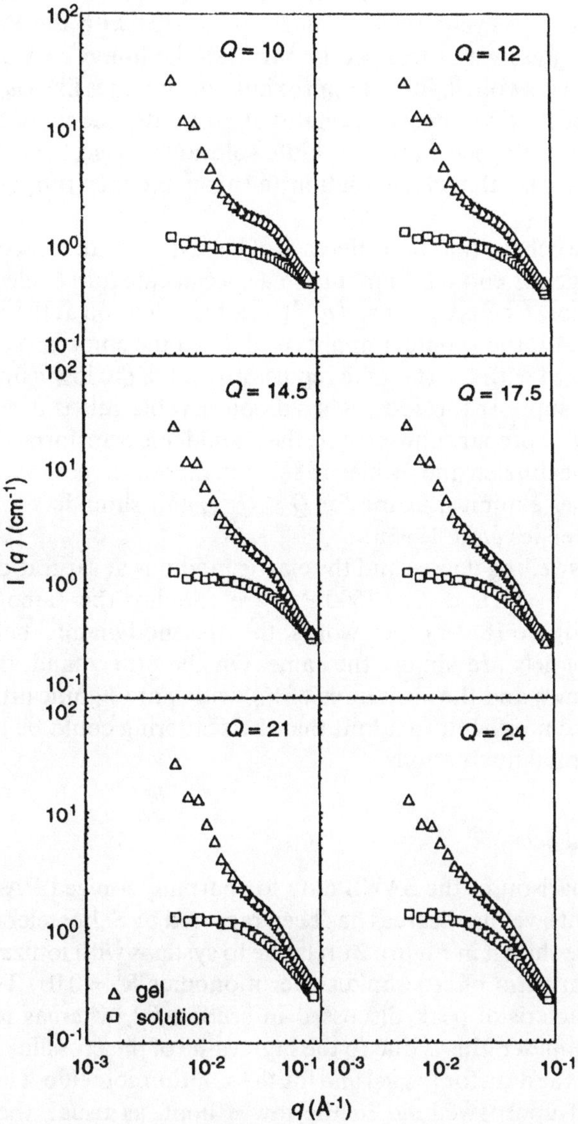

Figure 20. Comparison between intensity scattered by an 'anionic' gel ($M_c = 22\,000$; same as in section 5.4.1) at various swelling degrees and those of semi-dilute solutions of the same polymer concentration (swelling with respect to the state of preparation of the gel). From Mendes *et al* (1993a)

the range of q, it seems that there is a combination of three different behaviours. Contrary to what observed for most of the neutral gels (if one forgets about the 'stray' part of signal, at even lower q values, associated with the Picot–Benoît effect or with the presence of trapped impurities), the low q part of the signal *decreases* as the gel is swollen. In first approximation, it varies like ϕq^{-3}. Around the shoulder, the scattering intensity remains almost independent of the polymer concentration, like the spectra of semi-dilute solutions for $q\xi_{sd} \approx 1$. Finally, the signal of the gel is close to that of the solution and again becomes proportional to ϕ.

Comments We believe that the difference of behaviour with respect to most of the other neutral gels is enough to maintain a special category (called above '2') for such gels. The low q behaviour ($\approx \phi q^{-3}$) is not understood at the moment. As stated in section 5.4.1, the shoulder might result from the combination of a local maximum associated with the star-like repulsions and of the low q upturn. Some unpublished data support this idea: when a comparable gels is deswollen with respect to the state of preparation ($Q = 5$), the shoulder is transformed into a true maximum (Bastide Buzier and Boué, 1988). Furthermore, gels actually made from star molecules exhibit maxima for $Q \leqslant Q_{prep}$ and shoulders for $Q > Q_{prep}$ (Mendes, 1991; Mendes *et al*, 1995a).

The maximum swelling degree and the elastic modulus at saturation of the gel of Figure 20, $Q_{max} = 24$ and $G \approx 2500$ Pa) are roughly the same as for the statistical gel of Figure 18. In other words, the 'thermodynamic' behaviours of these two types of gels are almost the same. On the other hand, the absolute values of $I(q)$ at low q and the variations of $I(q)$ with q are significantly different. Therefore, it makes it difficult to admit that the scattering could be governed in both cases by thermal fluctuations.

5.4.3.2.4 Charged gels

A thorough comparison of the SANS data for partially ionized PAA solutions and gels at different swelling degrees has been reported by Schosseler *et al* (1994). Typical spectra are shown in Figure 21 relative to systems with ionization degree $\alpha = 0.05$ and molar ratio of crosslinker over monomer $R_c = 0.01$. The solution exhibits the characteristic peak discussed in section 5.2, whereas for gels one observes only a shoulder: this is due to the high value of the crosslinking density. In the high q limit, the data for the gel and for the solution coincide. The scattering is greatly reduced upon swelling. In the low q limit, as usual, the scattering intensity by the gel is larger than that of the solution, but surprisingly enough the excess of scattering varies only slightly with polymer concentration. This point will be discussed in more detail below in section 5.4.3.3.

Comment The analysis of the data for aqueous gels is complicated by the phenomenon of exchange between deuterons of heavy water and protons of the

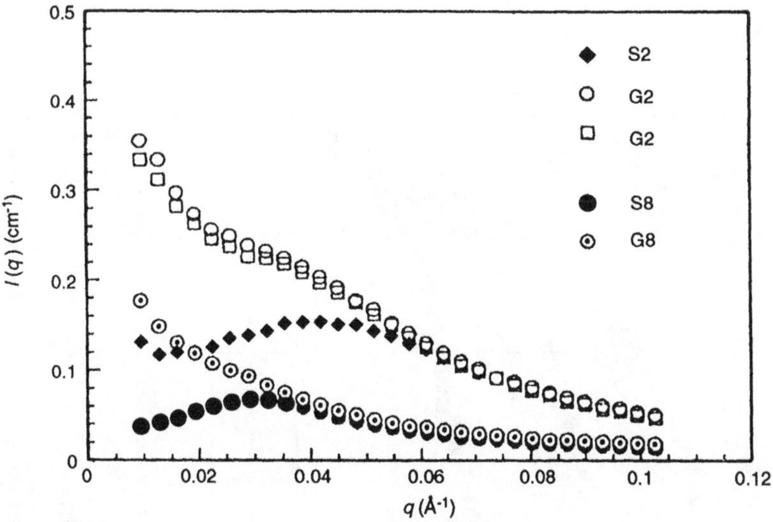

Figure 21. Comparison of the intensities scattered from swollen PAA gels ($\alpha = 0.05$; open symbols) and from corresponding solutions (closed symbols). S2(8) and G2(8) stand respectively for a solution and a gel with a swelling degree of 2(8) with respect to the state of preparation

gels. Furthermore, the deuterated water is inevitably exchanged, to some extent, with the light water of the atmosphere during the manipulation of the gels. A method to handle this problem is described in detail by Schosseler *et al* (1994).

5.4.3.2.5 *Normalized curves*

As stated in the introduction to this subsection 5.4.3, the normalization by the concentration of gel scattering curves is not accepted by all authors, and therefore not systematically performed. Nevertheless, we show in Figures 22 and 23 the data from Figures 19 and 20 which are replotted after normalization. In the case of the statistical gels (Figure 22), the curves separate in a fan-like way (i.e. strongly at low q, whereas they remain nearly superposed at large q). It can be seen that this effect is more pronounced that in the case of semi-dilute solution, shown on the same figure. In the case of the 'anionic' gels, the features are quite different; there is an approximate superposition both at very small q's and at large q's. In the intermediate regime there is a sort of fan-shaped separation, which is less than for statistical gels but comparable with that of semi-dilute solutions in the same range of q.

Comment Note that, since the superposition of curves at high q is satisfactory, the possible existence of a scattering invariant does not seem to have detectable effects, at least for such gels.

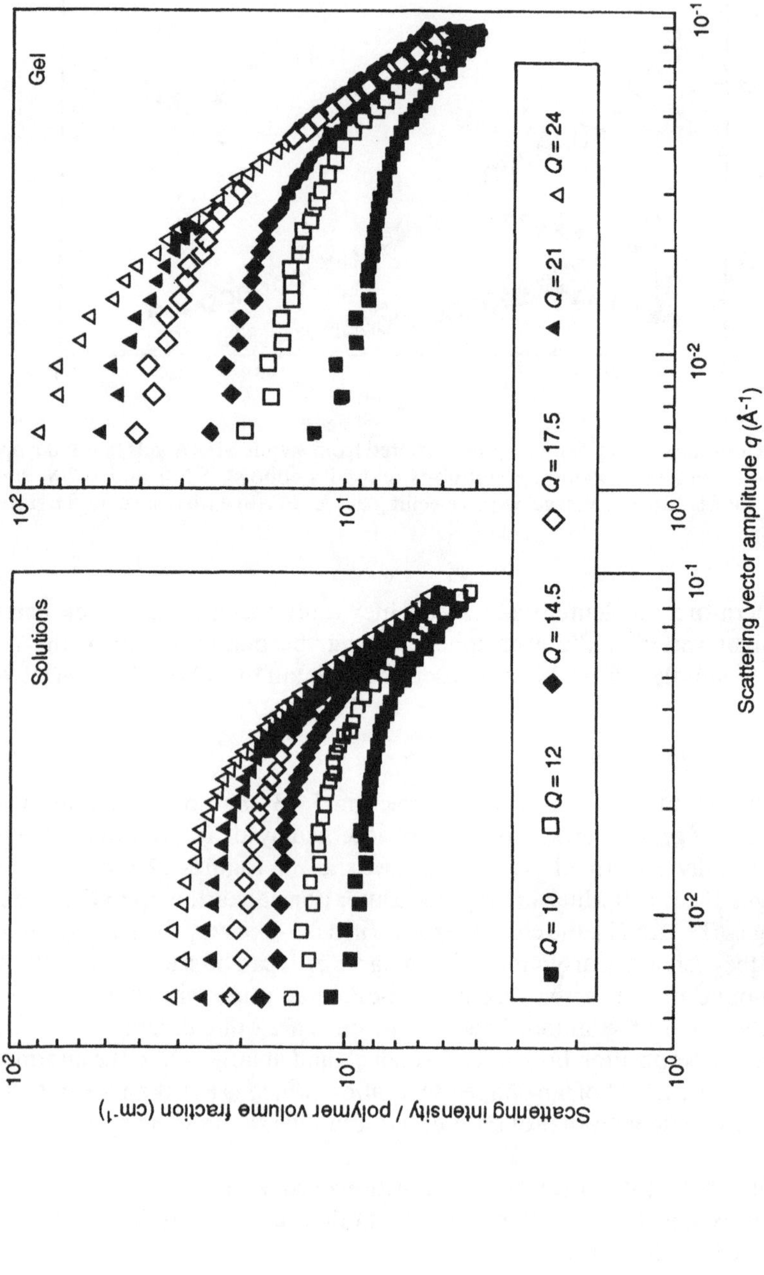

Figure 22. Normalized scattering intensities of a polystyrene 'statistical' gel (same as in section 5.4.1); $R_c = 0.8\%$) at different swelling ratios and a series of semi-dilute solutions of the same polymer concentrations. From Bastide *et al* (1992a)

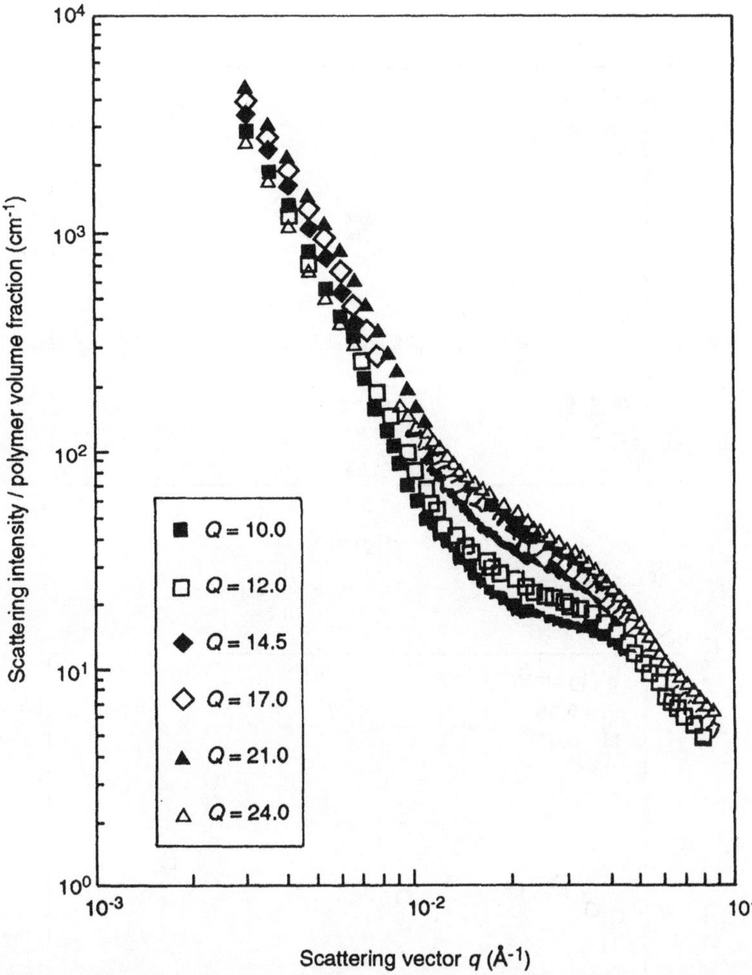

Figure 23. Normalized scattering intensities for an 'anionic' polystyrene gel ($M_c \approx 22\,000$, same as in section 5.4.1) at different swelling degrees Q. From Mendes *et al* (1993a)

5.4.3.3 Light scattering

PAA gels (as above) have also been studied by light scattering using the scanning procedure described in section 5.4.2.2. Two gels, with $\alpha = 0.1$ and $R_c = 0.01$, but prepared at two different concentrations (0.08 g cm^{-3} and 0.05 g cm^{-3}) were investigated. Both the fluctuating and the frozen-in component were measured as a function of the swelling degree. The results are reported in Figure 24(a) and (b) (from Skouri, 1994). It can be seen that the fluctuating component is proportional

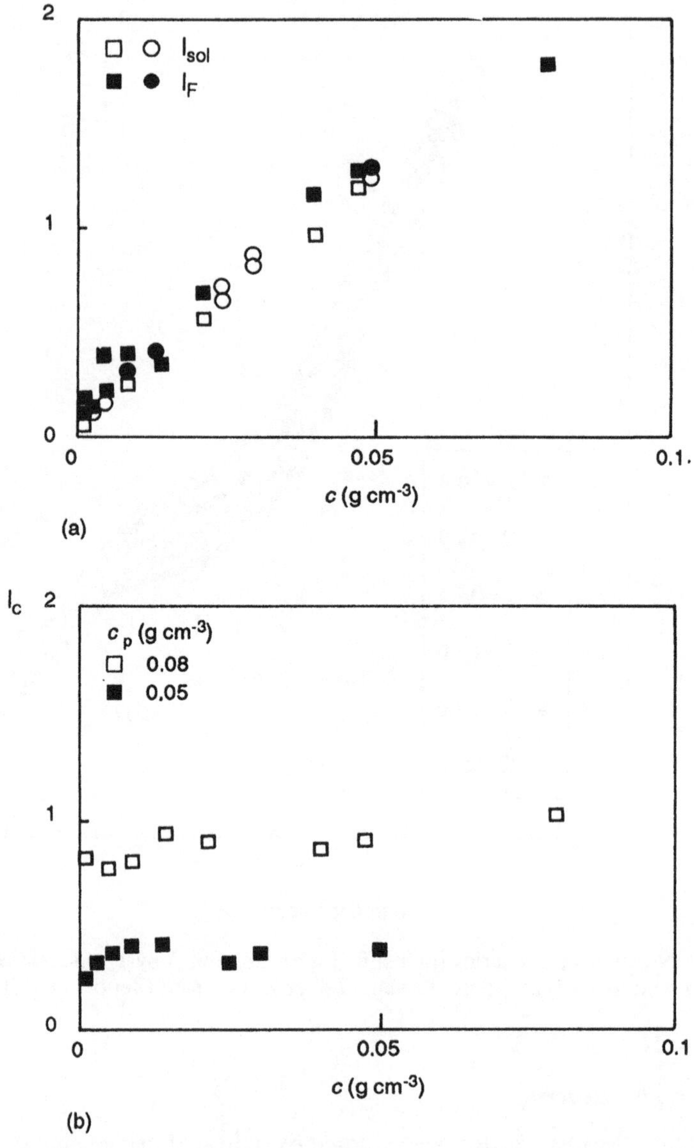

Figure 24. (a) Comparison of the fluctuating component of two PAA gels with $\alpha = 0.1$ ($c_p = 0.05$ and $0.08\ \mathrm{g\ cm^{-3}}$), at different swelling ratios, and of I_{sol} obtained from equivalent solutions. (From Skouri, 1994). (b) Evolution of the quenched part I_c of the scattering intensity with the polymer concentration. From Skouri (1994)

to the polymer concentration and coincides, within experimental accuracy, with the intensity scattered by a solution of same concentration.

Comment This shows that for the investigated charged gels, the excess of scattering intensity with respect to the solution is entirely due the static signal, whatever the swelling extent. A recent study performed on a neutral gel of PDMS leads to the same conclusion (Rouf *et al*, 1995). One must note, however, an important difference between neutral and charged gels: the fluctuating component increases with concentration for the charged gels, due to the leading contribution of the counter ions entropy. Conversely, for the neutral gels, I_F decreases upon increasing the concentration, reflecting the progressive screening of intra-chain correlations. The behaviour of the frozen-in component is also different. In the charged gels it is found to be almost constant with concentration (cr. Figure 24(b)). Such a concentration independence, confirmed by SANS experiments (Schosseler *et al*, 1994) is totally unexpected. If the heterogeneities could be considered as isolated islands in a matrix of polymer, their separation upon dilution should lead to a decrease of the intensity. Conversely, as will be discussed in section 5.6, some unscreening effects of large-scale heterogeneities can lead to an increase upon dilution. Obviously, we deal here with heterogeneities correlated in a non-trivial manner.

5.5 Stretched gels: experimental results

In this section we consider the effect of stretching on the scattering patterns. These experiments have been carried out mainly by means of SANS, but, more recently, results by light scattering have been reported. As will be seen, this type of experiment is of critical importance to discriminate between the different models that have been proposed to describe the spatial concentration fluctuations in both neutral and charged gels.

5.5.1 SMALL ANGLE NEUTRON SCATTERING

Most of the experiments have been performed by uniaxial mechanical stretching (sample clamped by its extremities) and at fixed polymer concentration. The scattering intensity has been recorded on a bidimensional detector whose plane was always set parallel to the stretching direction. This allows us first to draw iso-intensity lines. Then, by selecting only some narrow section (7–15°) in the detector, one obtains scattering curves for different orientations of the scattering vector **q** with respect to the stretching direction.

5.5.1.1 Neutral gels

We consider first a gel at swelling degree ($Q \approx 12$) only slightly larger than that of preparation (Bastide *et al*, 1990; Mendes *et al*, 1991). Figure 25 shows typical

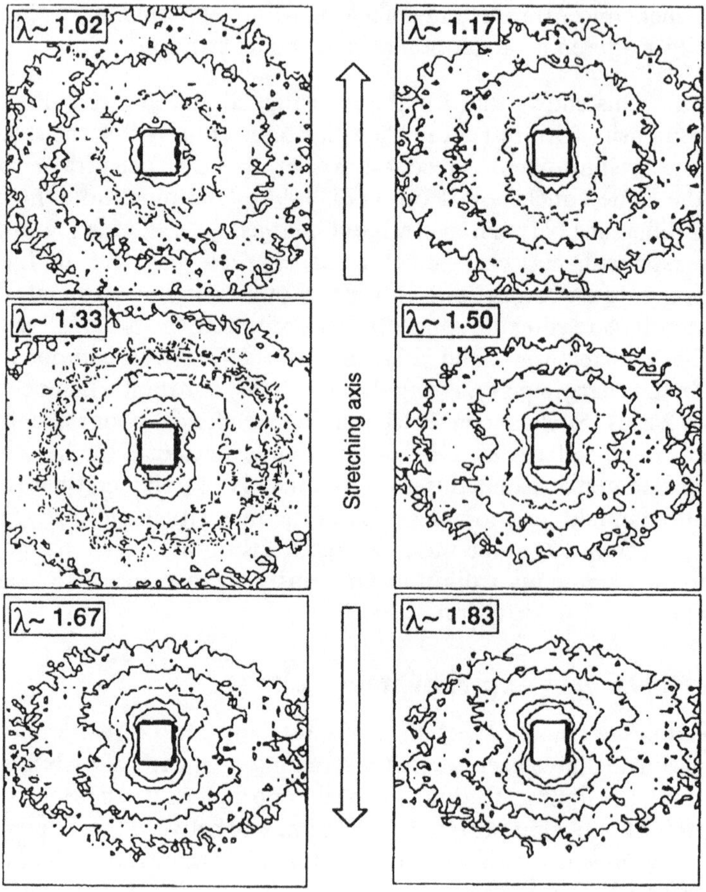

Figure 25. Iso-intensity lines (measured on a XY detector) for a statistical gel of polystyrene (with $R_c = 0.8\%$, same as in section 5.4.1) submitted to uniaxial elongations of different ratios λ. Swelling degree $Q \approx 12$; q range: $6 \times 10^{-3}\,\text{Å}^{-1} \leqslant q \leqslant 4 \times 10^{-2}\,\text{Å}^{-1}$. The rectangle in the centre indicates the limits of the beam stopper. From Bastide *et al* (1990b)

iso- intensity lines for a polystyrene statistical gel analogue to those discussed in section 5.4. The scattered intensities in the isotropic case for a comparable sample were shown on Figure 26, where a larger range of q's was covered (the crosslinking densities are the same ($R_c = 0.8\%$), but the stretched sample comes from another batch). The stretching ratio $\lambda = L/L_i$ is varied from 1 to 1.83.

At relatively large q values (close to the edges of the detector) the iso-intensity lines are elliptical, with a major axis perpendicular to the stretching direction: this is the anisotropy classically expected in the case of chains with end-to-end vectors oriented, on an average, along the direction of stretching (Boué *et al*, 1987 and

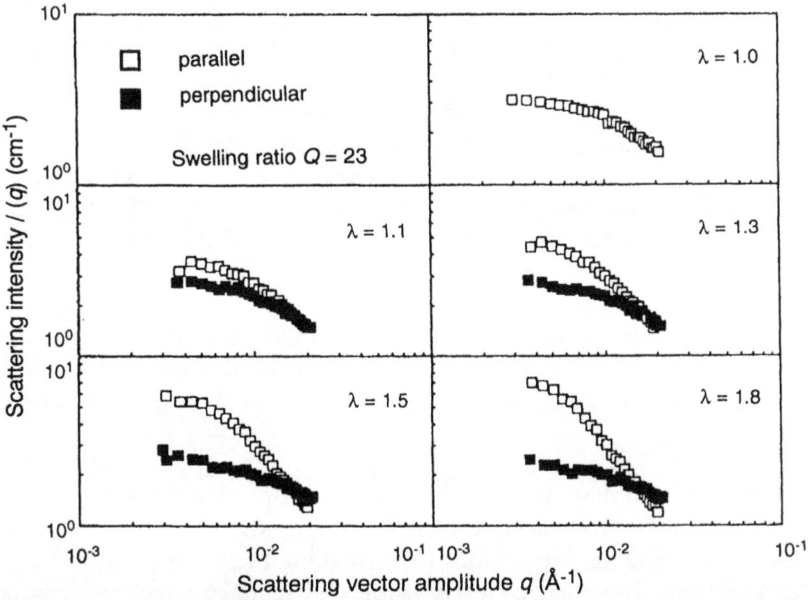

Figure 26. Scattering intensities of a uniaxially stretched polystyrene 'statistical' gel (same as in section 5.4.1; $R_c = 0.8\%$) for **q** parallel and **q** perpendicular to the elongation axis, at different stretching ratios λ. Swelling degree $Q = 23$. From E. Mendes (1991b)

references therein). Therefore, in this range of q, one probably detects, the individual orientation of the elementary chains of the network. But at low q's, one clearly sees the appearance of 'lobes' with a major axis parallel to the stretching direction. Thus the scattering anisotropy at low q is the opposite of what would be expected either on the basis of a chain orientation process, or from the classical expectation for the change, under strain, of the thermal concentration fluctuations. It is, in particular, opposite to that predicted for idealized gels by Onuki 1988 (see section 2). On the other hand, such double-winged curves, are similar to those experimentally observed by Oeser, Picot and Herz (1988) for PDMS rubbers containing mobile labelled chains. These are the so-called *butterfly patterns* (Oeser, Picot and Herz, 1988) or, according to Onuki's nomenclature (1992), *abnormal butterfly patterns.*

Double-winged patterns with an unexpected orientation are not specific to gels. As far as we know, comparable patterns were observed for the first time by Inoué *et al* (1971) in a light scattering experiment performed on tri-block copolymer forming microphases (the contrast being between the microphases and the matrix). The enhancement of scattering intensity was interpreted as resulting from a non-uniform displacement of the micro-phases. More recently, lobes more or less comparable to those of Figure 25 have been observed in a variety of systems submitted to a uniaxial strain or to a shear flow:

1. Labelled chains trapped in networks, without solvent (Oeser Picot and Herz, 1988; Boue *et al*, 1991; Zielinski *et al*, 1992; Bastide *et al*, 1992a; Ramzi, 1994; Ramzi *et al*, 1993, 1995).
2. Polymer blends, consisting of a matrix of long 'host' chains containing some smaller 'guest' chains. The latter can be deuterated homopolymer chains of the same chemical nature as the host chains (Bastide, Buzier and Boué, 1988; Boué *et al*, 1989), or of a different chemical nature (like e.g. PVME–PS, Kempkes and Picot, 1995), or even copolymers (e.g. PSH–PSD, Barea, Muller and Picot, 1988).
3. Latex films (Rharbi *et al*, 1995).
4. Semi-dilute solutions under shear (Wu, Pine and Dixon 1991, Dixon, Pine and Wu 1992, 1991; Hashimoto and Fujioka 1991; Hashimoto and Kume 1992, Kume *et al* (1995); van Egmond, Werner and Fuller, 1992; Boué and Lindner, 1994).

It must be stressed that the shape and the anisotropy of the butterfly patterns in elongated gels arise from a large increase, under stretching, of the difference between the extrapolated intensities $I(q \to 0)$ along the parallel and the perpendicular directions. This effect is clearly visible on Figure 26, which corresponds to the same type of gel as before but which is now swollen close to saturation ($Q = 23$). On this figure, one sees clearly the occurrence of a crossing of the scattering curves for q parallel and q perpendicular to the elongation axis. For scattering vector amplitudes beyond this crossing point, the scattering pattern is back to normal, that is again elliptical. Examination of Figure 26 reveals the existence of a crossover between an intermediate q regime where $I(q)$ decays rapidly upon increasing q, and a low q regime where the signal levels off. The crossover scattering vector $\mathbf{q}_c$ is a decreasing function of the stretching ratio in the parallel direction and an increasing function in the perpendicular one. Fits to the data at low q's to an Ornstein–Zernicke law provide a correlation length ξ_g and the extrapolated intensity $I(q \to 0)$ in the principal directions, which are found to vary strongly with elongation. If one force fits to scaling laws, one finds $\xi_g^{par} \sim \lambda^{1.6}$, $\xi_g^{per} \sim \lambda^{-0.5}$, $I^{par}(q \to 0) \sim \lambda^{1.8}$, $I^{per}(q \to 0) \sim \lambda^{-0.56}$ for $Q = 23$ and $\xi_g^{par} \sim \lambda^3$, $\xi_g^{per} \sim \lambda^{-1}$, $I^{par}(q \to 0) \sim \lambda^2$, $I^{per}(q \to 0) \sim \lambda^{-0.3}$ for $Q = 12$ (Mendes *et al*, 1991b, 1995b). The general features of the phenomenon do not change appreciably with the swelling degree.

Comments The above results show that, on spatial scales corresponding to the investigated q range, the gel does not deform homogeneously under stretching. The spatial fluctuations of polymer concentration are greatly enhanced along the direction of elongation and reduced in the perpendicular direction. This is consistent with the results of the isotropic swelling experiments which can also be viewed as an increase of the scattered intensity along the direction of dilation.

Obviously, as for the swelling properties, one may wonder again whether this amplification of concentration fluctuations concerns the *thermal fluctuations* (sometimes called *gel modes*) or, conversely, some *quenched fluctuations* associated with non-uniformities of the network structure. In the first case, it would mean that the classical description of the gel thermodynamics is incomplete, since it predicts an opposite effect. In the second, it would be necessary to understand the mechanism of this amplification. In section 5.6 we present and discuss the models with have been proposed from these two types of hypotheses. But the remainder of this section will give some clues for answering this question.

It must also be stressed that such behaviours were observed not only for statistical gels but also for other types of networks:

1. End-linked polystyrene gels ($M_c = 10000$ and $f = 3$, analogous to those mentioned in section 5.4; Mendes *et al*, 1993b).
2. PDMS gels prepared by various methods (end-linking and statistical cross-linking, both in the bulk) (Mendes *et al*, 1993a; Rouf *et al*, 1994; Rouf, 1995).

We will discuss the general character of the butterfly patterns in the framework of the different available models.

5.5.1.2 Charged gels

The case of charged gels is interesting since it has been shown in section 5.4.3 that the scattered intensity is a decreasing function of the swelling ratio. The results of

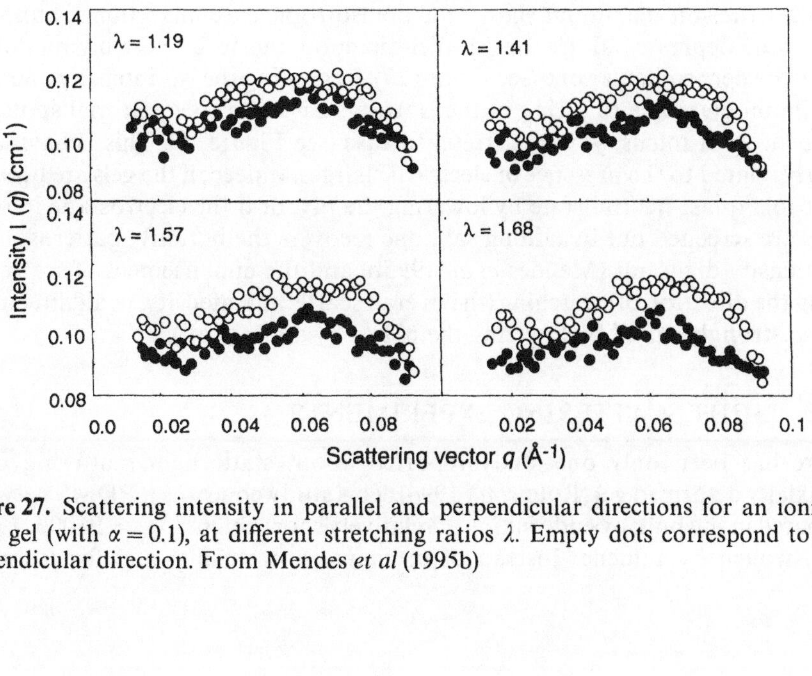

Figure 27. Scattering intensity in parallel and perpendicular directions for an ionized PAA gel (with $\alpha = 0.1$), at different stretching ratios λ. Empty dots correspond to the perpendicular direction. From Mendes *et al* (1995b)

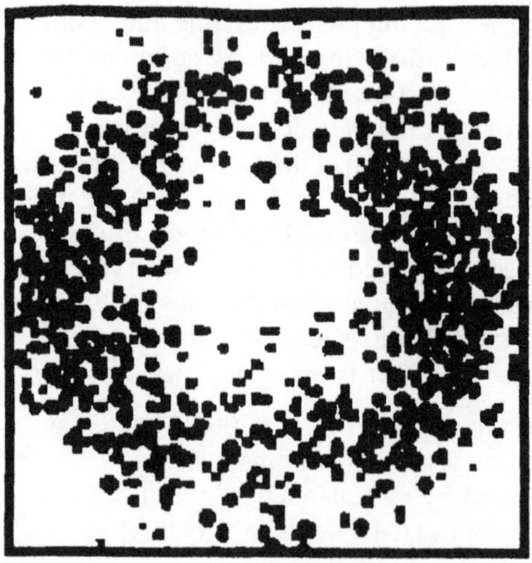

Figure 28. Typical iso-intensity patterns for an ionized PAA gel ($\alpha = 0.1$) under uniaxial stretching (the elongation axis is vertical) and polymer volume fraction $\phi \approx 0.07$. The stretching ratio is $\lambda = 1.57$. From Mendes *et al* (1995b)

a recent study on a partially charged polyacrylic acid indeed show a behaviour entirely different from that of neutral gels (Mendes *et al*, 1995b). On very low q's, the scattered intensity is very weak and is not changed upon stretching in a detectable manner. In particular, no butterfly patterns are observed. On larger q values the soft maximum shown for the isotropic case in section 5.4 appears now to be depressed along the parallel direction and to be possibly reinforced along the perpendicular one (see Figure 27). Regarding the iso-intensity lines, the maximum gives rise to a ring in the isotropic case and to equatorial spots, for some levels of intensity, in the stretched case (see Figure 28). This behaviour is clearly related to the presence of electrical charges. Indeed, if the gels are brought back to a quasi-neutral state by lowering the pH, or if the electrostatic interactions are screened out by adding salt, one recovers the butterfly patterns in the isointensity diagrams (Mendes *et al*, 1995b) and the enhancement of scattering along the direction of stretching (the overall scattering intensity, in all directions, is also strongly raised by lowering the pH or by adding salt).

5.5.2 LIGHT SCATTERING EXPERIMENTS

There has been only one study reported about static light scattering by a uniaxially deformed gel (Rouf *et al*, 1994). This study concerns a PDMS network prepared in the bulk by end-linking of relatively short chains ($M_c = 10\,000$, $f = 4$) and swollen by toluene. Instead of mechanical stretching, and for technical

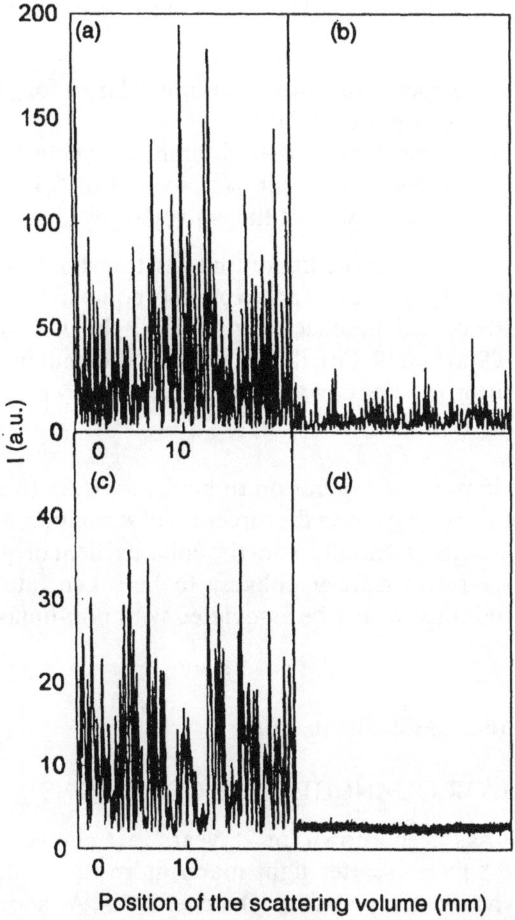

Figure 29. Variation of the scattered intensity (measured on a single coherence area) with the position of the scattering volume for a PDMS gel and semi-dilute solution with the same polymer volume fraction: $\phi = 0.25$. (a) Anisotropically swollen gel, **q** parallel to the stretching axis; (b) anisotropically swollen gel, **q** perpendicular to the stretching axis; (c) isotropically swollen gel; (d) semi-dilute solution. From Rouf *et al* (1994)

reasons, the uniaxial deformation of the gel was achieved by anisotropic swelling (a sample with an elliptical section was put in a cylindrical cell, according to a method already used by Geissler, Duplessix and Hecht, 1983). This method of deformation leads also to the appearance of butterfly patterns in the SANS profile. The light scattering experiments were performed using the scanning procedure described in section 5.1 and 5.4.2. Figure 29 shows the statistical fluctuations of light scattered intensity (in one coherence area, that is, in one diffraction spot of the scattering volume) for q parallel and perpendicular to the stretching direction, respectively. The following observations are relevant to Figure 29:

1. The ensemble average scattered intensity is much larger for q^{par} than for q^{per}, in agreement with the low q SANS data.
2. The threshold below which no scattered intensity is detected (corresponding roughly to the fluctuating component $I_F(q)$, see section 5.4.2) remains constant and approximately equal to that of the isotropic gel.

It can be observed on the same figure that the threshold associated with the fluctuating component $I_F(q)$ is also approximately equal in the deformed state to its value in the undeformed state (see Figure 29(a); the vertical scale is not the same as for Figure 29(a) and b). Finally it is not very different from the intensity of the semi-dilute solution of same concentration, measured with the same scanning device.

Comments the important conclusion to be drawn from these experiments is that the increase of scattering along the direction of stretching, at the origin of the butterfly patterns, results essentially from the enhancement of quasi-static spatial fluctuations. Such a result is very unlikely to be taken into account by the thermodynamics and must rather be associated with non-uniformities of the gel structure.

5.6 Recent models and discussion

5.6.1 THERMAL OR QUENCHED FLUCTUATIONS

As stated above, it has been known for 25 years that neutral gels at maximum swelling in a good solvent scatter light much more than calculated from the classical thermodynamical approaches (Bueche, 1970). As shown in section 5.4, an overscattering at low q, consistent with these pioneering experiments, is observed by SAXS, SANS as well as by light scattering, for a variety of systems, including charged gels when the crosslinking density is sufficiently large. These more recent studies indicate that, for a large family of gels, the overscattering depends strongly on the swelling degree. For neutral gels, it decreases as the concentration is increased and can be totally suppressed, for some gels, at a concentration larger than or equal to that of preparation. Furthermore, in several systems submitted to a uniaxial strain, the overscattering is found to be amplified along the direction of elongation. These behaviours upon swelling and stretching are key points for the determination of the 'structure' of gels.

The overscattering by gels indeed means that the spatial fluctuations of polymer concentration are larger than expected, and, in particular, larger than a semi-dilute solution of the same concentration. These 'excess' fluctuations are frequently attributed to the presence of non-uniformities in the network molecular architecture, which are expected to cause *quenched heterogeneities* in the polymer concentration profile (Stein, 1969). More generally, one may consider that these non-uniformities are intrinsic to the network formation, and one may look at them as a part of the structure (which can be called *superstructure*, with respect to the mesh size or microstructure, by contrast to nanostructure). Several models have been developed more recently on this basis. We discuss and review some of them, below.

But it is not obvious that the entire excess scattering originates in such quenched spatial fluctuations. The thermodynamics of gels may have been overlooked and the thermal fluctuations, whose profile is constantly being rebuilt, might have been underestimated. This is what was proposed, in particular, by Y. Rabin and R. Bruinsma in the theory which we consider first.

5.6.2 AMPLIFICATION OF THERMAL FLUCTUATIONS UNDER STRAIN

(Rabin and Bruinsma's theory 1992, Rabin, Pekarski and Bruinsma 1993a and b, Bruinsma and Rabin 1994):
The key ingredient of the present model is that, due to the twofold nature—solid and liquid—of gels, some non-trivial coupling might exist between the local fluctuations of polymer concentration and the local deformation. As in the theories summarized in section 5.3, the general framework is a Landau–Ginzburg formulation. In the present case, the following form of the free energy is proposed:

$$\Delta F = \int d^3r\, 1/2\{\lambda_l(\psi)\varepsilon_{ii}^2 + 2\mu(\psi)\varepsilon_{ij}^2 + L(\nabla\varepsilon_{ii})^2 + 2C_c\varepsilon_{ii}\psi + M(\nabla\psi)^2 + B\psi^2\} \quad (97)$$

where

$$\varepsilon_{ij} = \frac{1}{2}\left(\frac{\partial u_i}{\partial r_j} + \frac{\partial u_j}{\partial r_i}\right)$$

$\boldsymbol{u}(\mathbf{r})$ being, as before, the displacement field. Here $\psi(\mathbf{r})$ is a scalar field whose significance will be commented below. Equation (97) must be read with a sum over all indices repeated twice (Einstein's notation, e.g. $\varepsilon_{ii} = \sum_i \varepsilon_{ii}$).

Some similarities with Onuki's approach are immediately visible when converting the equations given in section 5.3.2 into the present notations. (Note that $\nabla\cdot\mathbf{u} = \varepsilon_{ii}$.) The average free energy $F = F_{\text{mix}} + F_{\text{el}}$ (neglecting the gradient term) in Eq. (76) could indeed have been written

$$F = \int_V \tfrac{1}{2}(\lambda_l\varepsilon_{ii}^2 + \mu\varepsilon_{ij}^2)\,d^3r \quad (98)$$

where λ_l and μ are the Lamé coefficients (see Landau and Lifshitz, 1973); μ, as before, is the shear modulus; λ_l can be related to the osmotic bulk modulus through $K_{\text{os}}^{\text{gel}} = \lambda_l + (2/3)\mu$ (Landau and Lifshitz, 1973).

There are also important differences with respect to Onuki's formulation. In Eq. (97), $\psi(\mathbf{r})$ is not the local concentration (by volume), but an *internal order parameter* assumed to represent the liquid-like part of the local concentration. Moreover, the strong coupling between the concentration and the displacement field $\delta\phi(r)/\phi = -\nabla\cdot\mathbf{u} = -\varepsilon_{ii}$, (see section 5.3.2) is believed no longer to be valid. Instead, a softer coupling involving Ψ is proposed:

$$\delta\phi(r)/\phi = -\varepsilon_{ii} + \Gamma'\nabla^2\psi + O(\nabla^4\psi) \tag{99}$$

Note that Eq. (99) allows for some deviations, essentially on short length scales, with respect to the usual affine assumption giving $\delta\phi(r)/\phi = -\varepsilon_{ii}$. This means that, according to the present formulation, the network disposes of a much larger freedom of reorganization. For example, ψ can fluctuate locally at fixed strain. This is why the two last terms, which give the free energy cost of such fluctuations, have been added. Note also that the elastic constants λ_1 and μ for a fixed ψ are no longer the measured ones. The latter are believed to be often much smaller, precisely because of the additional freedom given to the network by this model.

But the concentration ϕ and the displacements ε_{ij} cannot be completely independent. In the present model they are coupled in a somewhat indirect way through the combination of relation (99) and the term $2C_c\varepsilon_{ii}\psi$ in the expression of the free energy (Eq. (97)). Note that there is nothing analogous to $2C_c\varepsilon_{ii}$ in Onuki's formulation. This term introduces a free energy penalty for creating local concentration inhomogeneities.

Another key ingredient is that the elastic constants are allowed to depend locally on the internal order parameter ψ. Expanding them to the lowest order, one writes

$$\begin{aligned} \lambda_1(\psi) &= \lambda_{1g} + (\partial\lambda_1/\partial\psi)_g\psi \\ \mu(\psi) &= \mu_g + (\partial\mu/\partial\psi)_g\psi \end{aligned} \tag{100}$$

where $(\partial\lambda_1/\partial\psi)_g$ and $(\partial\mu/\partial\psi)_g$ are assumed to be negative.

Physically, the consequence of Eq. (100) is that under a spontaneous increase of the liquid-like character at some point (represented by an enhancement of ψ), the elastic constants λ_1 and μ are decreased. As a result, if the sample is submitted to a strain, such regions will be deformed more than the average gel. The structure of the free energy is such that this effect should not relax and, therefore, the thermal fluctuations of polymer concentration are expected to increase along the direction of elongation. Thus one can guess the consequence on the scattering intensity: it should be increased along the axis of extension.

In order to formulate this quantitatively, one follows the same route as in section 5.3.2. To obtain the structure factor from the free energy, one expresses F in the Fourier space and applies the equipartition theorem (see Section 5.3.2). The various constants appearing in Eq. (97) are estimated assuming that the gel can be described as a *close packing of blobs* (c^* theorem see section 5.3) and one obtains

$$S_{gel}(q) \sim k_B T \frac{\phi^2}{K_{osm}^{gel} + \frac{4}{3}\mu - (\lambda - 1)\gamma(3\cos^2\beta - 1) + L_s q^2} \tag{101}$$

with $\gamma = (C_c/B)\,(\partial\mu/\partial\psi)_g$, where $C_c/B \approx -1$, which means that, in general, γ is positive for a uniaxial extension ($\lambda > 0$).

As before, β denotes the azimuthal angle between $\mathbf{q}$ and the stretching axis. It is easy to see that $S_{gel}(q)$ given by Eq. (101) is increased along the elongation axis ($\beta = 0$), which is sufficient to give rise to lobes with a great axis parallel to the stretching direction (Figure 30). In other words, this theory predicts the appearance under strain of 'butterfly patterns' with the same basic features and the same orientation as those which are observed (see section 5.5). Moreover, it predicts the change of the anisotropy with the progressive relaxation of the order parameter ψ. At small times a *normal anisotropy* with the major axis of the iso-intensity curves perpendicular to the stretching axis is expected to show up. As the relaxation proceeds, an evolution towards lobes with a major axis parallel to the stretching direction is predicted to happen. Such an evolution has been observed in a slightly different system, namely an *asymmetrical polymer blend* (polystyrene protonated and deuterated) made of a matrix of long chains containing smaller 'guest' chains which are labelled (Bastide, Buzier and Boué, 1988; Boué *et al*, 1989, 1991). The former can be viewed as a temporary network and the latter, which remain essentially unentangled, as a kind of polymeric solvent. It must be stressed that, in Figure 30, one also observes the temporary appearance of a peak in the structure factor. Comparable maxima in the centre of the lobes have been observed for semi-dilute solutions under shear (Wu, Pine and Dixon, 1991; Dixon, Pine and Wu, 1992).

Incidentally, it can also be remarked that Eq. (101) is very similar to a random phase approximation (RPA, see de Gennes, 1979) expression with a directional dependence of χ, $(\lambda - 1)\gamma(3\cos^2\beta - 1)$ being comparable to χ. The idea of directional dependence of χ for reproducing the butterfly iso-intensity lines had been tested by Bastide, Buzier and Boué (1988) but no physical explanation could be found and the idea was abandoned.

Note that, for sufficiently large λ, a kind of anisotropic spinodal decomposition could easily be attained:

$$K_{osm}^{gel} + \frac{4}{3}\mu - (\lambda - 1)\gamma(3\cos^2\beta - 1)$$

may decrease down to zero and $S(q^{par} \to 0)$ may diverge. In practice, if one fits to this model, for small deformation ratios, some data partly shown in section 5.5, one gets a value of γ which should lead to a divergence of the intensity in the domain of study (say $\lambda_{spinodal} \approx 1.5$, whereas $\lambda_{max} \approx 1.8$). Such a divergence is not observed. According to Y. Rabin and R. Bruinsma, this disagreement might come from the fact that their theory is strictly valid for small deformations only. In the

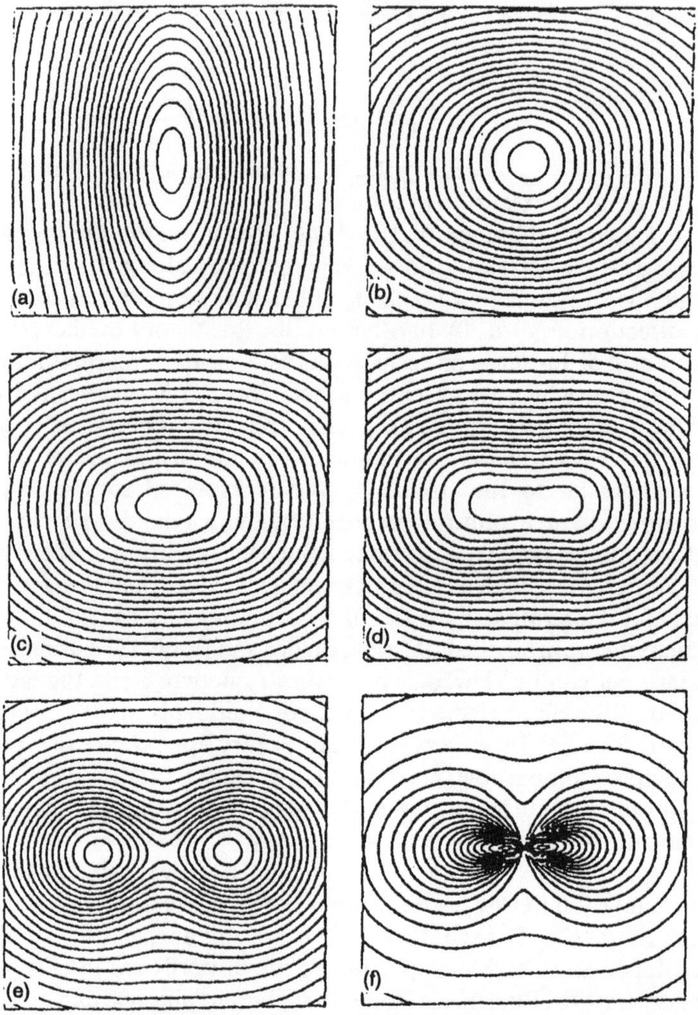

Figure 30. Iso-intensity lines calculated from Rabin–Bruinsma's theory for different times after the deformation. The stretching axis is horizontal. Time is measured in units of the relaxation time of the scalar order parameter ψ. (a) $t = 0$; (b) $t = 2$; (c) $t = 4$; (d) $t = 4$, (e) $t = 10$; (f) $t = 10^5$. The extension ratio is $\lambda = 1.3$ and the coupling constant is equal to 0.3. From Bruinsma and Rabin (1994)

case of relatively large elongation ratios some terms of higher order should be introduced in the calculation, which might soften the effects.

This theory was originally derived for the description of anisotropic cases (Rabin and Bruinsma, 1992). But it is clear that the same ideas should apply to an isotropic expansion with respect to the preparation state (i.e. what we have sometimes called an overswelling). One assumes that the deformation is identical in any direction, thus

$$(\lambda - 1)(C_c/B)(\partial\mu/\partial\psi)_g(3\cos^2\beta - 1)$$

should become a quantity of form $(\Lambda - 1).\,\tilde{\gamma}$, with $\Lambda = (Q/Q_O)^{1/3} = (\phi/\phi_O)^{-1/3}$ as in section 5.3.2. The swelling with respect to the preparation state would then be accompanied by the approach of a spinodal decomposition. The effective longitudinal modulus

$$M' = K_{\text{osm}}^{\text{gel}} + \tfrac{4}{3}\mu - [(\phi/\phi_0)^{-1/3} - 1]\tilde{\gamma}$$

is indeed expected to decrease upon swelling and might come close to zero. In other words, according to Bruinsma and Rabin (1994), due to the effect of the swelling, polymer networks put in contact in a good solvent of the constitutive chains might behave as if they were on the border of a spinodal decomposition, that is in very poor solvent conditions.

Comments and discussion For at least two reasons, this model is very appealing. First, it qualitatively predicts the overscattering of gels under swelling with respect to preparation state and the appearance of butterfly patterns under uniaxial strain without requiring the presence of heterogeneities. This fits well the celebrated razor-like recommendation: 'one should not use any unnecessary concept'. Second, its predictions are associated with a softening of the strong affine assumption, which has been found difficult to reconcile with a number of measurements of the deformation on molecular scales in networks (see e.g. Bastide, Herz and Boué, 1985; Bastide, Boué and Buzier, 1989; Bastide *et al*, 1984). Therefore, this theory is constructed in such a way that it could solve two problems at the same time.

Unfortunately, in view of the recent light scattering data, this uniformity of physical origin seems to be a two-edged sword. In sections 5.9 and 5.5, we have indeed seen that each time gels have been studied by a scanning light scattering device, the overscattering has been found to be almost entirely contained in the *quasi-static speckle pattern*, which is thought to be caused by almost quenched spatial fluctuations. Recall in particular that, in the case of the uniaxially deformed gel (Figure 29) which shows butterfly patterns in neutron scattering, the spectacular increase in parallel direction is entirely contained in the part of the intensity which varies with the scattering volume position. Contrary to Rabin–Bruinsma's predictions, the fast fluctuating part of the intensity is far lower. The amplitude and the dynamics of this fast fluctuating component remain essentially unchanged under strain. In short, the fluctuations of polymer concentration which are truly thermal (whose profile is entirely rebuilt very fast) do not seem to make a significant contribution to isotropic or anisotropic overscattering.

Thus, bearing in mind these experiments only, the simplest picture for gels is a modified idealized description: the thermal fluctuations would be those of an idealized gel of the same concentration but, in addition, there would exist quenched fluctuations which should be coupled to non-uniformities of network structure. The only point would then be to understand how these quenched fluctuations can be amplified in neutral gels, upon swelling or stretching the network. We discuss this question below (it can be explained by at least four models based on structural heterogeneities). However, the situation is probably not that simple. First it has been pointed out by Geissler, Horkay and Hecht that thermodynamical data do not agree entirely with such a description: the swelling pressure of gels is a little lower than expected (by, say, 20–30%; see section 5.3.1). Second, one has to remember that, in the case of neutral polyacrylamide gels at constant polymer concentration, the fluctuating part increases slightly, but in a detectable manner, with increasing crosslinking density (Joosten, McCarthy and Pusey, 1991). This enhancement of thermal fluctuations is not expected from the idealized description of section 5.3: the latter predicts a measurable decrease. Thus, since one observes at the same time a strong increase of the quasi-static part of the intensity, there must be at least some coupling between the quasi-static and the thermal fluctuations.

This leads us back to the model by Y. Rabin and R. Bruinsma. We have seen above that, according to these authors, gels swollen with respect to their state of preparation might be on the border of a spinodal decomposition. More recently Bruinsma and Rabin argued (1994) that, under further strain (e.g. a uniaxial elongation), the decomposition threshold could be passed. After reintroducing at this level the presence of moderate heterogeneities, they proposed that one should observe a heterogeneous nucleation pinned by the non-uniformities. This would prevent the divergence of $I(q \to 0)$ and, according to this picture, the quasi-static speckle pattern would then originate in the presence of micro-phases with rather high polymer concentration and whose profile would evolve very slowly in time. This explanation is meant, in particular, to explain the slow dynamics of the speckle pattern (section 5.4.3). This interesting idea does not come without posing problems. If the threshold of spinodal decomposition was truly passed, it would be difficult to understand why the gel does not expel solvent macroscopically (or in other words, the swelling pressure would be reduced much further than by 30%). Remember that gels usually swell upon a uniaxial stretching (this behaviour is explained semi-quantitatively by a Flory–Rehner approach; see Dušek and Prins, 1969; Horkay and Zrinyi, 1988) and there is apparently no exception for gels exhibiting butterflies. So, the explanation in terms of micro-phases does not seem very convincing to us.

To sum up, conventional ideas, those based on structural heterogeneities, seem to be more adapted for explaining the observed phenomena.

Although it goes beyond the scope of this book, we come back briefly to the case of non-gel systems which exhibit scattering anomalies under shear or elongation. It has been known for some time that shearing a semi-dilute polymer solution in poor solvent induces a shift of its cloud point (Ver Strate and

Phillippoff, 1974; Schmidt and Wolf 1979; Rangel-Nafaile, Metzner and Wissbrun, 1984). More recently, scattering experiments have revealed, in the same regime of concentrations and temperatures, an anisotropy characterized by lobes resembling those observed in the case of stretched gels (Wu, Pine and Dixon, 1991; Hashimoto and Fujioka (1991), Hashimoto and Kume (1992); van Egmond, Werner and Fuller, 1992). Such a behaviour is maintained when the quality of the solvent is raised, that is, further apart from the demixtion curve (Boué and Linder, 1994). As stated above, the lobes in gels are themselves looking like those observed in the case of asymmetrical polymer blends under uniaxial elongation (Bastide, Buzier and Boué, 1988). Therefore, one observes patterns which look surprisingly similar in three types of systems, two of them being liquids where truly quenched fluctuations cannot exist. If, in the case of gels, the phenomenon seems to be related to the presence of a static superstructure, such an explanation cannot hold for true liquids, even if viscoelastic. For such systems, 'thermodynamical' theories have been proposed, which seem to reproduce the observed effects satisfactorily (Helfand and Frederickson, 1989); Milner, 1991; Olmsted and Milner, 1994). It is, however, surprising that behaviours which are so similar have to be explained on the one hand with 'quenched' fluctuations and on the other hand with thermodynamical fluctuations. One may wonder whether the 'quenched' fluctuations in gels could not have a kind of equivalent in viscoelastic liquids, in the form of a slowly fluctuating quantity, decoupled from the fast fluctuating 'thermal' fluctuations. A comparable situation exists in some spin glasses made of blends of magnetic and non-magnetic atoms. The fast fluctuations of the spin orientations are superimposed on slow spatial rearrangements of positions of the magnetic atoms. In the case of semi-dilute solutions, the recent observations, by Dixon, Pine and Wu (1992), of a relatively slow decay of the effect after stopping the shear might perhaps be interpreted along theses lines. In the case of asymmetrical polymer melts, the 'butterfly' patterns could be viewed as resulting from an inhomogeneous deformation associated with non-uniform, though random, inter-chain entanglements (Bastide *et al*, 1993).

5.6.3 RECENT MODELS OF STRUCTURAL HETEROGENEITIES IN GELS

We turn now to explanations of overscattering in which the non-uniformity of the network structure plays a major role. As stated above, this idea seems at present to be more suitable for describing the existing experimental data. From this common basis different models have been developed. In the following, we summarize these models and try to stress their similarities and their differences.

5.6.3.1 Separation of clusters

The first approach that we will present in this category is twofold. There is first a model, which we call the *frozen-blobs model*, which deals with a special type of gels, made by almost instantaneous random crosslinking of a semi-dilute solution. Then, there is an attempt at generalizing it to a larger spectrum of systems. The main idea on which the whole approach is based is that uniformity and randomness are two separate things and that the difference between the two is especially important in quenched systems. In the field of gels, this distinction was noticed first by Bueche (1970). In the present model, it is introduced in the framework of the percolation model, which is used here not to describe the gelation process but the quenched disorder. In order to illustrate this idea, we focus first on a situation simpler than that of gels.

5.6.3.1.1 Basic idea illustrated by the lattice gas paradox

Imagine a *lattice gas* consisting of molecules stuck randomly on the nodes of a regular lattice (each site of the lattice being approximately the same size as the molecule). The structure of this 'quenched' lattice gas can be described as the result of a site percolation process (see e.g. Stauffer, Coniglio and Adam, 1982 or Stauffer, 1985): the occupancy of a site is drawn at random with a probability equal to the ratio p of the number of molecules over the total number of sites. It is well known that, in a site percolation process, clusters of first neighbours on the lattice are formed. These clusters are not compact; they are branched, their size distribution is very wide and the larger are self-similar with a fractal dimension equal to 5/2 (if the number of occupied sites is large enough). See Figure 31.

The average size of the clusters increases with p and, for p larger than a certain threshold p_c, the larger cluster becomes 'infinite', which means that it reaches the edges of the lattice. Since the molecules have been put at random on the lattice, there are no correlations between them. The correlation length ξ of the number density of particles is of the order of the size of the lattice site. And if there are no spatial correlations between the molecules, the ensemble-averaged small-angle scattering is independent of q for $qa \ll 1$ and its absolute value will be low (see Eqs (9) and (10)). Nevertheless, for p comparable with p_c, this system contains large clusters, which, if they were separated and studied in dilute state, would give rise to a considerable scattering exhibiting a strong dependence with q (see Figure 31). There is therefore some kind of paradox: objects possibly leading to a large scattering are 'invisible' in a scattering experiment provided they are studied in the state where they have been prepared. This apparent paradox arises from screening effects: the correlations between scattering centres of one given cluster (the *intra-cluster* correlations which would lead to a strong scattering in the case of well-separated objects) are exactly balanced by the inter-cluster correlations (i.e., the correlations between pairs of points carried by different clusters) (Daoud and Leibler, 1988).

A comparable effect obviously exists in semi-dilute solutions (see section 5.2): the structure of the chains on distances greater than ξ is undetectable for a scattering experiment, due to screening effects. The same phenomenon has even

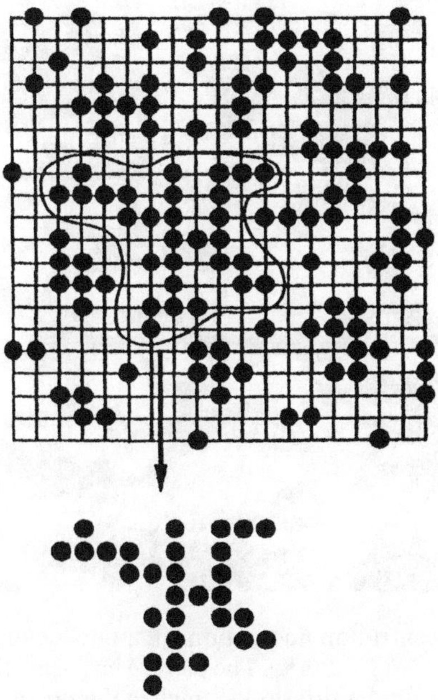

Figure 31. Schematic representation of a 'frozen' lattice gas in two dimensions (small molecules placed at random on a lattice). From Bastide *et al* (1992a)

been observed directly in critical binary mixtures. As shown by Guenoun–Perrot–Beysens (see Figure 32), in the vicinity of a critical point, some fractal clusters span length scales which are much larger than the correlation length ξ. The clusters represent regions where the concentration of one component is higher than the average one, or, which amounts to the same, fluctuations of the order parameter. There is therefore a complementary assembly of two types of clusters (black and white on Figure 32). As in the preceding example, the correlations carried by the larger clusters are screened out, on length scales larger than ξ, by inter-cluster correlations.

In short, randomness means non-uniformity (represented here by clusters), but these inhomogeneities are not detectable with a scattering experiment on length scales larger than ξ. Obviously, in the case of the molecules on a lattice, screening effects are thorough only in the state of complete randomness. Slight displacements of the clusters with respect to each other would result in the breakdown of the delicate balance of intra- and inter-cluster correlations. One would then observe a clear enhancement of the ensemble-averaged scattering intensity, which could be isotropic or anisotropic depending on whether the separation

Figure 32. Picture of concentration fluctuations in a binary fluid (isobutyric acid–water) close to its critical point ($T - T_c = 1$ mK). The picture has been digitized with two intensity levels (black and white) around a threshold which is the mean intensity. The correlation length of these fluctuations is much smaller than the average size of the clusters (significantly smaller than the bar visible in the right-hand corner). From Guenoun, Perrot and Beysens (1989)

itself is isotropic or anisotropic. In other words, the cluster structure would then be partly 'unscreened'.

5.6.3.1.2 Swelling in the frozen-blobs model

As stated above, this model addresses the somewhat idealized situation of a gel formed by an almost instantaneous and completely random introduction of crosslinks into a semi-dilute solution (Bastide and Leibler, 1988). The precursor chains are assumed to be very large compared to the correlation length ξ_r in the reaction bath, prior to the crosslinking. If the crosslinking molecules react very fast, one expects that inter-chain bridges will essentially be formed in the vicinity of the pre-existing inter-chain contact points. Crosslinking molecules reacting with chains far enough from the inter-chain contact points should essentially lead to the formation of small intra-chain loops, whose effect is not taken into account. Therefore, in this model, one considers the set of inter-chain contact points existing at a given time, which form a kind of irregular lattice, and one assumes that some of them are 'frozen-in', i.e. are transformed into permanent crosslinks. Under these assumptions, the analogy with a site percolation process is straightforward. Let us insist on the fact, again, that percolation is not used here in order to describe the vicinity of the gelation threshold, but is utilized for modelling the disorder in the spatial distribution of crosslinks. One focuses on well-formed gels, far above the true gelation threshold, with essentially no sol fraction.

On the 'lattice' of the inter-chain contact points (represented schematically in Figure 33(a)) a given junction may or may not be neighbour to another one. When junctions are located on neighbouring lattice sites, a 'frozen blob' is formed: the chains cannot slide any more with respect to each other. Obviously, if the fraction p of crosslinks over the number of interchain contact points is large enough, the 'frozen blobs' will often be interconnected. In other words, clusters of frozen blobs (with a very wide size distribution) should be formed, representing the large-scale crosslinking density fluctuations. The clusters can be very large. One may even reach, for $p = p_c$, the percolation threshold of these frozen blobs. But, for simplicity, only the cases for which $p \leqslant p_c$ were considered explicitly[†].

Whatever their size, the frozen-blob clusters should remain, in the reaction bath, undetectable in a scattering experiment. First, in such a blob picture one does not introduce any difference of polymer concentration, that is, no 'contrast', between the frozen blobs and the rest of the gel. This hypothesis is probably a bit unrealistic: the formation of crosslinks might induce, at least, a slight local shrinking of the frozen blobs and therefore some difference of the concentration with respect to the interstitial regions. But, even then, provided the frozen blobs remain distributed at random in space, the intra-cluster correlations should be screened out by the inter-cluster correlations, as explained above. Therefore the scattering of the gel should remain approximately identical to that of the pre-existing solution. In other words, the larger correlation distance ξ_g associated with the spatial fluctuations of polymer concentration should remain approximately equal, in the reaction bath, to that of the equivalent semi-dilute solution.

Imagine now that the gel is 'overswollen' with respect to the preparation conditions (by putting it in contact with an excess of 'good solvent' of the polymer chains). In a frozen blob, it will be difficult to move the crosslinks apart from each other since they are connected by a chain segment which is already in an optimal excluded volume condition. Most of the swelling will presumably be accounted for by the 'interstitial regions' (the less crosslinked regions between the clusters): there, some expansion can still be achieved by removal of inter-chain contact points. As a result, one expects the swelling of the network to be very in-

[†] For $p > pc$ the average crosslink density is expected to be sufficiently high for inducing concentration heterogeneities in the reaction bath. Moreover, since the "frozen blobs" percolate, one may guess that the gel will swell with difficulty with respect to the reaction bath.

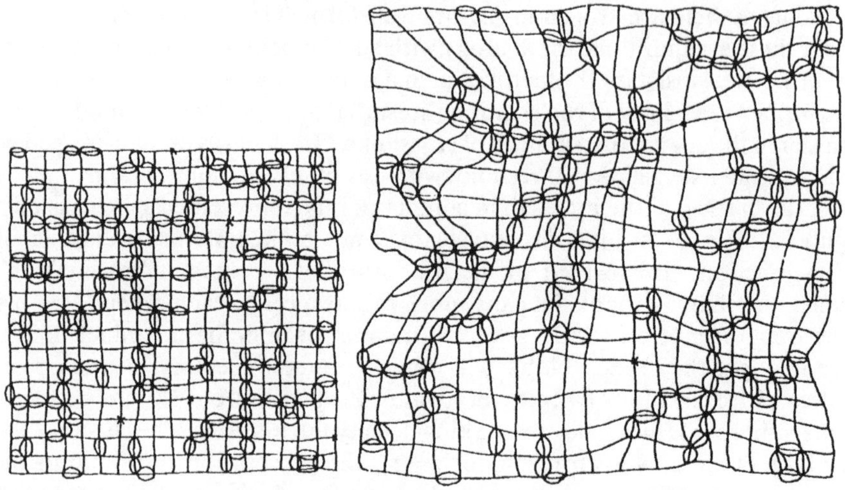

(a) (b)

Figure 33. Schematic representation of a gel prepared by random crosslinking of a semi-dilute solution. The 'frozen blobs', often associated in clusters, are represented by this sign: ⊖. (a) In the preparation state; (b) after inhomogeneous swelling with respect to the preparation state. From Bastide, Boué and Buzier (1989)

homogeneous, with distortions spanning over rather large length scales (compared to the average mesh size), due to the clustering of frozen blobs (which form regions 'harder' than average). This mechanism is represented schematically in Figure 33(b).

The distortion of the gel means naturally an amplification of quasi-static concentration fluctuations. It must therefore lead to an enhancement of scattering intensity. In order to make more specific predictions, the problem was approached in terms of separation of the frozen blob clusters: because of the unscreening of correlations mentioned above, a strong enhancement of $I(q)$ is also expected along this way of reasoning. Note that the separation should be accompanied by an increase of the polymer concentration 'contrast' between the clusters and the interstitial regions. In order to describe this process, one borrowed the conclusions of previous works by Daoud Family and Jannink (1984) and Daoud and Leibler (1988), dealing with the dilution of 'true' percolation clusters. Because the size distribution of the clusters is very large, one expects, upon dilution, the small clusters to be expelled from the inside of the larger ones. The, then inter-cluster correlation length ξ_c can be viewed as the typical size of clusters which are not entangled with the others. Focusing on the regime of crosslinking densities corresponding to the vicinity of the percolation threshold,[‡] Daoud and Leibler (1988) showed that ξ_c should scale with ϕ as

$$\xi_c \sim \xi_0(\phi/\phi_0)^{-m} \tag{102}$$

with $m = 5/3$; ϕ_0 and ξ_0 denote respectively the concentration prior to dilution and the correlation length prior to the chemical reaction. Whereas,

$$\mathrm{I}(q \to 0) \sim \phi^{-t} \quad \text{with} \quad t = 5/3 \tag{103}$$

Then, postulating that, in statistically crosslinked networks, the scattering is largely dominated by the signal from the frozen-blob clusters, one simply identifies the larger correlation distance ξ_g in the gel with ξ_c. Note that the equality for m and t is a characteristic of the vicinity of the percolation threshold. Outside this critical region, scaling laws for the variation with ϕ of ξ_c and $\mathrm{I}(q \to 0)$ are expected to be conserved, but the exponents m and t, which have not been calculated, should take different values, presumably smaller.

Another prediction of Daoud–Leibler's theory which can be transposed to statistical gels is the scattering in the 'intermediate regime'. In the case of dilution of percolation clusters, it was shown that, for $q \gg 1/\xi_c$:

$$I(q) \sim \frac{\phi}{q^{D_{\mathrm{eff}}}} \tag{104}$$

with $D_{\mathrm{eff}} = D_a(3 - \tau) = 8/5$. Here D_a is the fractal dimension of the larger clusters in the presence of excluded volume interactions; it is predicted to be equal to 2 (Daoud and Leibler, 1988) and experimental data are in good agreement with this value (Bouchaud *et al*, 1986). In the case of a semi-dilute solution of clusters, the excluded volume interactions are relevant on length scales $\leqslant \xi_c$. On larger length scales ($\geqslant \xi_c$), the percolation clusters are expected to recover their unperturbed fractal dimension $D_p = 5/2$. This means that if one single large cluster was labelled (with $R \gg \xi_c$), one should obtain $\mathrm{I}(q) \sim q^{-5/2}$ for $q \ll 1/\xi_c$.

The term $(3 - \tau)$ in the exponent D_{eff} reflects the effect of polydispersity on the scattering. The q domain probed corresponds to the 'intermediate regime' of the

[‡] Considering the practical synthesis of statistical gels (polystyrene, see section 5.4) at a polymer concentration $\Phi = 0.1$, the vicinity of the percolation threshold of frozen blobs is not an irrelevant assumption. For such a concentration $\xi_{sd} \approx 13.5$ Å, which means a blob radius of gyration of 23 Å and therefore a blob "molecular weight" $M_{\mathrm{blob}} \approx 7000$, that is, roughly 70 monomer units. Transforming all the blobs into frozen blobs ($p = 1$) would require 1 crosslink per 2 blobs, that is, 1 crosslink for 140 monomer units. Since, for a site percolation, p_c should be of the order of 0.3, the percolation threshold of the frozen blobs would be then reached for one effective crosslink density of about 0.2%, in moles of true tie-points per mole of monomer. But, as stated above, only an unknown fraction of crosslinking molecules are transformed into real interchain bridges (because of incomplete chemical reaction and also because of formation of small intrachain loops). Therefore, the percolation threshold of the frozen blobs might correspond to higher values of the density of crosslinking molecules which are introduced, comparable to those of statistical gels (0.5% or 0.8%) discussed in section 5.4.

larger clusters but to the Guinier regime of the smaller ones. Here τ is the exponent appearing in the expression of the number density n_s of clusters made of s elementary units and is written

$$n_s(p) = s^{-\tau} f(s/s^*) \tag{105}$$

where s^* is a 'typical size' related to the distance from the percolation threshold as

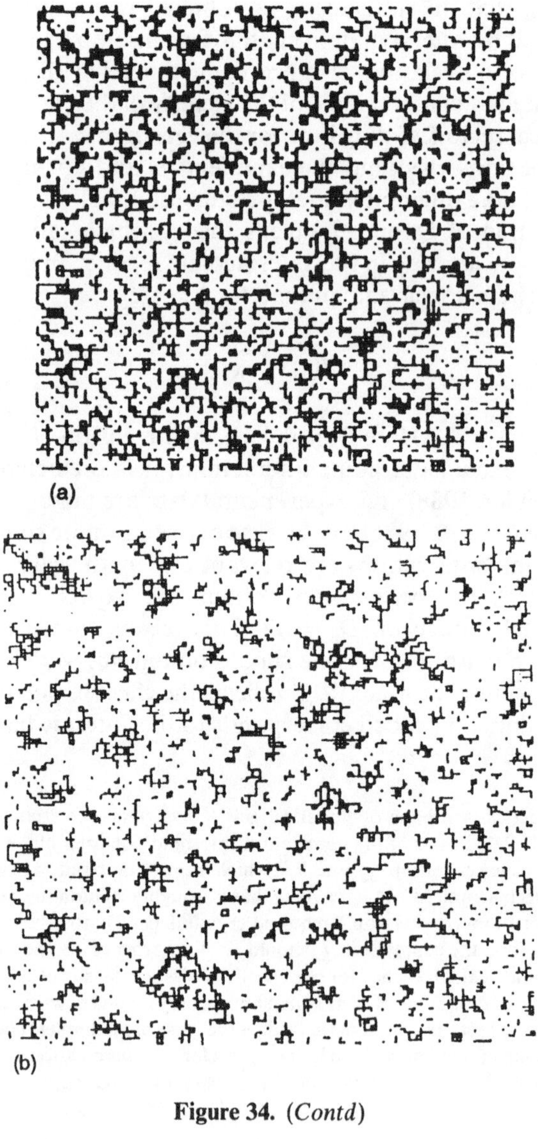

Figure 34. (*Contd*)

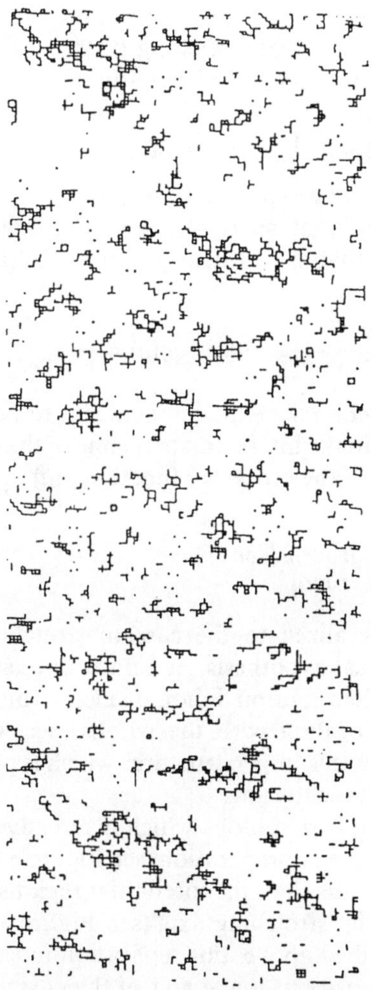

Figure 34. Schematic representation of crosslinking heterogeneities, according to the 'frozen blobs' model. The black clusters represent clusters of 'frozen blobs'. The white regions represent regions of lower crosslinking density. (a) In the preparation state (equivalent to Figure 33(a)); (b) after swelling with respect to the preparation state (equivalent to Figure 33(b)); (c) after uniaxial elongation with respect to swollen state (b). Note that fuzzy lines perpendicular to the stretching direction appear spontaneously. From Bastide, Leibler and Prost (1990)

follows: $s^* = (p_c - p)^{-1/\sigma}$. In the framework of the percolation theory the exponents τ and σ are respectively equal to 2.2 and 0.46.

In the model of gels with 'frozen blobs', one expects the same scattering behaviours to hold for $1/\xi_c < q < 1/\xi_{sd}$. For $q > 1/\xi_{sd}$, one should probe the correlations inside the blobs, frozen or not. Assuming, contrary to the classical picture (see appendix) that no real stretching of the chains is associated with the swelling, the intermediate regime of semi-dilute solutions should then be recovered:

$$I(q) \sim q^{-1/\nu} \sim q^{-1.7} \quad \text{for } q > 1/\xi_{sd} \tag{106}$$

where ν is the excluded volume exponent ($\nu = 0.588$).

Altogether, the scattering of such idealized non-homogeneous gels was expressed, by analogy with the Ornstein–Zernicke behaviour of semi-dilute solutions, as follows:

$$I(q) \sim \frac{\phi \xi_c^{D_{\text{eff}}}}{1 + g(q) q^2 \xi_c^2} \quad \text{with} \quad g(q) = \frac{1}{1 + [q\xi_c]^{2-D_{\text{eff}}}} \tag{107}$$

In order to obtain such a compact expression, meant to be valid in the Guinier regime ($q\xi_g < 1$) as well as in the intermediate regime of the clusters ($q\xi_g \gg 1$), one had to postulate the form of the crossover function $g(q)$.

5.6.3.1.3 Stretching in the frozen-blobs model
(Bastide, Leibler and Prost, 1990)

This model is easily generalized to the case of stretching by means of the introduction of an additional hypothesis. As before, one assumes that the frozen blob clusters make the deformation inhomogeneous on 'mesoscopic' length scales. Again, instead of dealing with the whole spectrum of concentration fluctuations, one focuses on this 'superstructure', which is postulated to carry the correlations dominating the scattering.

Consider then a gel with frozen blobs which is already overswollen to some extent with respect to its state of preparation (see Figure 34(b)). If the clusters of frozen blobs are deformed less than the interstitial regions, they separate along the direction parallel to the stretching axis (see Figure 34(c)). Because of the unscreening process described above, this separation must lead to an enhancement of the scattering intensity at low q and of the correlation length ξ_g^{par}, for q 'parallel' to the elongation axis. In the perpendicular direction, the gel is compressed. Therefore, the clusters should reinterpenetrate (Figure 34c)). Hence, the correlations they carry must be partly screened out again and one expects then a decrease of both the scattering intensity at low q and ξ_g^{per}.

Along the directions, represented by the azimuthal angle β, which are intermediate between the parallel ($\beta = 0$) and perpendicular ($\beta = \pi/2$) axes, one naturally expects a cross-over to take place between the enhancement and depression respectively. A simple function, that is an elliptic interpolation, was proposed for describing this transition:

$$\frac{1}{\xi_g^2(\beta)} = \frac{\cos^2 \beta}{\xi_g^{\text{par}}} + \frac{\sin^2 \beta}{\xi_g^{\text{per}}} \tag{108}$$

Then, to obtain the correlation length $\xi_g(\beta)$ in any direction, the variation law with the extension ratio λ for ξ_g^{par} and ξ_g^{per} must be known. To this end, a trick was used: it was postulated that the Flory–Rehner hypothesis can be used following a reverse path: the swelling of a gel has been described, on the basis of the model of rubbers, as a kind of three dimensional extension by a factor λ, so that one writes $\phi/\phi_0 = Q_0/Q = \lambda^{-3}$. Conversely, in the case of the 'frozen-blob' model, the swelling is described first and the hypothesis is made that an imposed extension λ, in the parallel direction, is analogous to a change of swelling degree Q:

$$\Lambda = (Q/Q_0)^{1/3} = (\phi/\phi_0)^{-1/3} \tag{109}$$

Hence by combining Eq. (109) with (102), one obtains

$$\begin{aligned} \xi_g^{par} &= \xi_0(\phi/\phi_0)^{-m} = \xi_0\lambda^{3m} \\ &= \xi_0\lambda^5 \quad \text{with } m = 5/3 \end{aligned} \tag{110}$$

In the perpendicular direction, the sample is compressed by a factor $\lambda^{-1/2}$, which can be compared to a deswelling, thus:

$$\begin{aligned} \xi_g^{per} &= \xi_0(\phi/\phi_0)^{-3m/2} = \xi_0\lambda^{-3m/2} \\ &= \xi_0\lambda^{-5/2} \quad \text{with } m = 5/3 \end{aligned} \tag{111}$$

Finally, after introducing Eqs (109) and (110) in Eq. (108) and then generalizing Eqs (107) as follows, one obtains the expression of the scattering intensity for any extension ratio λ, orientation β and scattering vector $\mathbf{q}$ (provided $q\xi_{sd} < 1$):

$$\mathrm{I}(q) \sim \frac{\phi\xi_c^D(\beta)}{1 + g(q,\beta)q^2\xi_g^2} \quad \text{with} \quad g(q) = \frac{1}{1 + [q\xi_g(\beta)]^{2-\mathrm{Deff}}} \tag{112}$$

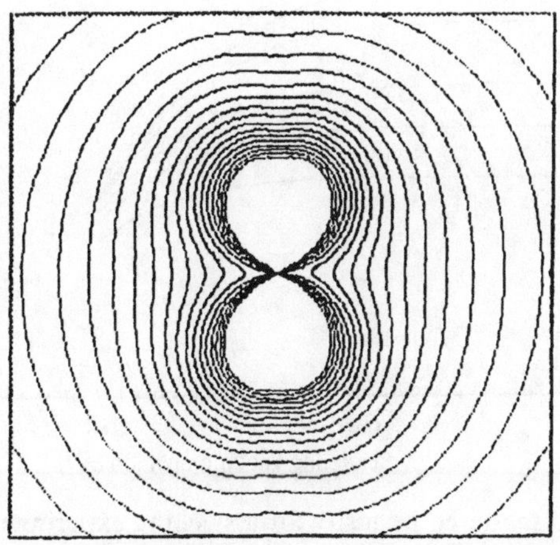

Figure 35. Example of 'butterfly-like' iso-intensity lines calculated from the 'frozen-blob' model. From Bastide, Leibler and Prost (1990)

As can be seen in the example displayed on Figure 35, this last expression (Eq. (112)) leads to lobes in the iso-intensity lines, which are oriented in the same direction as the observed ones.

Comments and discussion The main strength of the 'frozen blob model' is that, from rather crude assumptions, it is able to anticipate consistently the main trends of the scattering by a statistical gel which is swollen or is uniaxially stretched. As observed, it predicts indeed that

1. The scattering intensity is strongly enhanced upon swelling and, in the parallel direction, upon stretching (see section 5.5). The calculated iso-intensity lines (for a stretched gel) are qualitatively similar to the observed 'butterfly patterns'. It was the first model to predict such a behaviour.
2. The enhancement of the intensity is accompanied by the shift, towards the small values of q, of a crossover between a 'Guinier regime' (appearing like a 'saturation' on a log scale) and the development of an 'intermediate behaviour' (where the observed decay of the scattering intensity is grossly consistent with the predicted one, namely $I(q) \sim q^{-8/5}$ for $q\xi_g \gg 1$). Moreover, note the similarity in the aspect of the calculated curves (Figures 36 and 37) and the experimental curves given in Figures 22 and 38.

On Figure 36, the scattering intensity has been calculated, from Eq. (107), for decreasing polymer concentrations, starting from $\phi = 0.1$ with $\xi_g \approx \xi_{sd} = 14$ Å. The values of ϕ are very close to those of the statistical gel of Figure 22 (for which

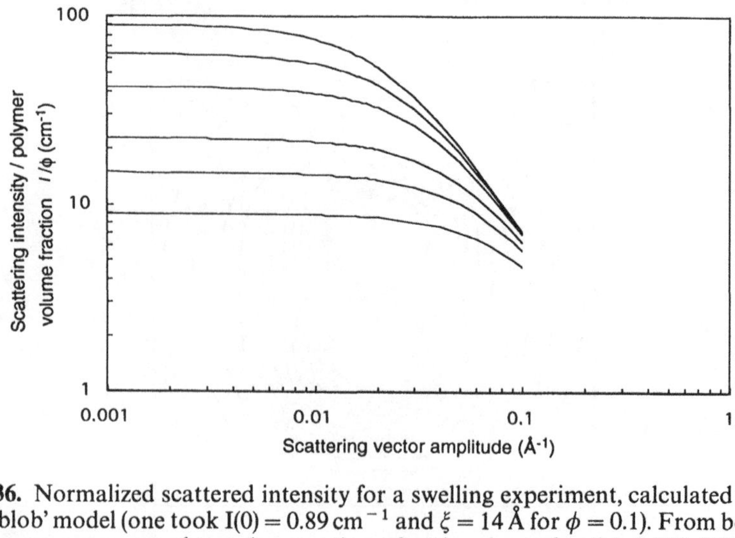

Figure 36. Normalized scattered intensity for a swelling experiment, calculated with the 'frozen-blob' model (one took $I(0) = 0.89\ \text{cm}^{-1}$ and $\xi = 14$ Å for $\phi = 0.1$). From bottom to top, the curves correspond to polymer volume fraction ϕ equal to 0.1, 0.083, 0.071, 0.056, 0.05, 0.042

$R_c = 0.8\%$) and one may observe that (1) the shapes, (2) the absolute values of the scattering intensities compare very well on the two sets of curves (in Figure 36, for $\phi = 0.1$, $I(q \to 0)$ and ξ_g have been taken as equal to the measured values, that is, respectively 0.89 $Å^{-1}$ and 14 Å, the values for the other concentrations have been calculated). When fitting the values found for ξ_g and $I(q \to 0)$ to scaling laws of the polymer concentration, the exponents found for the statistical gel with $R_c = 0.8\%$ are also very close to the values predicted by the model. Indeed, one obtains $\xi_g^{exp} \sim \phi^{-1.65}$ instead of $\xi_g^{th} \sim \phi^{-5/3}$ and $I^{exp}(q \to 0) \sim \phi^{-1.5}$ instead of $I^{th}(q \to 0) \sim \phi^{-5/3}$. For a lower crosslinking density, $R_c = 0.5\%$ the experimental exponents are lower, but this is not contradictory to the model prediction. It is not surprising that, decreasing the crosslinking density, the behaviour progressively returns closer to that of a semi-dilute solution.

A surprising fact is that gels prepared in a completely different manner (end-linking of short chains with tri-functional junctions) still exhibit a behaviour qualitatively—and sometimes quantitatively—similar to that predicted by the frozen-blob model. In principle, it should not apply to such samples. We have seen in section 5.4 that, for three different lengths of the chemical mesh size, one indeed finds the same apparent exponents for the larger correlation distance. Indeed one obtains $\xi_g^{exp} \sim \phi^{-1.5}$ and $I^{exp}(q \to 0) \sim \phi^{-1.5}$. The only clear difference between the three gels is the maximum swelling degree, and therefore the maximum values of ξ_g^{exp} and $I^{exp}(q \to 0)$ reached in the three cases. This result can be given with two opposite viewpoints. The pessimistic one is to discard the model for getting the right result only by chance. It can be argued that a really good model should apply *if and only if* a given set of assumptions are satisfied. It may also be observed that it is quite difficult to synthesize gels for which the assumptions of the frozen-blobs model strictly applies. The crosslinking reaction should have been performed instantaneously, which is not the case for the real statistical gels whose scattering properties are discussed above. The optimistic point of view is to seek for reasons why the model could work qualitatively outside the narrow range defined by the original assumptions. The idea is that cluster-like heterogeneities could be encountered in a much wider range of situations than the original one. Typical attempts at generalizing the model are given below.

As for the behaviour under strain, even for the statistical gel with $R_c = 0.8\%$ the agreement is only qualitative. The main success of the model is that it predicts the development under strain of a 'limit curve', indicated on Figure 37, and consistent with that observed for $q < 10^{-2}$ $Å^{-1}$ in Figure 38. The change of scattering intensity, under strain, in the parallel direction is less than predicted by Eqs (110) and (112). It must be stressed, however, that this discrepancy might be explained by a stretching of the clusters, which, for simplicity, was not incorporated in the model. If, as it is very likely, the clusters are deformed, they must separate less, in the parallel direction, than predicted by the model. As a result the unscreening effects and thus the enhancement of scattering intensity should be less pronounced than expected from Eq. (112). Note that a stretching of the chains,

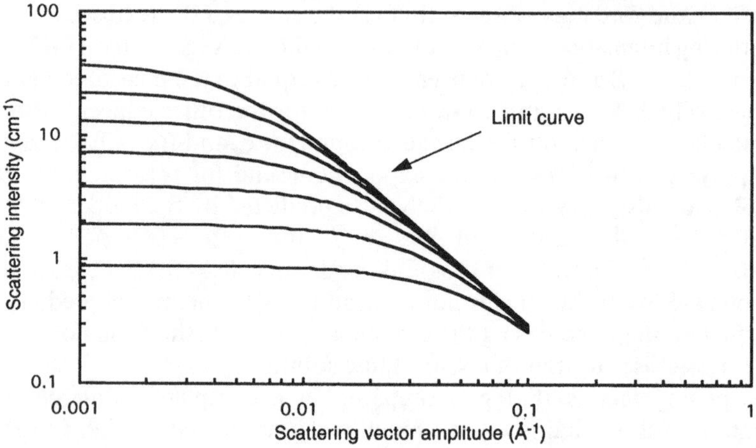

Figure 37. Scattered intensities in the direction parallel to the stretching axis, calculated with the 'frozen-blob model for different elongation ratios λ (one choose $I(0) = 0.89\,\text{cm}^{-1}$ and $\xi = 20\,\text{Å}$ for $\lambda = 1$). From bottom to top: $\lambda = 1, 1.1, 1.2, 1.3, 1.4, 1.5, 1.6$

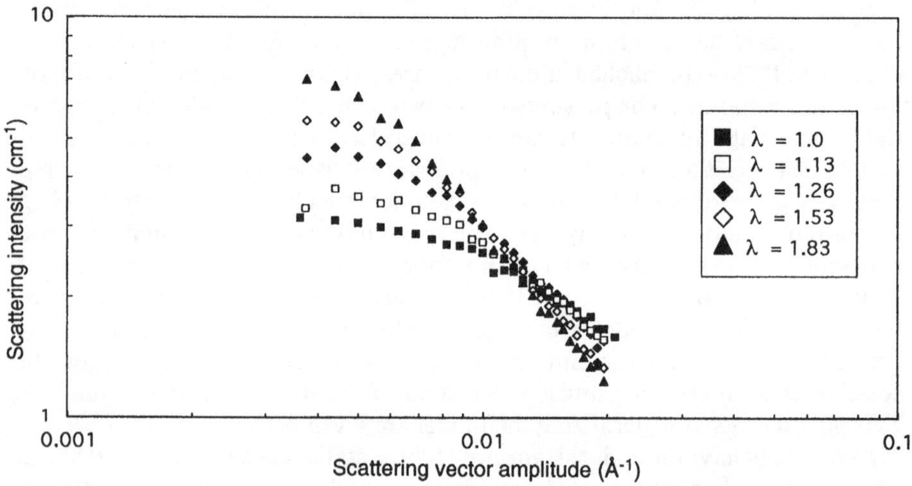

Figure 38. Experimental scattered intensities, in the direction parallel to the stretching axis, for a 'statistical gel' ($R_c = 0.8\%$, same as in Figure 26). The different stretching ratios λ are as indicated

belonging either to the clusters or to the interstitial regions (the deformation of the latter was also neglected for simplicity in the model) is clearly evidenced by the crossing of the scattering intensities in the parallel and perpendicular direction (see Figure 31). It also explains the departure, for $q > 10^{-2}\,\text{Å}^{-1}$, from a true limit curve in Figure 38.

In addition to the above-mentioned shortcomings, this model suffers from being purely geometrical. It does not deal with the thermodynamics: the scattering is assumed to be dominated by the static fluctuations. The larger correlation distance in the gel (that is associated with the crossover to a Guinier regime) is identified with the inter-cluster correlation length. No precise predictions on the dynamical fluctuations are derived from this approach and the question of the subtraction of $I_F(q)$ is left pending.

Finally, there is a feature which makes this model different from theories based on statistical thermodynamics: it can be viewed as a very rough 'evolution' picture: randomness produces a 'structure', that is, clusters where the crosslinking density is larger than average. Under addition of solvent, this structure is in a sense 'selected' by a swelling which deforms the clusters less than on an average. Because of the resulting separation of the clusters, the structure is indeed partly 'revealed' in the scattering through an unscreening process.

5.6.3.1.4 Representation of a gel in more general cases

The above model was developed on the basis of rather restrictive hypotheses (it was assumed that the crosslinking density was such that the frozen blobs were close to percolation). One might then be tempted to infer that, outside this particular regime, or for different crosslinking methods, both the structure of the superstructure and the response to a strain would be entirely different. However, it can be argued that the formation of 'cluster-like' non-uniformities is not restricted to the preparation method described above (Bastide *et al*, 1992a, b; Lal *et al*, 1993; Ramzi *et al*, 1993, 1995). This is the object of the next discussion, which could be entitled: what is the shape of a fluctuation?

What we have called above 'superstructure' corresponds to quenched fluctuations of crosslinking density. And we 'labelled' regions (the clusters of frozen blobs) where the 'local rigidity' had reached an arbitrary value; one crosslink per blob of the initial solution (see Figures 33 and 34). The 'less rigid' regions, whatever their crosslinking density, were left unlabelled. A comparable labelling is usually carried out on blackboards, when attempts are made to represent a fluctuation. There is a difference, however: fluctuations of concentration, for instance, are generally given as circles, or at least as convex islands. Such drawings may be misleading: a random fluctuation, whatever the quantity ζ which fluctuates (but depending on the level chosen to establish a 'contrast'—i.e. to label some regions) would be better represented, on length scales larger than the correlation length, as a branched object. This has nothing to do with the vicinity of a critical point. The 'size' of the cluster would be determined by the arbitrary level of labelling and, in many cases would be larger than the correlation length of the fluctuating quantity ζ (we believe that Figure 32 can be used as an illustration of this idea). In fact, a more satisfactory representation of a random fluctuation would be a landscape of hills and valleys. And, for random hills and valleys of ζ, islands for which ζ is larger than a given value presumably look like percolation clusters (Bastide *et al*, 1992b). The average size of these clusters would naturally depend on the level of labelling. Very generally, for polymer networks, and not only in the case of those formed by statistical crosslinking of a semi-dilute solution, one may expect the presence of random fluctuations in space of the local rigidity. Thus, large clusters, demarcating regions where the rigidity is larger than a certain level, are likely to exist in a vast class of networks.

The next point to consider is the relationship between these fluctuations of rigidity of the network and the fluctuations of polymer concentration in the gel. It probably depends on the quality of the solvent and on the presence of electrical charges. In the case of polyelectrolyte gels, the tendency to form micro-phases may often dominate the spatial fluctuations of polymer concentration (Moussaid, Candau and Joosten, 1994) when the ionization degree is low. Non-uniformities of the network structure would then appear only as rather weak corrections. This would also be the case for higher ionization degrees, when the swelling is incomplete. The variations of the local rigidity should induce very small variations of local concentration because of the high osmotic pressure of the counterions. However, in the case of neutral gels swollen by organic solvents, one may expect a direct coupling between the local rigidity and the local degree of swelling. Hills and valleys of crosslinking density should be transformed, respectively, into hills and valleys of polymer concentration, the strength of the coupling depending on osmotic properties (and thus on the degree of swelling). Therefore, in the case of random fluctuations of crosslinking density, regions where the local concentration is larger than a given level should look like clusters, possibly percolation-like, the average size of the clusters depending on the choice made for the level. In Figure 39, we show an example of such cluster-like fluctuations, in three states of a gel.

Therefore, it is not impossible for a scattering experiment to be especially sensitive to the level of fluctuations which are almost percolating. This would lead us back to the situations described in sections 5.6.3.1.2 and 5.6.3.1.3 and explain why comparable behaviours are often observed for different kinds of gels. Along these lines, one may expect the existence of a state of 'complete randomness' (not necessarily confounded with the state of preparation), in which cluster-like regions would remain 'invisible' for a scattering experiment (as in Figure 39(a)). As in the frozen-blob model, this would be due to screening effects, even if the polymer concentration in the clusters is larger than average. An increase of swelling degree, which should be larger in the valleys and should therefore separate the hills, would amount to unscreening the intra-hill correlations and increasing the scattering (Figure 39(b)). Such an effect would operate in the same way, but anisotropically, in a stretching experiment (Figure 39(c)), thus explaining the formation of butterfly patterns for the neutral gels studied under strain.

It must be stressed that an increase of swelling degree is not the only way to enhance the scattering by a gel. Qualitatively, from this type of description, one

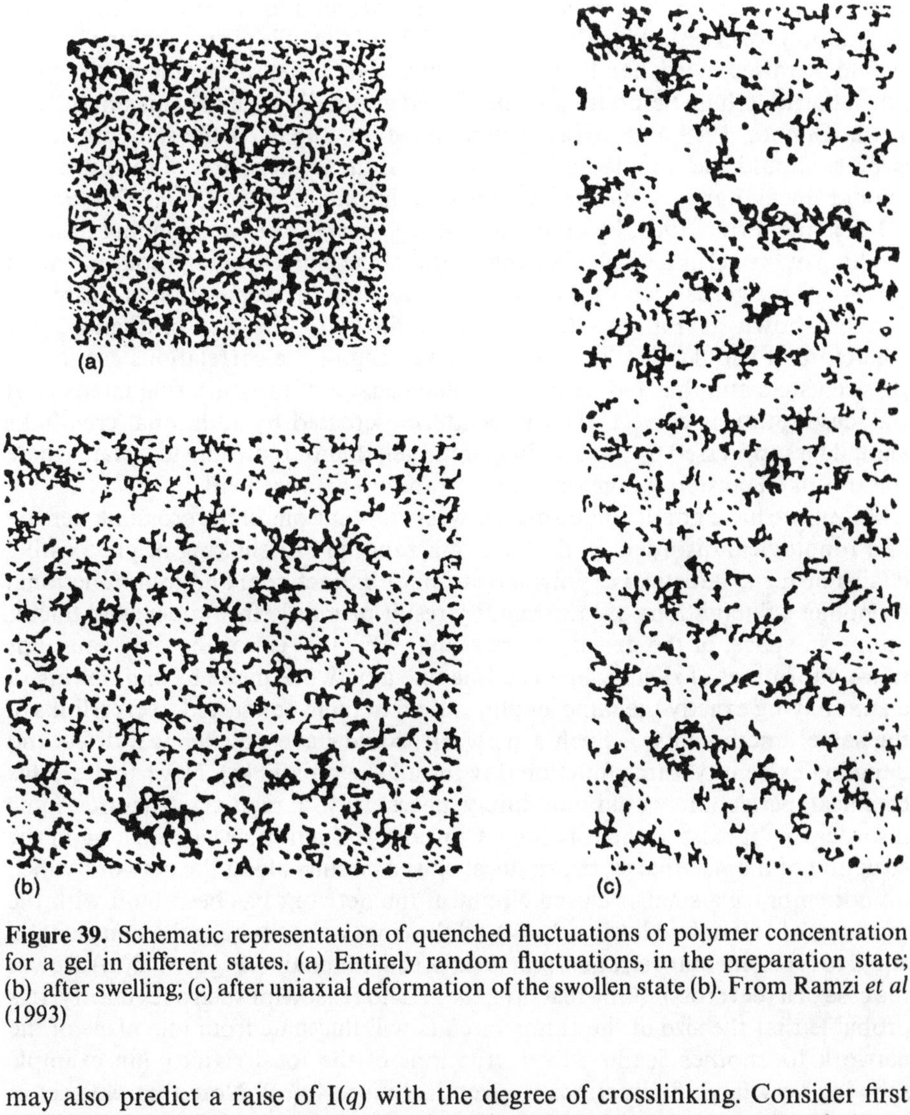

Figure 39. Schematic representation of quenched fluctuations of polymer concentration for a gel in different states. (a) Entirely random fluctuations, in the preparation state; (b) after swelling; (c) after uniaxial deformation of the swollen state (b). From Ramzi *et al* (1993)

may also predict a raise of I(q) with the degree of crosslinking. Consider first a statistical gel with a moderate crosslinking density, in such a way that the state of complete randomness coincides with the state of preparation (as in Figure 39(a)). Quenched fluctuations of polymer concentration, with cluster-like shapes, are believed to exist, but remain undetected by a scattering experiment. The scattering from the gel should not be greatly different from that of a semi-dilute solution of the same concentration. Assume now that new crosslinking molecules are introduced at random in the system. The efficiency of the reaction will presumably be larger in the regions (the clusters) where the concentration is higher than average. As a result, one may expect an amplification of the spatial fluctuations of crosslinking density: the height of the hills of crosslinking density would be raised more than that of the valleys). Note that this would also happen when the interchain tie-points are introduced gradually instead of instantaneously (de Gennes, 1979 and private communication, 1985). At the same time, this process should lead to a lateral contraction of the higher parts of the hills: they should shrink partly to ensure swelling equilibrium, achieving a *micro-syneresis* (Dušek and Prins, 1969, by contrast to *macro-syneresis* corresponding to a deswelling of the gel as a whole). A contraction of the clusters of high concentration but not of the gel as a whole is obviously equivalent to a separation of clusters: it can be shown as Figure 39(b) rescaled affinely to the size of Figure 39(a). Therefore, it should also lead to an unscreening of the correlations carried by these clusters and thus induce a clear enhancement of the scattering intensity. If this description is correct, the overscattering created by additional crosslinks should be suppressed by a deswelling of the gel, that is, returning to a state where the shrunken clusters recover an almost random arrangement.

As yet, we have not distinguished between fluctuations of the crosslink density and topological disorder. In the case of a random crosslinking of pre-existing chains or for a statistical copolymerization of bifunctional and multifunctional monomers, fluctuations of local rigidity result primarily from quenched fluctuations, in space, of the density of crosslinks. But this is not the only cause for non-uniformities of rigidity in a gel. Imagine a network made by end-linking of chains having exactly the same length, and all of them connected to crosslinks of the same functionality f. Such a network would have no chemical defect (no dangling extremity, same functionality for all the crosslinks). Nevertheless, this chemical perfection would not imply necessarily a regular structure: some disorder might exist in the topology. Consider the *empty circuits* existing in the structure of the net, that is, the smallest closed circuits along the network which do not embrace a smaller closed circuit. If the network has been built with the topology of a diamond lattice ($f = 4$), all the empty circuits have the same length, equal to six elementary strands, that is six precursor chains (Figure (40(a)). But, in the case of a real crosslinking reaction, there is no reason for such a regularity. It is probable that the size of the empty circuits will fluctuate from one place of the network to another, leading to fluctuations of the local rigidity (an example showing very large fluctuations is given in Figure 40(b)). Note that clusters of small empty circuits, spanning rather large length scales, can be formed even if the size distribution of the elementary strands is very narrow. Obviously, still in the idealized case of a 'perfect' chemical structure, different classes of topology, i.e. different families of networks, might exist, depending on the details of the synthesis (Bastide and Boué, 1986; Boué *et al*, 1987; Bastide, Boué and Buzier, 1989).

Finally, it must be stressed that the introduction at random of 'classical defects' in a network (i.e. pendant chains and variations of the functionality of

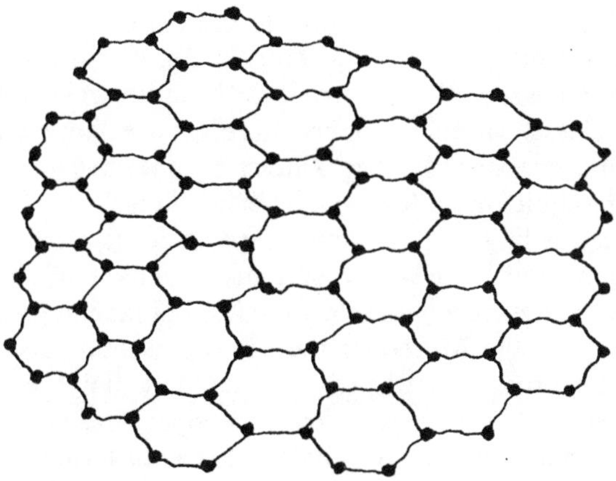

(a)

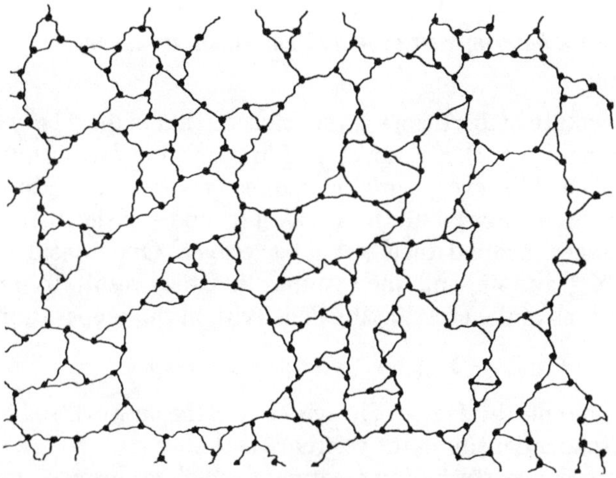

(b)

Figure 40. Schematic representation, in two dimensions, of networks with no chemical defect (except on the edges). Tridimensional networks with functionality $f = 4$ can be built easily from such pictures: imagine several layers of such two-dimensional networks connected by elementary strands, alternately upwards and downwards. (a) Regular topology: all the shortest closed loops have the same size; (b) irregular topology: the shortest closed circuits have different sizes

the junctions) can also lead to the formation of cluster-like heterogeneities of structure (Moussaid, Slot and Joosten, 1995). The reason for this is again the same: randomness and uniformity are different things in quenched systems.

Comments Clearly, the generalization of the frozen-blob model to various types of gels is not totally convincing. To sum up, it is proposed that, due to a freezing of disorder, a sort of 'superstructure' spanning relatively large scales is formed. Several classes of such 'superstructures' are liable to exist. To this superstructure, one associates clusters which may not have intrinsic boundaries: they depend on the choice made for the level of labelling. The concept of revealing such clusters by a swelling or a stretching experiment is consistent with experimental observations. But there is at least a missing link: one should understand precisely why the scattering would select the level of labelling corresponding to almost percolating clusters. Moreover, there is at present no quantitative prediction which can be proved or disproved. The only point which makes this approach refutable is the existence of this superstructure. The key point is that this superstructure may remain partly hidden for a scattering experiment. Thus, the existence or the non-existence of large clusters cannot be discriminated by a scattering experiment and the use of other techniques should be considered.

5.6.3.2 Quenched heterogeneities in a linear elasticity theory (Onuki No. 2)

The general structure of this theory is very close to that of the theory of section 5.3 (that we call Onuki No. 1): it is developed in the framework of the elasticity of continuous media. However, a major change has been introduced here: one postulates that the osmotic bulk modulus K_{os}^{gel} and the shear modulus μ vary randomly in space, around their spatial averages. One considers only small fluctuations δK_{os}^{gel} and $\delta\mu$, and one assumes that they result from uncorrelated spatial fluctuations of the crosslinking density v_0 in the preparation state:

$$\langle \delta v_0(\mathbf{r})\delta v_0(\mathbf{r}+\mathbf{r}')\rangle = \delta(\mathbf{r}')\bar{v}_0 p \tag{113}$$

where $p \ll 1$ represents the degree of irregularity of the network and $\bar{v}_0$ denotes the average crosslinking density in the reference state.

The next step is a perturbation treatment which is not easy to summarize. Following Boué (1995) (see also Ramzi *et al*, 1995) one can, however, state that it recalls the RPA procedure (de Gennes, 1979) where the structural non-uniformities play the role of the perturbation $\delta U_{\text{quench}}(r)$. Thus, the response to a deformation could be derived using the fluctuation–dissipation theorem:

$$\delta\phi_{\text{quench}}(\mathbf{q}) = \chi(\mathbf{q})\cdot\delta U_{\text{quench}}(\mathbf{q}) \tag{114}$$

with

$$\chi(\mathbf{q}) \cong S_{\text{therm}}(\mathbf{q})/k_B T \tag{115}$$

where $S_{\text{therm}}(\mathbf{q})$ is the structure factor of idealized gels (Eq. (79)) in which all the fluctuations are assumed to be thermal. Then, the total scattering is written as the sum of the thermal scattering and of the contribution of the quenched fluctuations:

$$S(\mathbf{q}) = \langle \delta\phi(\mathbf{q})\delta\phi(-\mathbf{q}) \rangle = S_{\text{therm}}(\mathbf{q}) + \langle \delta\phi_{\text{quench}}(\mathbf{q})\delta\phi_{\text{quench}}(-\mathbf{q}) \rangle \tag{116}$$

Precisely because they are frozen in, the modes corresponding to quenched fluctuations do not obey the equipartition. Here $\langle \delta\phi_{\text{quench}}(q)\delta\phi_{\text{quench}}(-q) \rangle$ is given simply by taking the square of Eq. (114). This is why there is a squared Lorentzian in the final result, which can be expressed as

$$S(\mathbf{q}) \propto \frac{\phi^2 k_B T}{\left\{K_{\text{osm}} + \dfrac{\mu}{3} + \mu J(\beta) + C\phi^2 q^2\right\}} + \phi^2\left(\frac{\mu}{\bar{v}_0}\right)^2 \frac{[J(\beta) - (\phi/\phi_0)^{2/3}]^2 \langle v_0(\mathbf{q}) v_0(-\mathbf{q}) \rangle}{\left\{K_{\text{osm}} + \dfrac{\mu}{3} + \mu J(\beta) + C\phi^2 q^2\right\}^2} \tag{117}$$

or

$$S(\mathbf{q}) \propto \frac{\phi^2 k_B T}{\left\{K_{\text{osm}} + \dfrac{\mu}{3} + \mu J(\beta) + C\phi^2 q^2\right\}} + \frac{[J(\beta) - (Q/Q_0)^{-2/3}]^2 (Q/Q_0)^{2/3} p\mu\phi^2 k_B T}{\left\{K_{\text{osm}} + \dfrac{\mu}{3} + \mu J(\beta) + C\phi^2 q^2\right\}^2}$$

where, as before,

$$J(\beta) = (\lambda^2 - 1/\lambda)\cos^2\beta + 1/\lambda = \lambda^2 \cos^2\beta + 1/\lambda \sin^2\beta \tag{117a}$$

with β denoting the angle between q and the stretching axis.

Here $\langle v_0(\mathbf{q}) v_0(-\mathbf{q}) \rangle$ is the Fourier transform of the fluctuations of the crosslinking density, which are assumed to be uncorrelated:

$$\langle \delta v_0(\mathbf{q}) \delta v_0(-\mathbf{q}) \rangle = \bar{v}_0 \cdot p \cdot (\phi_0/\phi) \tag{118}$$

Recall that Q_0 and Q are the swelling degrees of the gel in the reference state and in the state of study respectively. Therefore, the scattered intensity is expressed as a function of parameters which can all be estimated from independent thermodynamic measurements, except for the *degree of inhomogeneity*, *p*.

Comments and discussion Contrary to the preceding model, which is mostly geometrical, this approach is complete since it deals quantitatively with the quenched fluctuations as well as with the thermal ones. It gives an analytical expression of the intensity associated with each of these contributions (the first and second terms of Eq. (117) respectively). Depending on the degree p of inhomogeneity in the crosslinking density and on the degree of overswelling

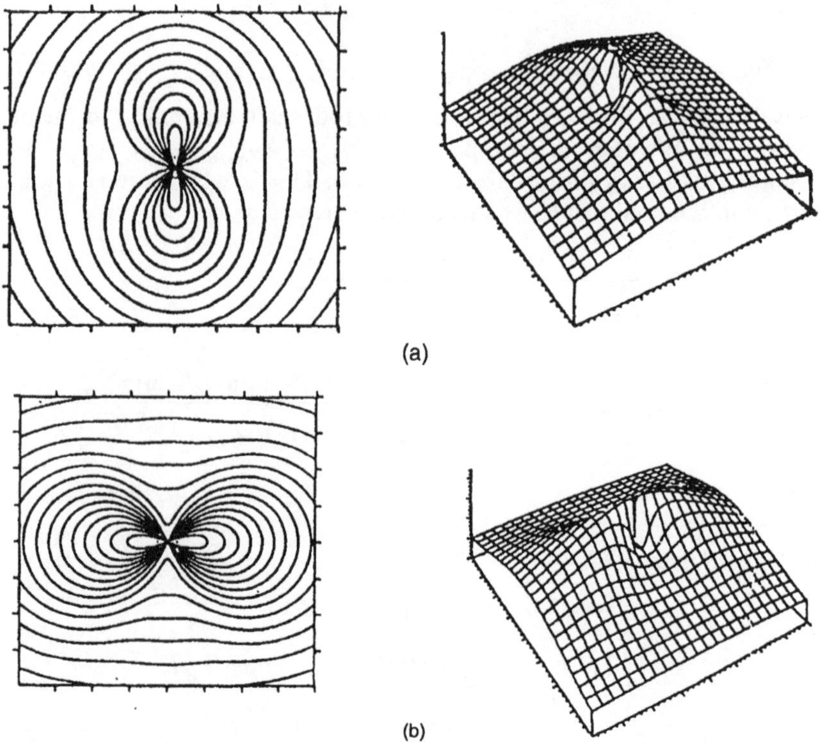

Figure 41. Calculated iso-intensity lines, in the framework of Onuki No. 2 theory. The stretching axis is horizontal. (a) 'Normal' butterfly pattern (perpendicular to the stretching axis) obtained for: $\varepsilon = 4$, $p^* = 0.1$ (small degree of heterogeneity) and $c = 2$; (b) 'abnormal' butterfly patterns (parallel to the stretching axis) obtained for: $\varepsilon = 4$, $p^* = 1.3$ (large degree of heterogeneity) and $c = 2$. From Onuki (1992)

$Q/Q_0 = \phi_0/\phi$, either the former or the latter type of fluctuation may dominate the scattering. More explicitly, under uniaxial strain, one expects, for $p(Q/Q_0)^{2/3} \ll 1$, *normal butterfly patterns* to show up under extension (Figure 41(a)). The reason for this is the expected decrease of the thermal fluctuations along the parallel direction (as for idealized gels; Figure 9), whereas the effect of the quenched fluctuations should remain negligible in this case. Recall that, according Onuki's nomenclature, normal butterfly patterns are double winged iso-intensity lines with a major axis perpendicular to the direction of stretching (see section 5.3.2). They must be associated with a decrease of $I(q \rightarrow 0)$ along the stretching axis. On the other hand, for $p(Q/Q_0)^{2/3} \geqslant 1$, one expects the second term to dominate and *abnormal butterfly patterns* to show up under uniaxial extension, because of the enhancement of the numerator of the second term (Figure 41(b)). The physical reason for this change is essentially the same as for the frozen-blob model of section 5.6.3: now, the quenched disorder in the crosslinking density is postulated to be sufficient to induce a noticeable distortion, upon strain, of the network. Then the quenched fluctuations of polymer concentration should be amplified along the direction of stretching, inducing an anisotropical enhancement of scattering.

Recall that in the case of neutral gels under uniaxial strain, only abnormal butterfly patterns have been observed as yet. However, a vanishing or, at least, a very strong reduction of the latter has been observed recently by Mendes *et al* (1995b) on introduction of electrical charges along the network chains. The investigated sample was first a gel of polyacrylic acid, almost in the state of preparation and containing D_2O as a diluent (which is poor solvent of the polymer, $\theta \approx 25\,°C$). In this weakly acidic form, the ionization degree of the chains is very low and the gel exhibits abnormal butterfly patterns under uniaxial strain. After partial neutralization of the weak acid with an appropriate amount of a strong base (NaOH), the ionization degree was raised to 10% approximately and the butterfly patterns vanished. Here $I(q \rightarrow 0)$ appeared to be drastically decreased in any direction with respect to the quasi-neutral case and anisotropy at very small q's could no longer be detected (but new experiments, involving larger counting times and therefore better precision are forecast). We also reported the reverse experiment in section 5.5: one starts with charged gels which do not show detectable butterfly patterns, then one adds salt or a strong acid and the great enhancement of scattering in the parallel direction, inducing the butterfly patterns, is re-established. The direct and reverse experiments are obviously consistent with each other and their results are qualitatively predicted by Onuki's theory. Equation (117) can indeed be rewritten as follows:

$$S(q \rightarrow 0) \cong \frac{\phi^2 k_B T}{K_{osm} + \dfrac{\mu}{3} + \mu J(\beta)} \left\{ 1 + \frac{[J(\beta) - (Q/Q_0)^{-2/3}]^2 (Q/Q_0)^{2/3} p}{\varepsilon + J(\beta)} \right\} \tag{119}$$

where $\varepsilon = (K_{os}^{gel} + \mu/3)/\mu$ (see Onuki, 1992, 1993).

As the ionization degree is raised to 10%, the osmotic pressure Π^{gel} and thus K_{os}^{gel} strongly increase, due to the entropy of the counterions. As a result, ε is neatly enhanced. The raise of ε is even amplified by the fact that μ decreases upon ionization at constant concentration, as found recently by Skouri *et al* (1994, 1995). Then, since $Q/Q_0 \approx 1$ in the present case, the term

$$\frac{[J(\beta) - 1]^2 p}{\varepsilon + J(\beta)}$$

can become almost negligible compared to 1 in the usual conditions of stretching ($\alpha \leqslant 1.5$): this is enough to make the abnormal butterfly patterns practically undetectable provided the degree of heterogeneity p is not too large.

Onuki's No. 2 theory is also complete in the sense that it makes predictions at the same time for the overswelling and the uniaxial deformation. Because of the term $[J(\beta) - (Q/Q_0)^{-2/3}]$ in the numerator of the second term of Eq. (117), no overscattering with respect to an idealized gel is expected to show up at the concentration of preparation, in the isotropic state. This prediction agrees with that made by the model for statistical gels (section 5.6.3.1) and with some of the experimental results. However, very often one observes an unpredicted enhancement of scattering intensity, in the reaction bath, upon increase of crosslinking density (see section 5.4.2). Qualitatively, this problem could be overcome by assuming that, upon increasing the crosslinking density, the reference state ceases to be confounded with the state of preparation. One might assume that the swelling degree Q_0 corresponding to the reference state becomes smaller than Q_{prep}. Although the language is different, such an assumption would be consistent with that made in section 5.6.3.1 when trying to generalize the frozen-blob model.

More serious is the question of deswelling. According to Eq. (117), $I(q \to 0)$ should increase again when the gel is deswollen with respect to the reference state. As stated above, the reference state—whose definition has long appeared problematic (see Dušek and Prins, 1969)—might not always be confounded with the preparation state. Nevertheless, for gels exhibiting no or little overscattering in the preparation state, in the framework of Onuki's theory, there is no reason to assume that the reference and preparation states are very different. Therefore, according to Eq. (117), a deswelling with respect to the state of preparation should shortly lead to a further increase of the scattering intensity associated with quenched fluctuations. No effect of this type has ever been observed. On the contrary, each time the spectra of the gel and the solution appeared to merge for all q's for a given swelling degree Q_m, the superposition was kept for $Q < Q_m$ (Mendes, 1991, see an example in Figure 19; section 5.4.3), with, in particular a decrease of $I(q \to 0)$. This is the main problem we have encountered in our attempts to check the qualitative predictions of Onuki's theory.

For a quantitative analysis of this theory, the key point is to estimate the value of the parameter p. To describe the variation of $I(q \to 0)$ for overswollen gels first, one has generally to assume a rather large value for p, for example $p \approx 25$–30 for the statistical gel of Figure 18, in section 5.4.3. Such a value in principle exceeds the limits considered by the theory, i.e. p small (say $p < 1$). If we ignore this difficulty, we meet another one: the calculated increase of $I(q \to 0)$ in the direction parallel to the stretching axis is found to be much too strong, when using the value of p obtained from the fit of the overswelling data. Therefore, there is a lack of consistency in the quantitative predictions made from this model. Possibly, as mentioned by Onuki, the structural heterogeneities do not act as a small perturbation, which is the basic assumption of this linear elasticity theory.

It can be also observed that a similar discrepancy between swelling and stretching behaviours was encountered with the frozen-blob model: for gels with $R_c = 0.8\%$, the agreement of the swelling data with the model predictions was quantitative, whereas the model overestimated the scattering enhancement in the parallel direction for the stretching experiment.

Another comment concerns the q dependence of the intensity. As observed above, the contribution of the quenched fluctuations term has the form of a squared Lorentzian, which means that it should decay like $1/q^4$ for sufficiently large values of q. As can be seen on the different figures of section 5.4, such a behaviour has not yet been observed in neutral gels. As long as the thermal contribution was not subtracted, the $1/q^4$ variation might be masked by the single Lorentzian term. However, in the case of systems for which Geissler, Horkay and Hecht (1991, 1993) did estimate and subtract the thermal contribution, softer decays were obtained, which were analysed with Gaussian functions or stretched exponentials. Systematic analyses for a wider range of systems are under way at present. Another delicate point concerns the crossover between the apparent saturation at low q and the decay at larger q's. We have seen in sections 5.4 and 5.5 that, experimentally, this crossover seems often to be shifted to lower q's, both under overswelling and under-stretching (for q parallel to the stretching axis in the latter case). But, according to Onuki's theory, the position of the crossover is not expected to change drastically upon swelling or stretching (see Mendes *et al*, 1995b). Moreover, in the stretching case, this model does not predict the appearance of a *limit curve* in the intermediate regime of scattering, which seems to be observed, at least in some cases (see, above, the comments about the cluster model). This is probably inevitable, given the choice made for approaching the problem: within a theory of continuum elasticity one cannot expect to describe precisely correlations on length scales where the molecular structures start to have an effect.

To sum up, the Onuki No. 2 model predicts many of the features observed in the scattering of gels submitted to a change of swelling degree or to a uniaxial stretching. This approach does not provide a picture for the gel 'structure' but allows a quantitative analysis of the data through the use of a semi-empirical heterogeneity parameter p.

5.6.3.3 Panyukov's theory

Independently of A. Onuki, S. Panyukov developed recently a theory of the scattering by heterogeneous polymer networks containing some labelled probes (chains or small solvent molecules) (1992; see also Panyukov, 1989, 1990a, b; Panyukov and Kuchanov, 1991). This theory addresses the case of networks made by statistical crosslinking of long chains and the problem is approached alongside that of free energy. The way of handling the heterogeneities is slightly different than in Onuki's No. 2 theory and is, in a sense, closer to the frozen-blob model. It is not assumed phenomenologically that the elastic constants fluctuate over the space. These fluctuations are presented as a consequence of the freezing-in of some topological disorder when making the network. This key ingredient of the theory is treated with the replica trick, introduced in polymer networks by S. F. Edwards (see e.g. Deam and Edwards, 1976; Warner and Edwards, 1978). In a quenched system, the free energy F must be averaged over all the frozen-in topologies:

$$F = \sum_m p_m F_m \tag{120}$$

where p_m and F_m represent respectively the probability of occurrence and the free energy of the topology m.

When the network is deformed, the probabilities p_m remain unchanged because the topologies are fixed. On the other hand, the free energy F_m of each topology m is modified (and can be written $\tilde{F}_m$. Since F_m can also be written $-k_B T \ln Z_n$, Z_n being the partition function of the mth topology (written $\tilde{Z}_m$ after deformation), one has to calculate

$$\tilde{F} = \sum_m p_m \tilde{F}_m = -k_B T \sum_m p_m \ln \tilde{Z}_m = -k_B T \langle \ln Z \rangle \tag{121}$$

Equation (121), which represents the average value of a logarithm, is difficult to calculate. In particular, it is known that, for any random variable x, $\langle \log x \rangle \leqslant \log \langle x \rangle$. The basic idea of the replica trick is that it is easier to calculate the moments $\langle Z \rangle$, $\langle Z^2 \rangle$, $\langle Z^3 \rangle, \ldots, \langle Z^n \rangle$. Then one tries to find an expression valid for any n and one calculates the average of $\ln Z$ by taking the limit

$$\langle \ln Z \rangle = \lim_{n \to 0} \frac{\langle Z^n \rangle - 1}{n} \tag{122}$$

This limit is taken simply from the series expansion:

$$\langle Z^n \rangle = \langle e^{n \ln Z} \rangle = 1 + n \langle \ln Z \rangle + (n^2/2) \langle \ln^2 Z \rangle + \ldots$$

But before reaching this step, when dealing with Z^n, one has, in a sense, generated a network (or any structure in general) consisting of n *replica* of the original system. It can be viewed as a hyper-network embedded in a space with a $3(n+1)$ dimension. In order to simplify the problem, some assumptions are made is this space, like for example neglecting the fluctuations (mean field approach). The rather complex expressions which are generated can be represented more compactly as functional integrals over two fields. The values of the latter are subsequently found by minimizing the Hamiltonian, which involves a spontaneous breaking of symmetry of the replicas. The expression obtained in this way for the elastic contribution to the free energy is consistent with the classical theory (see appendix):

$$F_{el} = \frac{k_B T \tau^2}{4\omega} V^{(0)} (\lambda_x^2 + \lambda_y^2 + \lambda_z^2) \tag{123}$$

where $V^{(0)}$ is the volume in which the network was synthesized, ω represents the activity of the crosslinks and τ is a localization parameter. Here τ is equal to the inverse of the average number N_c of monomer units between two crosslinks.

The next step is to consider the concentration fluctuations in the replica space. A key point is the presence of 'soft modes' in the spectrum of these fluctuations. These translational modes are associated with the loss of affineness of the deformation; they correspond to independent displacements u^k in the different replicas. After expressing the correlation functions, the final result is written

$$I(q) = \phi_D(1 - \phi_D)S_{1b}(q) + \phi_D^2\{G(q) + C(q)\} \tag{124}$$

where $S_{1b}(q)$ represents the form factor of the labelled molecules, whose volume fraction is ϕ_D.

$$S_{1b}(q) = \frac{\rho}{[l^{-1} + a^2(\lambda\mathbf{q})^2]} = \frac{\rho l}{[1 + l \cdot a^2(\lambda\mathbf{q})^2]} \tag{125}$$

where l represents here the average number of elementary links in the deuterated probes, $\rho = \rho^{(0)}/\lambda_x\lambda_y\lambda_z$ is the density of crosslinks in the network and $\lambda\mathbf{q}$ denotes the vectors with components $\lambda_i q_i (i = x, y, z)$. As previously, the terms λ_i are the components of the deformation along the principal directions of strain and a is the length of the monomer unit. Note that $S_{1b}(q)$ reduces essentially to the scattering of deformed chains in the Guinier regime and shows a *normal anisotropy*, that is an intensity decreased in the direction parallel to the elongation axis.

The second term, $G(q)$, represents the correlation function of the thermodynamic (temporal) fluctuations. As in Onuki's theory, it leads to butterfly-like iso-intensity lines, but with a normal anisotropy, that is with the major axis of the lobes perpendicular to the stretching direction.

$$G(q) = \frac{\rho}{[B\rho + \frac{3}{2}\tau(\lambda\mathbf{n})^2 + 6ca^3(\lambda\mathbf{q})^2]} \tag{126}$$

with $\mathbf{n} = \mathbf{q}/|\mathbf{q}|$, $\lambda\mathbf{n}$ has the components $\lambda_i q_i (i = x, y, z)$ and B is an effective virial coefficient.

Finally the third term, $C(q)$, is the correlation function of the spatial fluctuations; $C(q)$ leads to butterfly-like iso-intensity lines, with an *abnormal anisotropy*, that is, to lobes having their major axis parallel to the stretching direction (i.e. as in the experiments).

$$C(q) = \left[\frac{\frac{3}{2}\tau(\lambda\mathbf{n})^2 + 6ca^3(\lambda\mathbf{q})^2}{B\rho + \frac{3}{2}\tau(\lambda\mathbf{n})^2 + 6ca^3(\lambda\mathbf{q})^2}\right]^2 \cdot \left(\frac{2}{3}\rho\tau^{-1}B(q) + \frac{2\rho A^2(q)}{2B^{(0)}\rho^{(0)} - 2\tau + a^2(\lambda q)^2}\right) \tag{127}$$

where c is a constant, which is calculated in the theory and found equal to 1.44.

$$A_c(\mathbf{q}) = 1 + \frac{2 - a^2\tau^{-1}(\lambda\mathbf{q})^2}{3[1 + R^2(\lambda\mathbf{q})^2 + R^2\mathbf{q}^2]}$$

with

$$B_c(\mathbf{q}) = \frac{2}{[1 + R^2(\lambda\mathbf{q})^2 + R^2\mathbf{q}^2]} - \frac{1}{[1 + 2R^2\mathbf{q}^2]} + \frac{2 - a^2\tau^{-1}(\lambda\mathbf{q})^2}{3[1 + R^2(\lambda\mathbf{q})^2 + R^2\mathbf{q}^2]}$$

where $R^2 = \ll \overline{\delta \mathbf{r}_i^2} \gg$ with $\delta \mathbf{r}_i = \mathbf{r}_i - \langle \mathbf{r}_i \rangle$ describing the thermal fluctuations of the monomer units.

The double angular brackets represent a Gibbs average together with an average over the structural disorder.

Comments A direct confrontation of Panyukov's theory with the experimental observations is difficult because of the complexity of its mathematical analysis. However, one can note that the structure of the result is similar to that of Onuki No. 2 theory. The first term, $S_{sl}(q)$, has been added with respect to Onuki's theory, which is restricted to the case of gels swollen with small organic molecules; $G(q)$ is Lorentzian and its shape seems to be consistent with the first term of Eq. (117); $C(q)$ is rather complicated but, as the second term of Eq. (118), contains a squared Lorentzian term.

5.6.3.4 The Panyukov–Rabin et al theory

Recently S. Panyukov and Y. Rabin (1995; see also Panyukov, Rabin and Feigel, 1994) proposed a new theory dealing with the statistical physics of polymer gels. It considers the case of an instantaneous crosslinking of Gaussian chains. The chains are assumed to be 'phantom' in the sense that they are allowed to cross each other. In other words, the physical interactions between chains, like knots or trapped entanglements, are not taken into account. However, excluded volume between chain segments is described as a perturbation. As the one of the preceding paragraph, the theory is based on the replica trick. It is argued that, under given thermodynamic conditions (solvent, temperature, deformation state), the network has a single state of microscopic equilibrium defined by the average positions of all the monomers. Because of the random nature of the crosslinking process, this state is characterized by an inhomogeneous concentration profile in the gel.

The expression found for the free energy is consistent with the classical result by Kuhn (1936), Wall (1942, 1943), Flory and Rehner (1943), James and Guth (1943), Deam and Edwards (1976), Graessley (1975) and others. Contrary to these earlier theories, however, the present one predicts that the deformation of the elementary strands of the network depends on their length. Due to the randomness of the crosslinking, their size distribution is indeed very large. Strong deviations from affineness is obtained for the short chains (those which are shorter than the average mesh size). In other words, short chains are shown to react to deformation only through disinterpenetration (i.e., in a way similar to that proposed by Bastide, Picot and Candau, 1981). Affineness is predicted to be recovered for the longer chains of the distribution (those which are larger than the average mesh size).

The theory also deals with the scattering of networks either containing a labelled polymeric solvent or a 'good solvent' consisting of small molecules. In the latter case, the scattering intensity is written

$$I(q) \sim G(\mathbf{q}) + C(\mathbf{q}) \tag{128}$$

where $G(\mathbf{q})$ and $C(\mathbf{q})$ represent respectively the contributions of the *annealed* (or thermal) and of the *quenched* spatial fluctuations of polymer concentration.

Three q vector ranges are considered for these functions $G(\mathbf{q})$ and $C(\mathbf{q})$:

1. the microscopic regime, such as $q \gg \xi^{-1}$ (ξ denoting as before the correlation length of thermal fluctuations, which is found similar to that of a semi-dilute solution of the same polymer concentration). In this regime, the theory predicts $G(q) \sim \phi(aq)^{-5/3}$ (as for a semi-dilute solution of chains with excluded volume) and $C(\mathbf{q}) \to 0$ (which is satisfactory, since no quenched fluctuations are believed to exist on distances smaller than the mesh size).
2. The mesoscopic range, such as $q < \xi^{-1}$.
3. The long wavelength limit ($q \ll$ inverse average mesh size).

We refer to the original paper (1995) for the detailed expressions in the regimes (2) and (3) and report only the main qualitative conclusions:

(a) The frozen heterogeneities of network structure (resulting from the randomness of the network formation) lead to the dominant term of the scattering for gels prepared in conditions close to those of maximum swelling (rather large crosslinking density) or for gels with moderate crosslinking but swollen to equilibrium. In the latter case, the effect of the quenched fluctuations is shown to decrease when the gel is deswollen back to the conditions of preparation, where thermal fluctuations are predicted to dominate.

(b) When the gel is uniaxially stretched, the theory predicts that the angular dependence of the scattering will always (at small q's) be dominated by the quenched fluctuations and that butterfly patterns in the iso-intensity lines will be observed.

(c) Contrary to what is usually expected (see section 5.3.1), the thermal contribution to the scattering from gels is predicted to exhibit a maximum for a q value of about the inverse mesh size.

(d) The theory predicts the existence of a critical value of the crosslinking density to which is associated a divergence of the characteristic length scale associated with the heterogeneities.

Comments Since this theory has been completed very recently, it is too early to analyse its predictions in detail (in particular the shape of the scattering curves). On qualitative grounds, there is apparently no strong disagreement either with the existing experimental data or with the conclusions of the approaches summarized above. It is worth mentioning that this theory leads to the breakdown of the classical additivity assumption in the network free energy. Therefore, it is liable to explain the discrepancies between Π_{mix} and Π_{sol} reported by Geissler–Horkay–Hecht *et al* (see section 5.3).

Since the expressions of the structure factors are very detailed, they might be used, if they fit the data satisfactorily, to recover the statistical information about a gel in its state of preparation (like the density of crosslinks). One would then dispose of an independent estimate of the average chain length between crosslinks and it would be extremely interesting to compare it with those obtained from mechanical data in the framework of the classical theories.

5.6.3.5 *The Geissler–Horkay–Hecht approach*

Here we try to summarize the ideas that E. Geissler, A. M. Hecht, F. Horkay *et al*, recently introduced in the field of gels. These ideas, which may have either theoretical or experimental origins, give a coherent description of gels. Nevertheless, in order to facilitate the presentation, we divide this approach into several subsets which we discuss separately.

5.6.3.5.1 *Reduction of 'mixing pressure' under crosslinking*

In section 5.3, we have mentioned the careful measurements of the osmotic pressure Π^{gel} (or swelling pressure) together with the elastic modulus G, for gels at different states of swelling (Geissler *et al*, 1988; Horkay, Hecht and Geissler, 1989; Horkay *et al*, 1993c; Hecht *et al*, 1992; see also Nagy and Horkay, 1980). From Π^{gel} and G, one can calculate the mixing pressure Π^{mix} associated with the dilution of the polymer constituting the network (see appendix and section 5.3). Here, Π_{mix} was systematically found to be smaller than the corresponding value for a semi dilute solution of the same polymer (Horkay, Hecht and Geissler, 1989). This difference (typically 30%[†]) is large enough to attract attention but not to explain on the basis of thermodynamics only, the large overscattering exhibited by gels. From this analysis, it could be concluded, independently of the scanning light scattering experiments, that the overscattering by gels results essentially from the presence of structural heterogeneities. One has, however, to explain the difference between Π^{mix} and Π^{sol}. It is noteworthy that this difference was found to consist essentially of a *difference of prefactor*: both Π^{mix} and Π^{sol} are found to scale with concentration with the same apparent exponent, roughly 2.3. Therefore, it seems unlikely that the difference arises from an amplification of thermal fluctuations upon swelling, as proposed by Y. Rabin and R. Bruinsma (1992, and Bruinsma and Rabin 1994). Such a mechanism would probably lead to a coinci-

[†]The distinction between the network (Π^{mix}) and solution values (Π^{sol}) was found to vanish smoothly as the crosslinking density goes to zero (Horkay, Hecht and Geissler, 1989). Note also that an even larger difference of prefactor has sometimes been found by the same group, when comparing the values of the longitudinal modulus $K_{os}^{gel} + 4/3\mu$ obtained either by dynamic light scattering or by direct thermodynamic measurements (see Geissler *et al*, 1988; Horkay *et al*, 1993b). Apparently, the larger difference of prefactor was found in the case of polyvinyl acetate gels swollen by isopropyl alcohol (Horkay *et al*, 1993b).

dence of Π^{mix} and Π^{sol} in the state of preparation and to a steeper decay of Π^{mix} upon swelling.

The discrepancy between Π^{mix} and Π^{sol} is sometimes presented as originating from a difference of Flory–Huggins χ parameter in semi-dilute solutions and in networks (see Yen and Eichinger, 1978; McKenna, Flynn and Chen (1989, 1990)). Arguing about a change of χ is largely a matter of semantics, since, depending on the authors, χ is considered, or otherwise, as a combination of enthalpic and entropic contributions (recall that in the so-called 'scaling' descriptions there is no entropic part in χ). In the case of networks of rather large crosslinking density made with crosslinks which are less soluble than the polymer, an explanation based on a purely enthalpic change might make sense (Hecht, Duplessix and Geissler, 1985). But the phenomenon is apparently more general: networks made with soluble crosslinks seem to exhibit the same trends. Another option is to attribute the difference to a by-product of the quenched fluctuations of polymer concentration (Horkay, Hecht and Geissler, 1989). To put it simply, let us focus on a gel under incomplete swelling conditions. Some regions where the crosslinking density is higher than average will presumably stand much closer to maximum swelling than the 'softer' parts of the gel, where the concentration is less than average. Probably, both types of regions contribute less to the swelling pressure than the polymer in a homogeneous gel, the former because they are closer to swelling equilibrium, the latter because their polymer concentration is less than average. As a result the effective 'mixing pressure' may appear as reduced. In short, one could call the phenomenon a 'lump' effect: an imperfect mixture where 'lumps' are remaining is less effective, in terms of many properties, including osmotic ones, than a perfectly dispersed one. For those who do not accept 'scaling' pictures, this option can be expressed as a change of the entropic part of χ.

Discussion and comments The conclusion, due to Geissler *et al*, that Π^{mix} is smaller than Π^{sol} seems to be well documented. However, one should keep in mind that Π^{mix} is not a direct measurement but a quantity estimated in the framework of a model. Hence, there is some built-in uncertainty on its meaning and on its value.

Concerning the physical explanation of the difference between Π_{mix} and Π_{sol}, the 'lump effect' seems to us the more satisfactory. However, it is not easy to understand how it may lead to a difference of prefactor of the two quantities instead of a progressive merging of both quantities at high concentrations.

5.6.3.5.2 Decomposition of the structure factor: quenched and thermal fluctuations

Although the presence of static and dynamic contributions in the spectrum of concentration fluctuations had been proposed earlier by several groups, Geissler, Hecht, Horkay *et al* were the first to write explicitly the structure factor of a gel as a sum of two terms (a Lorentzian and a stretched exponential) involving two different types of correlation distances (see above, section 5.4 and Horkay *et al*, 1991a, 1993b; Horkay, Hecht and Geissler, 1994; Geissler, Horkay and Hecht, 1993):

$$\mathrm{I}(q) = \mathrm{I}_s + \mathrm{I}_F = I_s(0)\exp[-(q\Xi)^p] + \frac{\mathrm{I}_F(0)}{(1+q^2\xi_F^2)} \tag{129}$$

where $\mathrm{I}_s(q)$ and $\mathrm{I}_F(q)$ denote respectively the static and the fluctuating part of the scattered intensity.

The origin of the fluctuating part is natural: it is identified with the scattering by an idealized gel (Eq. (57):

$$\mathrm{I}_F(0) = k_B T b_v^2 \phi^2/(K_{os}^{gel} + 4/3G) \tag{130}$$

In other words, it is assumed that all the thermodynamic properties are contained in the Lorentzian component of the gel signal. Recall that, since Π^{mix} may be smaller than Π^{sol}, the longitudinal modulus $M_{os} = K_{os}^{gel} + 4/3G$ is not necessarily larger (although the elastic contribution G is present) than the compression modulus K_{os}^{sol} of a semi-dilute solution. According to Horkay *et al* (1993b), it can even be, in some particular cases, significantly smaller.

The choice of a shape for the static part was made in several steps. First, it was assumed that the permanent spatial concentration fluctuations (resulting from the presence of crosslinking non-uniformities) could be described by a Gaussian distribution (Geissler *et al*, 1988):

$$p(\phi) = \frac{1}{\sqrt{2\pi\langle\Delta\phi^2\rangle}}\exp\left[-\frac{1}{2}\frac{(\phi-\langle\phi\rangle)^2}{\langle\Delta\phi^2\rangle}\right] \tag{131}$$

where $\langle\phi\rangle$ represents the ensemble average of the local polymer concentration (above, $\langle\phi\rangle$ was in general written ϕ) and ϕ denotes the local average (in time) of the polymer concentration, and could also be written $\phi(r)$.

Later it was postulated that $\phi(r)$ varies slowly in space with a Gaussian probability of characteristic length $\Xi > \xi_F$ (Mallam *et al*, 1989b):

$$\langle\delta\phi(0)\delta\phi(r)\rangle = \langle\delta\phi^2\rangle^{loc}\exp\left(-\frac{r^2}{2\Xi^2}\right) \tag{132}$$

It is seen immediately that $\langle\delta\phi^2\rangle^{loc}$ is the limit, at $r \to 0$, of $\langle\delta\phi(0)\delta\phi(r)\rangle$. In section 5.4, we added the index loc (for local), because the value of $\langle\delta\phi^2\rangle$ depends on the scale at which this quantity is calculated (i.e., the size of the 'cell' characteristic of the 'local concentration'; see next section). As defined here, $\langle\delta\phi^2\rangle^{loc}$ represents the mean square amplitude of the concentration fluctuations in vanishingly small cells.

The spatial Fourier transform then leads to

$$I_s(q) \sim \langle\Delta\phi^2\rangle^{loc}\Xi^3\exp\left(-\frac{q^2\Xi^2}{2}\right) \tag{133}$$

As seen in section 5.4, such decomposition of I(q) was found satisfactory for fitting the scattering envelope of PDMS gels. In practice, Mallam, *et al* (1989a) calculated the following variance:

$$\mathrm{var} = \sum_i \left[\mathrm{I}_{\mathrm{exp}}(q_i) - \frac{A}{1 + q_i^2 \xi_{\mathrm{F}}^2} - B \exp(-\tfrac{1}{2} q_i^2 \Xi^2) \right]^2 \tag{134}$$

where $\mathrm{I}_{\mathrm{exp}}(q_i)$ denotes the experimentally measured value of the scattered intensity for the ith value q_i of the momentum transfer. The fitting procedure is as follows: the values of ξ_{F} and Ξ are incremented iteratively until a minimum is found for the variance. For any value of ξ_{F} and Ξ, the coefficients A and B are obtained from a linear regression.

Later, however, a Gaussian form was found inadequate for fitting the scattered intensity of polyvinylacetate gels (see section 5.4): the decay appeared to be smeared out in q. Empirically, it was found that $\mathrm{I_s}(q)$ could be better approximated by the stretched exponential appearing in Eq. (128), with an exponent $p \approx 0.7$ (Horkay *et al*, 1991a). As a result, Geissler, Horkay and Hecht (1993) adopted subsequently a fitting procedure analogue to that of Eq. (133), but with now five parameters; ξ_{F}, Ξ, p, $\mathrm{I_s}(0)$, $\mathrm{I_F}(0)$.

Comments and discussion The decomposition of the total structure factor as a sum of a fluctuating and a static part is a reasonable working assumption. It is consistent with the results of dynamic light scattering experiments (Candau *et al*, 1979; Geissler, 1993 and refs. therein) and also with coherent scattering data obtained by means of the scanning procedure described in sections 5.4 and 5.5. The fluctuating part has a clear physical meaning: it is identified with the Lorentzian function predicted for idealized gels (see section 5.3).

Regarding the proposed shapes for the static part, the stretched or simple exponential forms do not derive from any precise picture of the superstructure. Geissler *et al* used a Gaussian function to fit the data relative to the PDMS gels, whereas a stretched form seemed better adapted to PVAc gels. Note that, implicitly, this observation supports the idea of the existence of several classes of gel superstructures (see section 5.4).

The main argument in favour of such Gaussian or stretched exponential forms is that these functions decay very fast at large q. This property ensures the convergence of the 'second moment' $Q^* = \int q^2 \mathrm{I}(q) \mathrm{d}q$, discussed in the next section, from which Geissler *et al* calculate $\langle \delta\phi_s^2 \rangle$. A true Lorentzian or a pseudo-Lorentzian with a 'fractal' tail (i.e. $I_s(q) \sim q^{-D}$) would either not converge if $D \leqslant 2$ or converge too slowly if D is slightly larger than 2.

The drawback of the choice of such forms is the difficulty encountered when trying to interpret in a consistent way the behaviours observed upon swelling and stretching. We have seen in section 5.4 that both for PDMS and PVAc gels the characteristic length Ξ of the static fluctuations, measured by means of an exponential or a stretched exponential is found to change very little with the

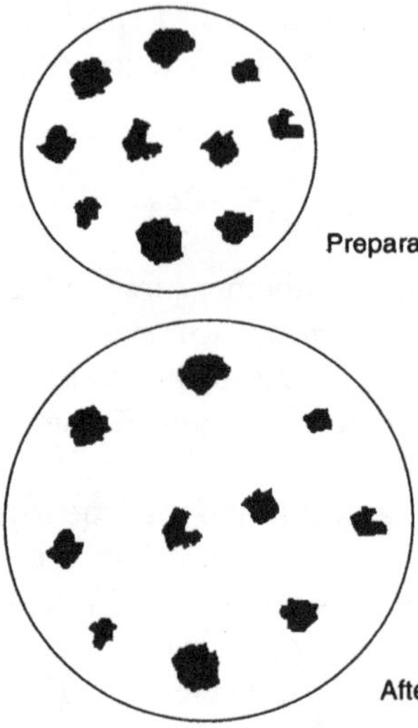

Figure 42. Schematic representation of 'island-like' heterogeneities in a gel

swelling degree of the gel. The variation of the low q scattering intensity is essentially attributed to the enhancement of the variance $\langle \delta\phi_s^2 \rangle$ of the quenched fluctuations of concentration (see below). Such a result would suggest the description of the gel heterogeneities as a set of 'islands' of crosslinking density higher than average which would became more dilute upon swelling (see Figure 42). There would be no unscreening effects associated with this process and Ξ would represent the 'mean size of the static non-uniformities' (Geissler *et al*, 1990; Geissler, Horkay and Hecht, 1993, 1994; Horkay *et al*, 1991a; Mallam *et al*, 1991), that is, the size of the islands. On the other hand, Geissler *et al* (1995b) showed recently that the same PDMS gels exhibit an strong enhancement of scattering intensity along the direction of maximum dilation, in an anisotropic swelling experiment (with butterfly iso-intensity lines). We have seen above that a similar behaviour is observed for all the neutral gels studied up to now (section 5.5). Using the same type of analysis, Geissler *et al* (private communication 1995) get then a clear increase of Ξ in the parallel direction associated with a decrease in the perpendicular direction, whereas $\langle \delta\phi_s^2 \rangle$ is found to remain approximately unchanged. Here, in this case, Ξ cannot represent a size of island-like heterogeneities. By contrast to the case of swelling, this result is in better agreement with the unscreening of a cluster-like structure, as proposed in the random-cluster model of section 5.6.2 and illustrated in Figures 34 of 39.

To sum up, the decomposition and the forms of scattering envelopes proposed by Geissler–Horkay–Hecht are shown to be efficient tools for fitting the scattering data of several types of gels. There is, however, an open question of representation: two different types of experiments on the same system seem, along this approach, to lead to two different pictures of the gel 'superstructure'.

5.6.3.5.3 'Second moment of scattering' and 'concentration polydispersity'

As frequently mentioned above, Geissler, Horkay and Hecht (1991, 1993), Hecht Duplessix and Geissler (1985), Hecht *et al* (1991, 1992), Horkay *et al* (1991a), Mallam *et al* (1989a and b) introduced the use of the second moment of scattering intensity for estimating the mean square amplitude of the concentration fluctuations. This quantity should characterize the distribution of local concentration in the system, which they call the 'concentration polydispersity'.

For a better understanding of this approach, it is useful to refer back to the basics of the scattering. We have seen in section 5.1 that

$$I(q) = b_v^2 \int_V \langle \delta\phi(0)\delta\phi(\mathbf{r})\rangle \exp(-\mathrm{i}\mathbf{qr})\mathrm{d}^3r \tag{135}$$

In principle, the correlation function $\langle \delta\phi(0)\delta\phi(r)\rangle$ of the fluctuations of polymer concentration may be obtained by the inverse Fourier transform of Eq. (134):

$$\langle \delta\phi(0)\delta\phi(\mathbf{r})\rangle = \frac{1}{b_v^2(2\pi)^3}\int \mathrm{I}(q)\exp(\mathrm{i}\mathbf{qr})\mathrm{d}^3q \tag{136}$$

We recall that

$$\langle \delta\phi(0)\delta\phi(\mathbf{r})\rangle = \frac{1}{V}\int_V \delta\phi(\mathbf{r}')\delta\phi(\mathbf{r}'+\mathbf{r})\mathrm{d}^3\mathbf{r}' \tag{137}$$

Thus the limit $r \to 0$ of Eq. (135) gives access to a quantity which can be written $\langle \delta\phi^2\rangle$, the mean square fluctuation of the local concentration:

$$\langle \delta\phi^2\rangle = \frac{1}{V}\int_V [\delta\phi(r')]^2\mathrm{d}^3r' = \frac{1}{b_v^2(2\pi)^3}\int I(q)\mathrm{d}^3q \tag{138}$$

In the case of an isotropic system, one finally obtains

$$\langle \delta\phi^2\rangle = \frac{1}{b_v^2 2\pi^2}\int q^2\mathrm{I}(q)\mathrm{d}q \tag{139}$$

where $Q^* = \int q^2\mathrm{I}(q)\mathrm{d}q$ was called the second moment of the scattering intensity.

As a result, a 'mean-square amplitude of concentration fluctuations' $\langle \delta\phi^2\rangle$ can, in principle, be estimated by calculating the integral (Eq. (138)). Obviously, some conditions must be satisfied to this end: the integral has to converge and, in order to approach the limit of Q^*, $I(q)$ has to be measured over a large enough range of q.

Such a condition is clearly fulfilled when the system is composed of two media having well-defined boundaries, like colloids in solution (see Porod, 1982 or Debye, Anderson and Brumberger, 1957). Furthermore, one can then decompose the integrating space in two subspaces. Subspace 1 can be defined as the set of all the colloids, representing an overall volume fraction ϕ, and where the local volume fraction $\phi_1 = 1$. Subspace 2 is then composed of the solvent; its overall volume fraction is $(1 - \phi)$ and the local volume fraction ϕ_2 of colloids is obviously zero in this medium. In such conditions, from Eq. (138), one can write

$$\langle \delta\phi^2 \rangle = \frac{1}{V} \int_V [\delta\phi(\mathbf{r}')]^2 d^3r' = \frac{1}{V}\{(1-\phi)^2 V_1 + (-\phi)^2 V_2\} \tag{140}$$

since in subspace 1, of volume V_1, $\delta\phi(\mathbf{r}') = (1-\phi)$, and in subspace 2, of volume V_2, $\delta\phi(\mathbf{r}') = (-\phi)$. Moreover, since $\phi = V_1/V$, $(1-\phi) = V_2/V$, one obtains eventually a very concise result:

$$\langle \delta\phi^2 \rangle = \phi(1-\phi) \tag{141}$$

Thus, the second moment Q^* can be expressed as function of the sole contrast factor b_v^2 and the volume fraction ϕ of scattering objects:

$$Q^* = \int q^2 I(q) dq = b_v^2 2\pi^2 \phi(1-\phi) \tag{142}$$

Therefore, in such conditions, Q^* must represent a 'scattering invariant', independent of the spatial arrangement of the scatterers, that is, independent of the structure of the system.

Let us postulate, as a working assumption, that this treatment can be applied to the present case of polymer chains and solvent. Then, assuming that Eq. (141) can be used, the superposition at high q of the spectra of various systems of same ϕ, but with different small q signals (a situation often encountered when comparing gels and semi-dilute solutions), might break down. For instance, it is expected that a gel spectrum with a very strong signal at low q must become significantly smaller at high q than the signal of a solution with same ϕ: because the second moment Q^* should be the same in both cases, the price of an enhanced intensity at small angle would indeed be a reduced one at large angle (with, however, some significant 'damping' of the effect due to the multiplication by q^2 in the integral (141)). Hecht *et al* (1991) and Geissler, Horkay and Hecht (1991) report examples of such crossings for pairs of gels and solutions of the same concentration. They interpret these crossings as consequences of the applicability of the scattering invariant to polymers. Since the signal from the gel is then comparable at large q with the spectrum of a semi-dilute solution of significantly smaller polymer concentration, they introduce the notion of the *effective corresponding solution.*

To sum up this first part, if polymers obey the same laws as colloids, the *overall* second moment Q^* is an invariant and does not carry any information about the structure. The key idea of Geissler–Hecht–Horkay is that some information about the structure is, however, contained in the second moment of the static part of the scattering. These authors exploit this property to estimate a mean-squared 'static' fluctuation $\langle \delta\phi^2 \rangle_s$ in the case of gels. First, consistently with Eq. (129), they assume that the total mean squared fluctuation $\langle \delta\phi^2 \rangle$ can be expressed as the following sum:

$$\langle \delta\phi^2 \rangle_{\text{total}} = \langle \delta\phi^2 \rangle_s + \langle \delta\phi^2 \rangle_F \quad (143)$$

where $\langle \delta\phi^2 \rangle_F$ is a 'liquid-like' or 'fluctuating' part associated with the 'thermal' fluctuations.

Then Geissler–Hecht–Horkay *et al* decompose the scattering of the gel into two parts, as shown above. Thus, the difference $I(q) - I_F(q)$ can be evaluated. In practice, as explained above, these authors fit $I(q) - I_F(q)$ by a Gaussian or a stretched exponential. Its integral, after multiplying by q^2, converges rapidly and can be easily calculated. Then, they write

$$\langle \delta\phi^2 \rangle_s = \frac{1}{b_v^2 2\pi^2} \int q^2 \{I(q) - I_F(q)\} dq \quad (144)$$

Therefore, using this method (and accepting the underlying assumptions), one obtains a quantity, $\langle \delta\phi^2 \rangle_s$, which can indeed be called a mean-squared amplitude of the 'static' fluctuations of polymer concentration. As stated in section 5.4 and above in this section, Geissler *et al* found $\langle \delta\phi^2 \rangle_s$ to be clearly increased, as the swelling degree of PDMS or PVAc gels was raised. On the other hand, Ξ, the 'mean radius the static concentration fluctuations' was found to be only marginally changed.

Comments and discussion As observed above, the use of Porod's invariant for compact objects with sharp boundaries seems to lie beyond any criticism. However, in the case of polymer gels, there are different issues which need to be discussed.

The first difficulty arises from the fact that the current models of idealized chains with Gaussian or excluded volume statistics (leading respectively to $I(q) \sim q^{-2}$ and $q^{-5/3}$ lead to a divergence of the integral (142)). Obviously such models cease to be valid on very short length scales, where the 'structure' of the monomer unit becomes important. Then, to overcome this difficulty, one may be tempted to compare the monomer unit with an elementary 'grain' with sharp boundaries, leading to a q^{-4} decay for the $I(q)$: this would ensure the convergence for $\int q^2 I(q) dq$). However, it is not clear whether this new idealization makes sense in the case of neutron scattering: neutrons interact with nuclei only and a monomer unit is not really a compact grain. It can be viewed as well as a rather 'cloudy' object with a thickness comparable with the average distance between nuclei (which are the scattering centres for SANS). Therefore, even if the integral of the overall intensity converges, it is not obvious that it gives exactly $b_v^2 2\pi^2 \phi(1-\phi)$.

A second point is to know whether arguments based on Porod's invariants may explain a crossing at high q of scattering curves corresponding to systems with the same number of elementary scatterers but with different large-scale-structures. This question is discussed at the beginning of section 5.4.3 and remains pending. For us, it is difficult to view the clear crossings which are reported as manifestations of scattering invariants.

Then there is the question of the physical meaning of $\langle\delta\phi^2\rangle$ (or $\langle\delta\phi^2\rangle_s$) determined through Q^*. Assume that $\int_0^\infty q^2\mathrm{I}(q)\mathrm{d}q$ converges and that $\mathrm{I}(q)$ is actually measured in a range of q wide enough to approach the limit of this integral. We can rewrite Eq. (137) as follows:

$$\langle\delta\phi^2\rangle = \frac{1}{V}\int_V \delta\phi^2(\mathbf{r}')\mathrm{d}^3r' = \lim_{N\to\infty}\frac{1}{N}\sum_i^N \delta\phi_i^2 \tag{145}$$

where $\delta\phi_i^2 = (\phi_i - \phi)^2$ denotes the mean squared fluctuation of polymer concentration in the infinitesimal element i of volume V/N, with N tending towards infinity. Therefore, it can be seen that $\langle\delta\phi^2\rangle$ calculated by integration of $q^2\mathrm{I}(q)$ over all the q range gives an estimate of the mean squared fluctuation of polymer concentration on very local length scales.

This is the reason why we have introduced in section 5.4 the notation $\langle\delta\phi^2\rangle^{\mathrm{loc}}$ for $\langle\delta\phi^2\rangle$ estimated from the second moment: we wanted to make the distinction with other values of $\langle\delta\phi^2\rangle$, which could be calculated by different means. To us, it is not clear whether $\langle\delta\phi^2\rangle^{\mathrm{loc}}$ is the relevant quantity, or at least the only one, to characterize the permanent departures of polymer concentration from uniformity in gels. In order to illustrate this point, let us imagine two different types of static non-uniformities, inspired from the island and cluster pictures presented above (Figures 43(a) and (b) resembling respectively Figures 42 and 39). In order to avoid any problem with the calculation of the second moment Q^*, we assume here that these 'heterogeneities' are built with elementary 'grains' having sharp boundaries, with a size of a few nanometres and a well-defined polymer concentration, higher than that of the gel on an average. Obviously, we do not claim that such simplifying conditions are satisfied in reality. In Figures 43(a) and (b) are represented two different arrangements in space of the same 'grains', the circle denoting a region in the gel of, say, 300 Å radius: for us, it cannot be said that these gels have the same degree of non-uniformity. It is also obvious that, under further swelling, the gel with clustering heterogeneities of Figure 43(b) will be distorted much more on mesoscopic length scales than the gel with well-separated hetero-

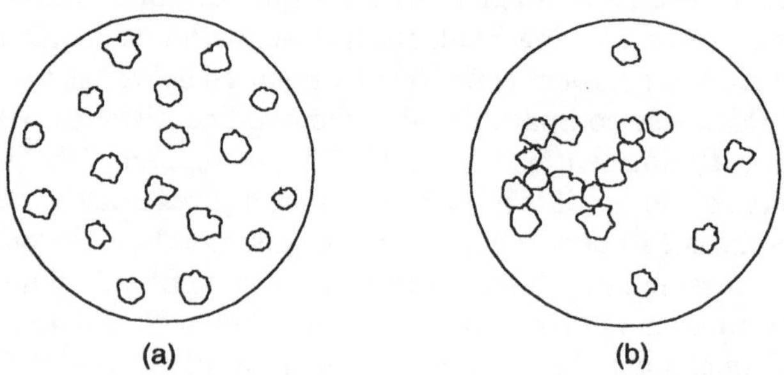

Figure 43. Schematic representation of (a) 'island-like' and (b) 'cluster-like' heterogeneities formed with identical elementary grains

geneities of Figure 43(a). This amplification of the gel heterogeneity on rather large length scales in the (b) case will, however, remain undetected by the second moment method: since the number and the size of the elementary grains are the same in both cases, $\langle \delta\phi^2 \rangle_s^{loc}$ will be found equal.

Finally, we are wondering about a possible dependence of the value found for $\langle \delta\phi^2 \rangle^{loc}$ with the forms chosen for $I_s(q)$ and $I_F(q)$. Because of the presence of the q^2 factor, the integral

$$Q_s^* = \int_0^{\infty} q^2 I_s(q) dq$$

depends strongly on the intensity at rather large q's. It is true that no quenched fluctuations are believed to exist on length scales smaller man the mesh size (there must be crosslinks to give rise to crosslinking heterogeneities). However, since the Fourier transform intrinsic to scattering experiments smooths the variations, it might remain a small contribution of $I_s(q)$ for $q \approx 0.1$ Å^{-1}, possibly not negligible after multiplication by q^2. Therefore, the precision of the subtraction of $I_F(q)$ in this q-range is of prime importance for the value of Q_s^*: a small uncertainty on $I_F(q)$ might give rise to a very large one on $\langle \delta\phi^2 \rangle_s$. A similar problem may arise because of the fit procedure. The integral is indeed calculated from the theoretical curve proposed for $I_s(q)$ and not from a curve joining the data points. It must be stressed that two different forms of the theoretical curves may be practically indistinguishable at rather high q's, whereas they lead to very different values of Q^*. For instance, a stretched exponential will lead to a small value of Q^*. On the other hand, a curve with a fractal tail and a cut-off might lead to a different and large value.

5.6.3.5.4 *'Concentration polydispersity' and thermodynamics*

In this subset, two ideas are exploited. The first, introduced first in Bueche's model (1970) and met also in the 'cluster model' given above, is that complete randomness in the crosslinking process may produce static non-uniformities of concentration via spatial fluctuations of the crosslinking density. The second is that the various regions of the gel, where the local values of the crosslinking density are different from the average, must be in thermodynamical equilibrium with each other.

We have seen in section 5.3 (Eq. (62)) that the osmotic pressure Π_{gel} of a gel (called swelling pressure ω in the GHH approach) can be written

$$\Pi_{gel}(\phi) = \Pi_{mix}(\phi) - G(\phi),$$

provided the elastic part of the free energy has a simple scaling form $F_{el} = a'(N_c/V_d)k_B T\phi^{1/3}$, where N_c is the number of crosslinks in a volume V_d of dry network and a' is a numerical prefactor independent of N, V and ϕ, see appendix. Hecht, Duplessix and Geissler (1985), Horkay, Hecht and Geissler

(1989), Geissler, Horkay and Hecht (1991) showed that this form of F_{el} is consistent with the mechanical data of many types of gels (those for which $G \sim \phi^{1/3}$). In order to deal with the spatial fluctuations of polymer concentration (more precisely its time-average value) in the *swollen state*, small cells with identical volume v in the *dry state* are considered. Due to randomness in the crosslinking process, there should be some fluctuations δn of the number of crosslinks in these cells (around the ensemble-average $\langle n \rangle$ calculated over the whole set of cells). After swelling of the network, the polymer concentration ϕ will also fluctuate from one cell to another because the random fluctuations of crosslinking density must obviously result in variations of local swelling degree. Therefore, a local deviation $\delta\phi$ with respect to the ensemble-averaged value $\langle \phi \rangle$ can be assigned to each cell. Note, incidentally, that the volume v' of the swollen cells will also vary from one cell to another one: $v' = v/\phi = v/(\langle \phi \rangle + \delta\phi)$.

However, since all the swollen cells are assumed to be in thermodynamical equilibrium, the swelling pressure $\Pi_{gel}(\phi)$ should not fluctuate from one to another. Therefore, one may write:

$$\delta\Pi_{gel} = (\partial\Pi_{mix}(\phi)/\partial\phi)_{\langle\phi\rangle}\delta\phi - \delta G(\phi, n) = 0 \tag{146}$$

Then, from the proportionality $G \sim (n/v)\phi^{1/3}$, Geissler, Horkay and Hecht (1993) obtain

$$\{\langle\phi\rangle \cdot (\partial\Pi_{mix}/\partial\phi)_{\langle\phi\rangle} - \tfrac{1}{3}G(\langle\phi\rangle)\}\left(\frac{\delta\phi}{\langle\phi\rangle}\right) = G(\langle\phi\rangle)\cdot\left(\frac{\delta n}{\langle n\rangle}\right) \tag{147}$$

where

$$\{\langle\phi\rangle \cdot (\partial\Pi_{mix}/\partial\phi)_{\langle\phi\rangle} - \tfrac{1}{3}G(\langle\phi\rangle)\}$$

is nothing less than the osmotic bulk modulus K_{os}^{gel} of the gel—recall that

$$K_{os}^{gel} = \langle\phi\rangle \cdot (\partial\Pi_{gel}/\partial\phi)_{\langle\phi\rangle} = K^{mix} - G/3$$

Taking the mean square of Eq. (146) (over the set of swollen cells), one then obtains

$$\frac{\langle\delta\phi^2\rangle}{\langle\phi\rangle^2} = \left(\frac{\langle\delta n^2\rangle}{\langle n\rangle^2}\right)\frac{[G(\langle\phi\rangle)]^2}{[K_{os}^{gel}]^2} \tag{148}$$

Assuming a random distribution of crosslink positions in space, in the dry state, leads to $\langle\delta n^2\rangle = \langle n\rangle$. Therefore, Geissler, Horkay and Hecht (1993) obtain finally:

$$\frac{\langle\delta\phi^2\rangle}{\langle\phi\rangle^2} = \left(\frac{1}{\langle n\rangle}\right)\frac{[G(\langle\phi\rangle)]^2}{[K_{os}^{gel}]^2} \tag{149}$$

This model therefore establishes a relation between some quenched fluctuations of polymer concentration in the gel and two measurable thermodynamic quantities, $G(\langle\phi\rangle)$ and K_{os}^{gel}. In practice, in order to test this model, Geissler *et al* identified $\langle\delta\phi^2\rangle$ with $\langle\delta\phi^2\rangle^{loc}$ defined above.

Comments and discussion Referring back to the preceding paragraph, the identification of $\langle \delta\phi^2 \rangle$ appearing in this thermodynamical model with $\langle \delta\phi^2 \rangle^{loc}$ has no clear physical justification. To obtain a significant ensemble average, the characteristic cell size has to be large with respect to the correlation length of the thermal fluctuations, which is of the order of 20 Å; $\langle \delta\phi^2 \rangle^{loc}$ obtained from Eq. (138) is presumably a more local quantity. However, the experiments of Geissler, Horkay and Hecht (1993, 1994) show a linear dependence of $\langle \delta\phi^2 \rangle^{loc}$ with the submicroscopic $\langle \delta\phi^2 \rangle$ obtained from measured values of G and K_{os}, which is an argument in favour of the approach. In addition, this method leads, in principle, to a determination of $\langle n \rangle$ (Eq. (48)). In fact, the results of Geissler *et al* provide a value of $\langle n \rangle = 1$ in the case of PVAc gels. Such a value, even if the distribution of structural non-uniformities is very wide, seems not to be physical.

5.6.3.6 The Falcao–Mortensen–Pedersen approach

In connection with experiments performed on PDMS gels (which are briefly mentioned in section 5.4), Falcao, Skov Pedersen and Mortensen (1993) developed a model addressing the case of networks prepared by random crosslinking of polymer melts. Some features of this approach are close to those of frozen-blobs model and of its generalization. The system is seen as a fractal arrangement of blobs similar to those of a semi-dilute solution. The scattering intensity is decomposed as the following product:

$$I(q) = \phi_N P(q) S_t(q) \tag{150}$$

where $P(q)$ is the square modulus of a 'particle' form factor averaged over the ensemble of particles (for centro-symmetric particles, $P(q)$ is simply the form factor; see Higgins and Benoît, 1994). In the present case, the 'particle' is identified with the 'blob' of the semi-dilute solution with the same concentration as in the gel. In practice, Falcao *et al* postulate

$$P(q) \sim \frac{(1 + q\xi_s)^{1/3}}{1 + q^2\xi_s^2} \tag{151}$$

For $q\xi_s \gg 1$ and $q\xi_s \ll 1$, this function tends to the limits predicted by the scaling approach of semi-dilute solution (see section 5.1). Moreover, it gives an expression for the crossover between these two limit behaviours. This function was found to fit satisfactorily the scattering of the semi-dilute solutions overall the q range accessed; subscript s denotes a 'small correlation length' or the size of the blob. Although defined differently, it should not be very different from ξ_F, the correlation length of the thermal fluctuation in the approach by Geissler, Hecht, Horkay *et al.*

Here $S_t(q)$ is the 'structure factor' describing the spatial distribution of the 'particles' (not to be confounded with $S(q)$ of section 5.1, which corresponds to another type of normalization). Falcao *et al*, used the expression derived by Freltoft, Kjems and Sinha (1986) or by Texeira (1988) consistent with the basic assumption of fractal distribution of blob particles:

$$S_f = 1 + \frac{1}{(qr_0)^D} \frac{D\Gamma(D-1)}{[1 + 1/(q\xi_L)^{(D-1)/2}]} \sin[(D-1)\arctan(q\xi_L)] \qquad (152)$$

where ξ_L denotes a 'large correlation length', representing either the size of the fractal clusters when these are well separated, or a screening length comparable to some extent to that appearing in the 'frozen-blobs' model (ξ_c).

Such an approach allows one to fit scattering curves relative to PDMS gels within a large range of crosslink density despite the fact that the shape of the scattering curves differ widely from one gel to the other. In particular, some of these curves exhibit a 'double-shoulder' profile similar to that observed in particular, by Geissler–Horkay–Hecht *et al.*

Comments It is not obvious that the factorisation given in Eq. (150) is permitted for polymer systems because polymer chains are not centro-symmetric.

The correlation ξ_L deduced from the analysis of Falcao *et al* is at first sight very similar to the inter-cluster correlation length ξ_c of the 'frozen-blobs' model, but due to the factorization of Eq. (149), ξ_L and ξ_c do not strictly coincide.

An interesting conclusion to be drawn from the Falcao *et al* study is that the double shoulder in a scattering curve is not necessarily the signature of an island-like superstructure.

5.7 Conclusion

When examining the gel structure factors which are gathered in this review, the most striking feature is the excess of scattering intensity with respect, at the same time, to a semi-dilute solution of the same polymer concentration and to the predictions of classical thermodynamical approaches. The excess of scattering is widely dependent on the preparation method, the crosslink density, the swelling ratio relative to the state of preparation and, for charged gels, the ionization degree. In some specific cases, the excess of scattering is very weak: that is what is observed for gels in the state of preparation, either made from semi-dilute solutions or electrically charged. In both cases the perturbation resulting from the crosslinking appears to be weak compared to the osmotic pressure of the comparable polymer solution. On the other hand, gels prepared by end-linking of chains with crosslinks of high functionality exhibit a huge scattering at low q, even in the preparation state.

The main difficulty encountered in trying to understand the observed behaviours is that none of the proposed models can describe accurately the totality of experimental features. However, one can classify the proposed models into two main families with fundamentally different basic assumptions. The first rests on the idea that there is a coupling between strain and thermal concentration fluctuations, which results in an amplification of these thermal fluctuations under swelling or stretching. This model does not consider the possibility of structural changes induced by the crosslinking, the spatial fluctuations are characterized by a single correlation length. In the second family, it is assumed that the crosslinking intrinsically produces structural heterogeneities. In some of the models, they are described as a sort of 'superstructure' consisting either of 'islands' or of 'clusters', with different characteristic size distributions depending on the approach. As mentioned above, some experimental data can be qualitatively explained by either kind of model. As an example, one of the most spectacular observation, i.e. the butterfly patterns obtained under stretching of the gels, has been used by the supporters of either type of approach as a key feature in favour of their model. There are, however, two kinds of experiments that suggest that the interpretation in terms of a structural heterogeneities is more appropriate. The first concerns ionized gels. In these materials, long-range concentrations are suppressed due to the high entropy cost of counterions. In fact, under uniaxial strain no enhancement of fluctuations is observed along the direction of elongation. The butterfly patterns reappear upon screening of electric interactions, by adding salt or upon lowering the pH. The second experiment refers to the separation of static and dynamic fluctuations by means of the scanning light scattering procedure. This reveals that, under stretching, the enhancement of the scattering intensity is almost entirely associated with the static (or quasi-static) profile of polymer concentration fluctuations.

The large variety of structure factors depending on the synthesis method is also in favour of crosslinking-induced superstructure. At present it is not clear whether this superstructure can be described as separated islands of crosslinking density larger than average, or as branched clusters, more or less interspersed and possibly connected on relatively large length scales. In fact, either of these pictures might apply, depending on the preparation method. By combining swelling and stretching experiments, one can obtain information that allows us to be more affirmative. For instance, in the case of statistical gels, a model based on the unscreening of fractal (or at least branched) clusters takes into account both the butterfly patterns and the increase of static scattering under swelling. On the other hand, we do not see how an island-like superstructure could lead to butterfly patterns. Some other gels, for example those prepared by end-linking of chains with crosslinks of rather high functionality exhibit a scattered intensity at very low q which, under swelling is roughly proportional to the polymer concentration, still remaining much larger than that of the corresponding semi-dilute solution. Obviously, such behaviour would be well described with an island-like superstructure model. Some gels present an intermediate behaviour, for instance, for ionized gels, one observes a decrease of the total intensity upon swelling, but the static contribution of the scattering intensity at zero q is roughly independent of concentration. SANS data suggest the presence of small compact aggregates, but obviously they must be correlated to lead to an intensity independence on the swelling ratio.

In this chapter, we have focused on static scattering. There have been recent developments of dynamic light scattering by gels that allow us to separate the dynamic and static contributions of the concentration fluctuations. The results obtained by means of this technique are consistent with the conclusions reported above. Furthermore, these experiments provide information on the dynamics of the gels, which, in particular, controls the kinetics of swelling and of volume phase transition.

From the large number of data available, one can make a rough empirical classification of the gels and predict empirically what kind of scattering behaviour will be observed for a given synthesis method. However, at the present stage, it is not possible to give a precise picture of the different superstructures that may be encountered in the gels. One can definitely forget about the naive description of a gel in terms of a frozen-in uniform semi-dilute solution. Clearly, some new experiments are needed, perhaps using fluorescent probes. Also, there have been very few studies aiming to correlate the scattering features with the mechanical properties of the gels. It is likely that future investigations will deal with the effect of the superstructure on the viscoelastic properties in the high-frequency range and the fracture behaviour. Also themes where the superstructure might be relevant merge, as with the properties of gels, with guest molecules (surfactants, linear chains, neutral or charged and/or colloidal particles).

List of symbols

$A(\mathbf{q})$	scattered field
α	ionization degree
α_p	polarizability
a	length of monomeric unit (or lateral size of a lattice site)
a^3	volume of the lattice site, in lattice theories
a_K	length of the Kuhn segment (or statistical unit)
β	azimuthal angle in a stretching experiment, that is the angle between the elongation axis and the direction of $\mathbf{q}$
b_i	coherent scattering length of atom *i*, or of 'scattering' molecule *i*. Often expressed in cm
b	average scattering length
b_v	'contrast factor' in the case of a binary system composed of scattering molecules *a* and *b*; $b_v = [(b_a/v_a) - (b_b/v_b)]$
c	polymer concentration in a gel expressed in $g\,cm^{-3}$
c_p	polymer concentration in a gel expressed in $g\,g^{-1}$ or mol. mol.$^{-1}$
C	crosslink content in a gel expressed in $g\,g^{-1}$ of monomer
$\delta\phi(\mathbf{r})$	local deviation (at point $\mathbf{r}$) of the polymer volume fraction with respect to its average value

ϕ_0 polymer volume fraction of a gel in the reference state
ϕ polymer volume fraction
ϕ_e polymer volume fraction of a gel at maximum swelling
ϕ_s salt concentration in polyelectrolytic systems
$\phi(\mathbf{r})$ local polymer volume fraction at point $\mathbf{r}$ ($\phi(\mathbf{r}) = \phi + \delta\phi(\mathbf{r})$)
$\phi(\mathbf{q})$ Fourier transform of $\phi(\mathbf{r})$ (over the scattering volume V) often written ϕ_q
f functionality
f_x elastic retractive force along the Ox direction
F free energy (we do not discriminate between Helmholtz and Gibbs free energy)
F_{mix} mixing contribution to the free energy
F_{el} elastic contribution to the free energy
G elastic modulus measured by uniaxial deformation ($G = \sigma_0/(\lambda - 1/\lambda^2)$. If the deformation is performed at constant volume $G = \mu$.
h Planck's constant ($h \approx 6.62 \times 10^{-34}$ J s)
$\hbar$ $= h/2\pi$
$\mathrm{I}(\mathbf{q})$ differential scattering cross-section per unit volume ($= (\partial\Sigma/\partial\Omega)/V$), often simply quoted as 'scattered intensity'. In general, $I(\mathbf{q})$ is expressed in cm^{-1}
$I(\mathbf{q})$ differential scattering cross-section ($= (\partial\Sigma/\partial\Omega)$), represents a non-normalized 'scattered intensity'. In general, $I(\mathbf{q})$ is expressed in cm^2
κ^{-1} Debye screening length of polyelectrolytes
$\mathbf{k}$ wave vector $\|\mathbf{k}\| = 2\pi/\lambda$
k_B Boltzmann constant ($k_B \approx 1.38 \times 10^{-23}\,\mathrm{J\,k^{-1}}$; at room temperature $k_B T \approx 4 \times 10^{-21}\,\mathrm{J} \approx 1/40\,\mathrm{eV}$)
$\mathbf{k}_i$ incident wave vector
$\mathbf{k}_s$ scattered wave vector
K_{os} isothermal osmotic bulk modulus ($K_{os} = \phi[\partial\Pi_{os}/\partial\phi]_T$)
K_{os}^{gel} isothermal osmotic bulk modulus of a gel
K^{mix} mixing part of the isothermal osmotic bulk modulus of a gel
χ_T isothermal osmotic compressibility ($\chi_T = 1/K_{os} = \phi^{-1}[\partial\phi/\partial\Pi_{os}]_T$)
l exponent governing the variation of the elastic free energy F_{el} with the polymer volume fraction, in a scaling description; $l = 1/3$ in the classical description of gels
l_B Bjerrum length (by definition $(1/4\pi\varepsilon).e^2/l_B = k_B T$, where ε is the dielectric constant of the solution)
λ deformation ratio along the direction of the imposed strain, with respect to an initial state which is not necessarily confounded with the reference state of the network ($\lambda = L/L_i$)
λ_i component of the deformation along the principal direction i ($i = x, y, z$) with respect to an initial state not confounded with reference state
λ_l first Lamé coefficient
λ_ω wavelength of the considered radiation in the scattering medium

Λ_i deformation ratio in the principal direction i ($i = x, y, z$) with respect to the reference state of the network (idealized state in which the elementary chains of the network are 'relaxed', the reference state is often assumed to be confounded with the preparation state)

L, L_i length and length in the initial state of a sample of gel

μ shear modulus (or second Lamé coefficient)

M longitudinal modulus

M_c weight average molecular weight of the precursor chains (in an end-linking reticulation)

M_n number average molecular weight

M_w weight average molecular weight

ν excluded volume exponent

ν_e number of chemical meshes or elastic elementary chains

ν_0 number of chemical meshes (or elastic elementary chains) per unit volume, in the reference state

n refraction index

$n(\mathbf{r})$ number density of scattering molecules (or monomer units) at point $\mathbf{r}$

N_c number of crosslinks

Π_{os} osmotic pressure

$\mathbf{q}$ scattering vector, $= \mathbf{k}_i - \mathbf{k}_s$

q scattering vector amplitude, in the case of an elastic scattering process $q = (4\pi/\lambda_\omega) \sin \theta/2$

Ω solid angle

r' 'reduction factor' of the mixing part of the osmotic pressure in a gel, with respect to the osmotic pressure of a semi-dilute solution of the same concentration. It is an empirical parameter introduced after the works of Geissler, Hecht and Horkay. A typical value for r' is 0.7

R_c ratio of the number of crosslinking molecules at the time of the reaction over the number of monomer units. Sometimes called 'crosslinking density'. This name is slightly misleading because the crosslinking reaction is liable to be incomplete

$\mathbf{R}$ instantaneous position of a point, whose mean position is $\mathbf{r}$ ($\mathbf{R} = \mathbf{r} + \mathbf{u}$), in Onuki's theory

σ stress (force per unit area of the deformed system)

σ_0 'nominal' stress (force per unit undeformed area)

Σ scattering cross-section (not to be confused with the summation sign)

$\mathbf{u}(\mathbf{r})$ instantaneous displacement of a point with respect to its average position $\mathbf{r}$

v_a volume of the scattering molecule (or monomer unit) of type a.

v_e electrostatic excluded volume of polyelectrolytes

v_1 volume of a solvent molecule (in general identified with a^3

V scattering volume

V_d volume of a 'dry' network (before swelling, or after complete deswelling of a gel)

w third virial coefficient
ξ correlation length of concentration fluctuations
Ξ correlation distance associated with quenched fluctuations in the Geissler Geissler *et al* approach.

Acknowledgements

We have benefited of many fruitful discussions with H. Benoît, F. Boué, J. F. Joanny, L. Leibler, E. Mendes, J. P. Munch, Y. Rabin, C. Rouf. We would like to thank them very much.

References

Alexander, S. (1984) *J. Phys. France* **45**, 1939
Alexander, S. (1985) in *Physics of Finely Divided Matter*, Springer Series in Pysics #5, Daoud, M, and Boccara, N, Eds. Springer Verlag Berlin
Antonietti, M., Sillescu, H. (1985) *Macromolecules* **18**, 1162
Bacon, G. E. (1955) *Neutron Diffraction*, Clarendon Press, Oxford
Barea, J. L., Muller, R., Picot, C. (1988) in *Polymer Motions in Dense Systems*, Proceedings in Physics, Vol. 29, Richter, D. and Springer, T. Springer-Verlag, Berlin; p. 86
Bastide, J., Candau, S. J., Leibler, L. (1980) *Macromolecules* **14**, 719
Bastide, J., Picot, C., Candau, S. J. (1981) *J. Macromol. Sci.:Phys. Ed.* **B19**, 13
Bastide, J., Duplessix, R., Picot, C., Candau, S. J. (1984) *Macromolecules* **17**, 83
Bastide, J. (1985a) in *Physics of Finely Divided Matter*, Springer Series in Physics Vol. 5, eds Daoud, M. and Boccara, N., Springer-Verlag, Berlin, p. 148
Bastide, J., Herz, J., Boué, F. (1985) *J. Physique* **46**, 1967
Bastide, J. (1985c) Thèse d'Etat, Université Louis Pasteur, Strasbourg
Bastide, J., Boué, F. (1986) *Physica* **140A**, 251
Bastide, J., Leibler, L. (1988) *Macromolecules* **21**, 2647
Bastide, J., Buzier, M., Boué, F. (1988) in *Polymer Motions in Dense Systems*, Proceedings in Physics **29**, eds Richter, D. and Springer, T. (Springer-Verlag, Berlin); p. 112
Bastide, J., Boué, F., Buzier, M. (1989) in *Molecular Basis of Polymer Networks*, Baumgärtner, A, and Picot, C, Eds. *Proceedings in Physics* **42**, Springer, Berlin p. 48
Bastide, J., Leibler, L., Prost, J. (1990a) *Macromolecules*, **23**, 1821
Bastide, J., Mendes, E., Boué, F., Buzier, M. (1990b) *Makromol. Chem., Makromol. Symp.*, **40**, 81
Bastide, J., Boué, F., Mendes, E., Zielinski, F., Buzier, M., Beinert, G., Oeser, R. and Lartigue, C. (1992a) *Polymer Networks '91*, Dušek and Kuchanov Eds., p. 119 VSP, The Netherlands.
Bastide, J., Boué, F., Oeser, F., Mendes, E., Zielinski, F., Buzier, M., Lartigue, C., Lindner, P. (1992b) *Mat. Res. Soc. Symp.* **248**, 313
Bastide, J., Boué, F., Mendes, E., Zielinski, F., Buzier, M., Lartigue, C., Oeser, R., Lindner, P. (1993) *Prog. Coll. Polym. Sci.* **91**, 105
Belkebir-Mrani, A., Beinert, G., Herz, J., Mathis, A. (1976) *Eur. Polym. J.* **12**, 243
Belkebir-Mrani, A., Beinert, G., Herz, J., Rempp, P. (1977) *Eur. Polym. J.* **13**, 277
Beltzung, M., Herz, J., Picot, C., (1983) *Macromolecules* **16**, 580
Beltzung, M., Herz, J., Picot, C., (1984) *Macromolecules* **17**, 663
Benoît, H., Picot, C. (1966) *Pure Appl. Chem.* **12**, 545
Benoît, H., Decker, D., Duplessix, R., Picot, C., Rempp, P., Cotton, J. P., Farnoux, B., Jannink, G., Ober, R. (1976) *J. Polym. Sci. Phys. Ed.* **14**, 2119

Benoît, H., Benmouna, M. (1984) *Polymer* **25**, 1059
Bonse, V., Hart, M. (1966) *Zeit. Phys.* **189**, 151
Borue, V., Erukhimovich, I. (1988) *Macromolecules* **21**, 3240
Bouchaud, E., Delsanti, M., Adam, M., Daoud, M., Durand, D. (1986) *J. Physique (France)* **47**, 1273
Boué, F. (1987a) *Adv. Polym. Sci.* **82**, 421
Boué, F., Bastide, J., Buzier, M., Collette, C. Lapp, A., Herz, (1987b) *J. Prog. Col. Polym. Sci.* **75**, 152
Boué, F., Edwards, S. F., Vilgis, T. A. (1988) *J. Phys. (France)* **49**, 1635
Boué, F., Bastide, J., Boué, F., Bastide, J., Buzier, M. (1989) in *Molecular Basis of Polymer Networks*, Baumgärtner, A., and Picot, C., Eds. *Proceedings in Physics* **42**, Springer-Verlag, Berlin p. 65
Boué, F., Bastide, J., Buzier, M., Lapp, A., Herz, J., Vilgis, T. A. (1991) *Colloid Polym. Sci.* **269**, 195
Boué, F., Lindner, P. (1994) *Europhysics Letters* **25**, 421
Boué, F. (1995) in *Molecular Characterization of Polymer Networks*, Ed., Galiatsatos V Oxford University Press, to be published
Brochard F., de Gennes, P. G. (1977) *Macromolecules* **10**, 1157
Brochard, F. (1981) *J. Physique (Paris)* **42**, 505
Broseta, D., Leibler, L., Lapp, A., Strazielle, C. (1986) *Europhysics Letters* **2**, 733
Brotzmann, R. W., Eichinger, B. E. (1981) *Macromolecules* **14**, 1445
Brotzmann, R. W., Eichinger, B. E. (1982) *Macromolecules* **15**, 531
Brotzmann, R. W., Eichinger, B. E. (1983) *Macromolecules* **16**, 1131
Bruinsma, R., Rabin, Y. (1994) *Phys. Rev. E* **49**, 554
Brumberger, H. (1966) in *Small Angle Scattering*. Gordon and Breach, New York, p. 121
Bueche, F. (1970) *J. Col. Interf. Sci.* **33**, 61
Cabane, B. (1986) in *Surfactant Solutions: New Methods of Investigation*, Ed. Zana R. Surfactant Series Vol. 2, M. Dekker N.Y.
Candau, S. J., Young, C. Y., Tanaka, T., Lemarechal, P., Bastide, J. (1979) *J. Chem. Phys.* **70**, 4694
Candau, S. J., Bastide, J., Delsanti, M. (1982) *Adv. Polym. Sci.* **44**, 27
Candau, S. J., Ilmain, F., Schosseler, F., Bastide, J. (1990) *Mat. Res. Soc. Symp.* **177**, 3
des Cloizeaux, J. (1975) *J. Phys. Paris* **36**, 281
des Cloizeaux, J., Jannink, G. (1987) in *Les Polymères en solution*, Les Editions de Physique, Les Ulis
Clough, S., Machonnachie, A., Allen, G. (1980) *Macromolecules* **13**, 774
Cohen, Y., Ramon, O., Kipelman, I. J., Misrahi, S. (1992) *J. Polym. Sci., Part B*, **30**, 1055
Collette, C., Lafuma, F., Audebert, R., Leibler, L. (1988) in *Biological and Synthetic Polymer Networks* O. Kramer Ed. Elsevier Applied Science, Amsterdam, p. 277
Cotton, J. P. (1991) in *Neutron, X-ray Light Scattering*, Lindner P., and Zemb, Th. (Eds) Delta Series North-Holland Elsevier Science Publishers, B. V. Amsterdam
Daoud, M., Cottom, J. P., Farnoux, B., Jannink, G., Sarma, G., Benoît, H., Duplessix, R., Picot, C., de Gennes, P. G. (1975) *Macromolecules* **8**, 804
Daoud, M., Jannink, G. (1976) *J. Phys. (Les Ulis, Fr.)* **37**, 973
Daoud, M. (1979) *J. Phys. Lett. (Les Ulis Fr)* **40**, L201
Daoud, M., Family, F., Jannink, G. (1984) *J. Phys. Lett. (Les Ulis Fr)* **45**, 199
Daoud, M., Bouchaud, E., Jannink, G. (1986) *Macromolecules* **23**, 1955
Daoud, M., Leibler, L. (1988) *Macromolecules* **21**, 1497
Dautzenberg, H. (1972) *J. Polym. Sci. Part C* **39**, 123
Davidson, N. S., Richards, W. R. (1986a) *Macromolecules* **19**, 2576
Davidson, N. S., Richards, W. R., Maconnachie, A. (1986b) *Macromolecules* **19**, 434

Deam, R. T., Edwards, S. F. (1976) *Phil. Trans. R. Soc. London Ser. A* **280**, 317
Debye, P., Bueche, A. M. (1949) *J. Appl. Phys.* **20**, 518
Debye, P., Anderson, H. R., Brumberger, H. (1957) *J. Appl. Phys.* **28**, 679
Dixon, P. K., Pine, D. J., Wu, X.-L. (1992) *Mat. Res. Soc. Symp.* **248**, 127
Doi, M., Edwards, S. F. (1986) *Theory of Polymer Dynamics*, Clarendon Press, Oxford
Duplessix, R. (1975) Thèse d'Etat, Université Louis Pasteur, Strasbourg France
Dušek, K., Prins, W. (1969) *Adv. Polym. Sci.* **6**, 1
Dušek, K. (1971) in *Polymer Networks: Structural and Mechanical Properties*, Eds. Chompff, A. J. and Newman, S, Plenum Press New York, p. 245
Edwards, S. F. (1966) *Proc. Phys. Soc., London* **88**, 265
Erman, N., Flory, P. (1978) *J. Chem. Phys.* **68**, 5363
Eliassaf, J., Silberberg, A. (1962) *Polymer* **3**, 555
Falcao, A. N., Skov Pedersen, J., Mortensen, K. (1993) *Macromolecules* **26**, 5350
Feigin, L. A., Svergun, D. I. (1987) in *Structure Analysis by Small Angle X-Ray and Neutron Scattering*, Plenum Press, N.Y.
Fernandez, A. M., Sperling, L. H., Wignall, G. D. (1986) *Macromolecules* **19**, 2572
Flory, P. J., Rehner, J. (1943) *J. Chem. Phys.* **11**, 521
Flory, P. J. (1953) *Principles of Polymer Chemistry*, Cornell Univ. Press, Ithaca
Flory, P. J. (1976) *Proc. R. Soc. London* **A351**, 351
Flory, P. J., Erman, B. (1982) *Macromolecules* **15**, 800
Frenkel, J. (1938) *Acta Physicochimica USSR* **9**, 235
Frenkel, J. (1940) *Rubber Chem. Tech.* **13**, 264
Freltoft, T., Kjems, J. K., Sinha, S. K. (1986) *Phys. Rev. B* **33**, 269
Geissler, E., Hecht, A. M. (1976) *J. Chem. Phys.* **65**, 103
Geissler, E., Hecht, A. M., Duplessix, R. (1982) *J. Polym. Sci., Polym. Phys. Ed.* **20**, 225
Geissler, E., Duplessix, R., Hecht, A. M. (1983) *Macromolecules* **16**, 712
Geissler, E., Hecht, A. M., Horkay, F., Zrinyi, M. (1988) *Macromolecules* **21**, 2594
Geissler, E., Hecht, A. M., Mallam, S., Horkay, F., Zrinyi, M. (1990) *Makromol. Chem., Macromol. Symp.* **40**, 101
Geissler, E., Horkay, F., Hecht, A. M. (1991) *Macromolecules* **24**, 6006
Geissler, E., Hecht, A. M., Horkay, F., Legrand, J. F. (1992) *Polymer* **33**, 3083 and refs there in.
Geissler, E., Horkay, F., Hecht, A. M. (1993a) *Phys. Rev. Lett.* **71**, 645
Geissler, E. (1993b) in *Dynamic Light Scattering* Monographs on the Physics and Chemistry of Materials-49, W. Brown Ed. Oxford Science Publication
Geissler, E., Horkay, F., Hecht, A. M. (1994) *J. Chem. Phys.* **100**, 8418
de Gennes, P. G. (1976) *Macromolecules* **9**, 594
de Gennes, P. G. (1979) *Scaling Concepts in Polymer Physics*, Cornell Univ. Press, Ithaca
Gillies, R., Kazmerczak, M., Keyes, T., Wostowicz, A. (1987) *Phys. Rev. B*, **36**, 9413
Goodman, J. W. (1964) in *Laser Speckle and Related Phenomena*, ed. by Dainty J. C. (Springer-Verlag, Berlin)
Graessley, W. (1975) *Macromolecules* **8**, 186 and **8**, 865
Guenet, J. M., Willmoth, N. F., Ellsmore, P. A. (1983) *Polym. Commun.*, **24**, 230
Guenoun, P., Perrot, F., Beysens, D. (1989) *Physical Review Letters*, **63**, 1152
Guinier, A., Fournet, G. (1957) in *Small Angle Scattering of X-rays*, Wiley N.Y.
Hakiki, A. (1991) Thesis, Université Louis Pasteur, Strasbourg, France
Hashimoto, T., Fujioka, K. (1991) *J. Phys. Soc. Japan* **60**, 356
Hashimoto, T., Kume, T. (1992) *J. Phys. Soc. Japan* **61**, 1839
Hecht, A. M., Geissler, E. (1983) *Polym. Commun.* **24**, 98
Hecht, A. M., Duplessix, R., Geissler, E. (1985) *Macromolecules* **18**, 2167
Hecht, A. M., Horkay, F., Geissler, E., Benoît, J. P. (1991) *Macromolecules* **24**, 4183

Hecht, A. M., Guillermo, A., Horkay, F., Mallam, S., Legrand, J. F., Geissler, E. (1992) *Macromolecules* **25**, 3677

Hecht, A. M., Horkay, F., Mallam, S., Geissler, E. (1993) *Makromol. Chem., Makromol. Symp.* **76**, 103–112

Helfand, E., Frederickson, G. H. (1989) *Phys. Rev. Lett.* **62**, 2468

Herz, J., Weiss, P., Rempp, P. (1968) *C.R.A.S.* **266**, 681

Higgins, J. S., Benoît, H. C. (1994) *Polymers and Neutron Scattering*, Oxford Series on Neutron Scattering in Condensed Matter, Clarendon Press, Oxford

Hinckley, J. A., Han, C. C., Moxer, B., Yu, H. (1978) *Macromolecules* **11**, 836

Horkay, F., Zrinyi, M. (1988) *Macromolecules* **21**, 3260

Horkay, F., Hecht, A. M., Geissler, E. (1989) *J. Chem. Phys.* **91**, 2706

Horkay, F., Hecht, A. M., Mallam, S., Geissler, E., Rennie, A. R. (1991a) *Macromolecules* **24**, 2896

Horkay, F., Zrinyi, M., Hecht, A. M., Geissler, E., Pruvost, P. (1991b) *Polymer* **32**, 835

Horkay, F., Geissler, E., Hecht, A. M. (1992) in *Synthesis, Characterization and Theory of Polymeric networks and Gels* SM Aharoni Ed. Plenum Press, New York and London, p. 303

Horkay, F., Burchard, W., Geissler, E., Hecht, A. M. (1993a) *Macromolecules* **26**, 1296

Horkay, F., Burchard, W., Hecht, A. M., Geissler, E. (1993b) *Macromolecules* **26**, 4203

Horkay, F., Burchard, W., Hecht, A. M., Geissler, E. (1993c) *Macromolecules* **26**, 3375

Horkay, F., Hecht, A. M., Geissler, E. (1994) *Macromolecules* **27**, 1795

Ilavski, M., Bouchal, K., Dušek, K. (1989) *Makromol. Chem.* **190**, 883

Ilmain, F. (1991) Thesis, Université Louis Pasteur

Inoue, T., Moritani, M., Hashimoto, T., Kaway, H. (1971) *Macromolecules* **4**, 500

James, H., Guth, E. (1943) *J. Chem. Phys.* **11**, 455

James, H., Guth, E. (1947) *J. Chem. Phys.* **15**, 669

James, H. Guth, E. (1949) *J. Polym. Sci.* **4**, 153

James, H., Guth, E. (1953) *J. Chem. Phys.* **21**, 1039

Joanny, J. F., Pincus, P. (1980) *Polymer* **21**, 274

Joanny, J. F., Leibler, L. (1990) *J. Phys. France* **51**, 545

Joosten, J. G. H., McCarthy, J. L., Pusey, P. N. (1991) *Macromolecules* **24**, 6690

Joosten, J. G. H., Geladé, E. T. H., Pusey, P. N. (1990) *Phys. Rev. A*, **42**, 2161

Kempkes (1995) Thesis, Université Louis Pasteur, Strasbourg, France

Koberstein, J. T., Picot, C., Benoît, H. (1985) *Polymer 1985* **26**, 673

Kuhn, W. (1936) *Kolloid Zeitschrift* **76**, 258

Kuhn, W. (1939) *Kolloid Zeitschrift* **87**, 3

Kume, T., Asakawa, K., Moses, E., Matsuzaka, K., Hashimoto, T. (1995) *Acta Polymer* **46**, 79

Lal, J., Bastide, J., Bansil, R., Boué, F. (1993) *Macromolecules* **26**, 6092

Landau, L. D., Lifshitz, E. M. (1973) Theory of Elasticity, Pergamon New York

Lapp, A., Picot, C., Strazielle, C. (1985) *J. Physique Lett.* (*Les Ulis, Fr.*) **46**, 1031

Lutz, P., Picot, C., Hild, G., Rempp, P. (1976) *British Polym. J.* **9**, 151

Maconnachie, A., Richards, R. W. (1978) *Polymer* **19**, 739

Mallam, S., Hecht, A. M., Geissler, E., Pruvost, P. (1989a) *J. Chem. Phys.* **91**, 6447

Mallam, S., Horkay, F., Hecht, A. M., Geissler, E. (1989b) *Macromolecules* **22**, 3356

Mallam, S., Horkay, F., Hecht, F., Rennie, A., Geissler, E. (1991) *Macromolecules* **24**, 543

Mandelbrot, B. (1977) *Fractals, Form, Chance and Dimensions* Freeman, San Franciso

Mark, J., Erman, B. (1988) *Rubberlike Elasticity, a Molecular Primer*, John Wiley, New York

McKenna, G. B., Flynn, K. M., Chen, Y. (1989) in *Molecular Basis of Polymer Networks*, Eds Baumgärtner, A. and Picot, C., *Proceedings in Physics* **42**, Springer Berlin, p. 127

McKenna, G. B., Flynn, K. M., Chen, Y. (1990) *Polymer* **31**, 1937
Medjahdi, G., Sarazin, D., François, J. (1991) *Macromolecules* **24**, 4138
Mendes, E., Lindner, P., Buzier, M., Boué, F. and Bastide, J. (1991a) *Phys. Rev. Lett.* **66**, 1595
Mendes, E. (1991b) PhD Thesis, Université Louis Pasteur Strasbourg (available on demand, contact E. Mendes, same address as S. J. Candau)
Mendes, E., Boué, F., Bastide, J. (1991c) unpublished data
Mendes, E., Hakiki, A., Ramzi, A., Herz, J., Schosseler, F., Munch, J. P., Boué, F., Bastide, J. (1993a) in *Polymeric Materials Science and Engineering* (*PMSE*) **68**, 231
Mendes, E., Girard, B., Picot, C., Buzier, M., Boué, F., Bastide, J. (1993b) *Macromolecules* **26**, 6873
Mendes, E., Lutz, P., Bastide, J., Boué, F. (1995a) *Macromolecules* **28**, 174
Mendes, E., Schosseler, F., Isel, F., Boué, F., Bastide, J. Candau, S. J. (1995b) *Europhysics Letters*, **32**, 273
Mendes, E. et al (1995c) in preparation
Milner, S. T. (1991) *Phys. Rev. Lett.* **66**, 1477
Moussaid, A., Munch, J. P., Schosseler, F., Candau, S. J. (1991) *J. Phys. II* **1**, 637
Moussaid, A., Schosseler, F., Munch, J. P., Candau, S. J. (1993) *J. Phys. II France* **3**, 573
Moussaid, A., Candau, S. J., Joosten, J. G. H. (1994) *Macromolecules* **27**, 2102
Moussaïd, A., Slot, H. J. M., Joosten, J. G. H. (1995) *Int. J. Pol. Anal. Char.*, in press
Munch, J. P., Candau, S. J., Duplessix, R., Picot, C., Benoît, H. (1974) *J. Phys. Lett. Fr.* **35**, 239
Munch, J. P., Candau, S., Herz, J., Hild, G. (1977) *J. Physique* (*France*) **38**, 971
Munch, J. P., Lemarechal, P., Candau, S. J., Herz, J. (1977) *J. Physique* (*France*) **38**, 1499
Nagy, M., Horkay, F. (1980) *Acta Chim. Acad. Sci. Hung.* **104**, 4
Obukhov, S. P., Rubinstein, M., Colby, R. H. (1994) *Macromolecules* **27**, 3191
Oeser, R., Picot, C., Herz, J. (1988) in *Polymer Motions in Dense Systems*, Proceedings in Physics **29**, Eds Richter D. and Springer T.; (Springer-Verlag Berlin); p. 104
Olmsted, P. D., Milner, S. T. (1994) *Macromolecules* **27**, 6648
Onuki, A. (1988) in *Space Time Organization in Macromolecular Fluids* eds Tanaka F., Doi, M. Chemical Physics **51**, Springer Berlin Heidelberg New-York, p. 94
Onuki, A. (1992) *J. Phys. II France* **2**, 45
Onuki, A. (1993) *Adv. Polym. Sci.* **109**, 63
Panyukov, S. V. (1989) *Zh. Eksp. Teor. Fiz.* **96**, 604 [*Sov. Phys. JETP* **69**, 342 (1989)]
Panyukov, S. V. (1990a) *Zh. Eksp. Teor. Fiz.* **98**, 668 (*Sov. Phys. JETP* **71**, 372 (1990)]
Panyukov, S. V. (1990b) *Pis'ma Zh. Eksp. Teor. Fiz.* **51**, 223 [*JETP Lett.* **71**, 253 (1990)]
Panyukov, S. V., Kuchanov, S. I. (1991) *Zh. Eksp. Teor. Fiz.* **99**, 643 [*Sov. Phys. JETP* **72**, 368 (1991)]
Panyukov, S. V. (1992a) *Sov. Phys. JETP* **75**, 347
Panyukov, S., Rabin, Y., Feigel, A. (1994) *Europhys. Lett.* **28**, 149
Panyukov, S., Rabin, Y. (1995) to be published in Physics Reports
Pines, E., Prins, W. (1972) *J. Polym. Sci. B* **10**, 719
Prins, W. (1971) in *Polymer Networks: Structural and Mechanical Properties*, Eds Chompff, A. J. and Newman, S. Plenum Press New-York, p. 323
Porod, G. (1982) in *Small Angle Scattering of X-rays* Eds Glatter, O. and Kratky, O., Academic Press, London
Pusey, P. N. (1974) in *Photon Correlation and Light Beating Spectroscopy*, Eds Cummins, H. Z., Pike, E. R., Plenum, N.Y.
Pusey, P. N. Van Megen, W. (1989) *Physica A.* **157**, 705
Rabin, Y., Bruinsma, R. (1992) *Europhys. Lett.* **20**, 79
Rabin, Y., Pekarski, P., Bruinsma, R. (1993a) in *Polymeric Materials Science and Engineering* (*PMSE*) **68**, 229

Rabin, Y., Pekarski, P., Bruinsma, R. (1993b) *Europhys. Lett.* **24**, 145
Ragnetti, M., Geiser, D., Höcker, H., Oberthur, R. C. (1985) *Makromol. Chem.* **186**, 1701
Ramzi, A., Mendes, E., Zielinski, F., Rouf, C., Hakiki, A., Herz, J., Oeser, R., Boué, F., Bastide, J. (1993) *J. Phys. IV France* **3**, 91
Ramzi, A., (1994) Thesis Université Pierre et Marie Curie Paris VI,
Ramzi, A., Zielinski, F. Bastide, J., Boué, F. (1995) *Macromolecules* **28**, 3570
Rangel-Nafaile, C., Metzner, A. B., Wissbrun, K. F. (1984) *Macromolecules* **17**, 187
Rawiso, M., Duplessix, R., Picot, C. (1987) *Macromolecules* **20**, 631
Reed, W., Gosh, S., Medjahdi, G., François, J. (1991) *Macromolecules* **24**, 6189
Rharbi, Y., Joannicot, M., Cabane, B., Boué, F. (1995) submitted
Richter, D., Jucknischke, O., Willner, L., Fetters, L. J., Lin, M., Huang, J. S., Roovers, J. Toporowski, P., Zhou, L. L. (1992) ACS-PMSE Symposium Neutron Scattering from Polymers. American Chemical Society, Washington D.C.
Rouf, C., Bastide, J., Pujol, J. M., Schosseler, F., Munch, J. P. (1994) *Phys. Rev. Lett.* **73**, 830
Rouf, C. (1995) Thesis, Université Louis Pasteur, Strasbourg, France
Rubinstein, M., Leibler, L., Bastide, J. (1992) *Phys. Rev. Lett.* **68**, 405
Rubinstein, M., Ajdari, A., Bastide, J., Leibler, L. (1992) *Makromol. Chem., Makromol. Symp.* **62**, 61
Shibayama, M., Tanaka, T., Han, C. C. (1992a) *J. Chem. Phys.* **97**, 6829
Shibayama, M., Tanaka, T., Han, C. C. (1992b) *J. Chem. Phys.* **97**, 6842
Shibayama, M., Muthukumar, M., Kurokawa, H. Nomura, S. Stein, R. S. (1992) *Polymer* **33**, 2883
Shibayama, M., Tanaka, T. (1993a) *Adv. Polym. Sci.* **109**, 1–62
Shibayama, M., Tanaka, T., Han, C. C. (1993b) *J. Phys. IV France Colloque C8* **3**, 25
Shibayama, M., Takahashi, H., Nomura, S. (1995) preprint
Schmidt, R., Wolf, B. A. (1979) *Colloid Polym. Sci.* **257**, 1187
Schosseler, F., Ilmain, F., Candau, S. J. (1991) *Macromolecules* **24**, 225
Schosseler, F., Moussaid, A., Munch, J. P., Candau, S. J. (1991) *J. Phys. II France* **1**, 1197
Schosseler, F., Skouri, R., Munch, J. P., Candau, S. J. (1994) *J. Phys. II France* **4**, 1221
Sellen, D. B. (1987) *J. Polym. Sci., Polym. Phys. Ed.*, **25**, 699
Silberberg, A., Eliassaf, J., Katchlasky, A. (1957) *J. Polym. Sci.* **23**, 259
Silberberg, A. (1988) in *Biological and Synthetic Polymer Networks* Ed. Kramer, O. Elsevier Applied Science, Amsterdam
Silberberg, A. (1992) in *Polyelectrolyte Gels*, eds Harland, R. S. and Prud'homme, R. K. ACS Symposium Series 480, Washington D.C., p. 159
Skouri, R., Munch, J. P., Schosseler, F., Candau, S. J. (1993) *Europhys. Lett.* **23**, 635
Skouri, R. (1994) PhD Thesis Université Louis Pasteur Strasbourg
Skouri, R., Schosseler, F., Munch, J. P., Candau, S. J. (1995) *Macromolecules* **28**, 197
Soni, V. K., Stein, R. S. (1990) *Macromolecules* **23**, 5257
Stauffer, D., Coniglio, A., Adam, M. (1982) *Adv. Polym. Sci.* **44**, 105
Stauffer, D. (1985) Introduction to Percolation Theory, Taylor and Francis, London and Philadelphia
Stavermann, A. J. (1982) *Adv. Polym. Sci.* **44**, 73
Stein, R. S. (1969) *J. Polym. Sci. B* **7**, 657
Takebe, T., Nawa, K., Suehiro, S., Hashimoto, T. (1989) *J. Chem. Phys.* **91**, 4360
Tanaka, T., Hocker, L. O., Benedek, G. B. (1973) *J. Chem. Phys.* **59**, 5151
Texeira, J. (1985) in *On Growth and Form* Eds, Stanley, H. E., Ostrowski, N. Martinus Nijhoff, Boston
Texeira, J. (1988) *J. Appl. Cristallogr.* **21**, 781
Treloar, L. R. G. (1943) *Trans. Faraday Soc.* **34**, 36

Van Egmond, J. W., Werner, D. E., Fuller, G. G. (1992) *J. Chem. Phys.* **96**, 7742
Ver Strate, G., Phillippoff, W. (1974) *J. Polym. Sci. Polym. Lett. Ed.* **12**, 267
Vilgis, T. A., Boué, F., Edwards, S. F. (1989) in *Molecular Basis of Polymer Networks*, Eds. Baumgärtner and Picot *Proceedings in Physics* **42**, Springer Berlin, p. 170
Wall, F. T. (1942) *J. Chem. Phys.* **10**, 132 and **10, 185**
Wall, F. T. (1943) *J. Chem. Phys.* **11**, 527
Wall, F. T., Flory, P. J. (1951) *J. Chem. Phys.* **19**, 1435
Warner, M., Edwards, S. F. (1978) *J. Phys.* **A11**, 1649
Weiss, P., Hild, G., Herz, J., Rempp, P. (1970) *Makromol. Chem.* **135**, 249
Weiss, N., Silberberg, A. (1977) *Br. Polym. J* **9**, 144
Weiss, N., Van Vliet, T., Silberberg, A. (1979) *J. Polym. Sci. Pol. Phys. Ed.* **17**, 2229
Witten, T. A., Pincus, P. A., Cates, M. E. (1986) *Europhysics Lett.* **2**, 137
Wu, X.-L., Pine, D. J., Dixon, P. K. (1991) *Phys. Rev. Lett.* **66**, 2408
Wun, K. L., Prins, W. (1974) *J. Pol. Sci. Pol. Phys. Ed.* **12**, 533
Xue, J. Z., Wu, X. L., Milner, S., Cheakin, P. (1991) *Phys. Rev. A.* **46**, 6550
Yen, L. Y., Eichinger, B. E. (1978) *J. Polym. Sci. Polym. Phys. Ed.* **16**, 121
Yu, H., Kitano, T., Kim, C. Y., Amis, E. J., Chang, T., Landry, M. R., Wesson, J. A. Han, C. C., Lodge, T. P., Glinka, C. J. (1986) in *Advances in Elastomers and Rubber Elasticity*, Eds Lal, J., Mark, J. E. Plenum Press, New York
Zielinski, F. (1991) PhD Thesis Université Pierre et Marie Curie Paris VI
Zielinski, F., Buzier, M., Lartigue, C., Bastide, J., Boué, F. (1992) *Prog. Coll. Sci.* **90**, 115
Zrinyi, F., Horkay, F. (1980) *Polymer Bulletin* **3**, 665

Appendix to Chapter 5

A.1 Swelling and elasticity of gels: a summary of the classical description

As stated in the introduction, the physics of gels was originally approached by considering gels as swollen rubbers (Frenkel, 1938, 1940; Flory and Rehner 1943, see also: Flory 1953, Dušek and Prins, 1969 and Mark and Erman, 1988). The basic assumption of the description, introduced originally by Frenkel, is that the swelling equilibrium of a gel results from a balance between the 'osmotic' interactions between chains and a tridimensional elastic retraction having the same 'entropic' origin as the uniaxial modulus of a rubber. As pointed out in particular by Deam and Edwards (1976), these opposite effects cannot be completely independent: for instance, the chain deformation in a network should necessarily modify the repulsions between chains. However, for simplicity, such couplings, or more sophisticated ones, are frequently ignored. Most authors assume a statistical independence of mixing and deformation effects and express it in the form of an *additivity assumption* for the total free energy F of a gel (for a more detailed discussion see also McKenna *et al*, 1989 and refs therein). In other words, F is generally written as follows:

$$F = F_{\text{mix}}(\phi) + F_{\text{el}}(\Lambda_x, \Lambda_y, \Lambda_z) \tag{A.1}$$

where F_{mix} corresponds to the mixing of the polymer chains constituting the network with a swelling agent, in general a low molecular weight solvent. As before, ϕ is the polymer volume fraction. In the original theories, regarding the mixing properties, a network was identified with a semi-dilute solution of infinite chains and F_{mix} was expressed through the Flory–Huggins theory. Since the latter is a lattice model, it is often convenient to deal with a free energy density F' per site. The size of the lattice site, of volume a^3, is generally identified with the size of a solvent molecule, of volume v_1 (strictly, one should speak of a free enthalpy). We recall the classical result by Huggins and Flory:

$$\frac{F'_{\text{FH}}}{k_B T} = (\phi/N)\ln\phi + (1-\phi)\ln(1-\phi) + \chi\phi(1-\phi) \tag{A.2}$$

The Physical Properties of Polymeric Gels. Edited by J. P. Cohen Addad.

where χ is the Flory–Huggins interaction parameter and N denotes the polymerization index of the chains constituting the solution. Thus, when $N \to \infty$, as assumed in the case of a network, Eq. (A.2) reduces to

$$\frac{F'_{\mathrm{FH}}}{k_{\mathrm{B}}T} = (1-\phi)\ln(1-\phi) + \chi\phi(1-\phi) \tag{A.3}$$

As for $F_{\mathrm{el}}(\Lambda_x, \Lambda_y, \Lambda_z)$, the elastic contribution, it is a function of the deformation ratios along the principal axes of the strain. $\Lambda_x, \Lambda_y, \Lambda_z$ are expressed with respect to a *reference state* (of volume fraction ϕ_0) in which the network is assumed to be *relaxed* (i.e. in which the chains exert no tension on the crosslinks). When the 'strain' is an isotropic swelling, one assumes that there is no difference between the corresponding deformation and that arising from a true mechanical stretching. In this case, one may therefore write

$$\Lambda_x = L_x/L_x^0 \tag{A.4}$$

where L_x and L_x^0 denote the length of the gel along Ox, in the deformed state and in the reference state respectively. Obviously, $\Lambda_x\Lambda_y\Lambda_z = V/V_0 = Q/Q_0 = (\phi/\phi_0)^{-1}$ (V, Q and V_0, Q_0 denoting respectively the volume of the gel and its swelling degree first in the state of study and then in the state of preparation) and taking into account the isotropy, one obtains

$$\lambda_y = \lambda_y = \lambda_z = (\phi/\phi_0)^{-1/3} = (Q/Q_0)^{1/3} \tag{A.5}$$

If the strain is a uniaxial deformation at constant volume, one writes

$$\begin{aligned}\Lambda_x &= L_x/L_x^0 = (L_x/L_x^{\mathrm{i}}).(L_x^{\mathrm{i}}/L_x^0) \\ &= \lambda_x(Q/Q_0)^{1/3} = \lambda_x(\phi/\phi_0)^{-1/3}\end{aligned} \tag{A.6}$$

where L_x^{i} represents the length of the sample, along direction Ox, in the isotropically swollen state, that is after a change of volume of the network with respect to the reference state but before uniaxial deformation.

Let us focus more closely on the expressions which have been proposed for the elastic part of the free energy F_{el}. They are always derived from Gaussian assumptions for the chain statistics and can be put in the following form (see Dušek and Prins, 1969):

$$F_{\mathrm{el}} = A(\nu_e/2)k_{\mathrm{B}}T(\Lambda_x^2 + \Lambda_y^2 + \Lambda_z^2) + B\nu k_{\mathrm{B}}T\ln\Lambda_x\Lambda_y\Lambda_z + \text{constant} \tag{A.7}$$

where ν_e is the number of elastically effective strands in the network, that is, the number of elementary meshes effectively connected to two first neighbour crosslinks.

The calculation of F_{el} has been at the centre of vigorous controversies that we will not enter into (see James and Guth, 1953, Flory, 1976). We restrict ourselves to an oversimplified scheme, taken from Dušek and Prins (1969) and centred on the values which can be given operationally to A and B.

In the framework of the theories by Kuhn (the simplest one, 1936, 1939) or by Wall 1943, Wall and Flory (1951) and Treloar (1943), $A = 1$. Then the term

$$A(\nu_e/2)k_B T(\Lambda_x^2 + \Lambda_y^2 + \Lambda_z^2)$$

represents exactly the free energy change of a collection of statistically independent chains whose end-to-end vectors are transformed affinely in the macroscopic strain (that is, e.g. $R_x = \Lambda_x R_x^0, R_x^0$ and R_x denoting respectively the projection of a end-to-end vector along the Ox direction, before and after deformation).

According to James and Guth (1943, 1947), Deam and Edwards (1976), Graessley (1976), Staverman (1982, see refs therein), A is reduced to 1/2 for tetrafunctional networks (for $f \neq 4$, A was found to be equal to $(f-2)/f$ or $2/f$, depending on the theory). In the latter case, some additional freedom is assumed to be given to the chain: only the mean positions of the crosslinks are assumed to be affinely transformed. But because of temporal fluctuations around the mean positions, the distribution of end-to-end vectors is no longer affinely transformed. More recently Erman and Flory, 1978, Flory and Erman (1977, 1982, see also Mark and Erman 1988) argued that, depending on the degree of elementary chain interspersion, values of A intermediate between 1 and 1/2 could be found.

The situation is not simpler regarding constant B. $B\nu_e k_B T \ln \Lambda_x\Lambda_y\Lambda_z$ was introduced in order to describe the localization in space of the crosslinks, or—which is almost the same thing—the entropy loss associated with the formation of the network. A slightly different formulation can be found in Brochard 1981. According to Flory and his followers, B should be maximum (and equal to $2/f$) when the crosslinks are strongly embedded in the chain matrix and is expected to diminish when some additional freedom is given to them (if the network is swollen, or if the crosslinking density is reduced). But such a localization term does not exist in the theories of James and Guth and Deam and Edwards. According to these approaches, B should then be taken as equal to zero.

For symmetry with the treatment of the 'mixing' term, we turn again to a free energy density per site. Then, one may write $\phi = \nu_e N_c a^3/V$, thus $\nu_e/V = \phi/N_c a^3$, and since $F' = Fa^3/V$, one obtains

$$F'_{el} = A(k_B T/2)(\phi/N_c).(\Lambda_x^2 + \Lambda_y^2 + \Lambda_z^2) + B(\phi/N_c)k_B T \ln \phi/\phi_0 + \text{constant} \quad (A.8)$$

In this notation, N_c is the number of sites occupied, on average, by an elementary elastic chain of the network. Note that, in general, N_c is not equal to the number N_{ck} of Kuhn's statistical segments of the chain, but is only proportional to this quantity. In most cases N_c is not even strictly equal to the average number N_m of monomer units of the average elementary mesh since the volume of the latter rarely coincides exactly with the volume of a solvent molecule. Conversion from N_m to N_c is straightforward, whereas conversion from N_m to N_{ck} is often controversial.

Note also that, because of Eq. (A.5), $\Lambda_x \Lambda_y \Lambda_z$ is equal to ϕ/ϕ_0. It is often convenient to rewrite Eq. (A.8) after combination with Eq. (A.6) and one obtains

$$F'_{\rm el} = A(k_B T/2)(\phi^{1/3}/N_c).\phi_0^{2/3}(\lambda_x^2 + \lambda_y^2 + \lambda_z^2) + B(\phi/N)k_B T \ln \phi/\phi_0 + \text{constant} \tag{A.8$'$}$$

The non-linear dependence $\phi^{1/3}$ arises from a combination of the concentration of chains ($\sim \phi/N_c$) with the effect of their assumed extension because of the swelling (Eq. (A.6)).

A.1.1 SWELLING EQUILIBRIUM

The equations governing the swelling equilibrium are obtained by minimizing the total free energy with respect to total volume, or, this is equivalent, by equating the excess chemical potential $\Delta\mu_s$ of the solvent in the gel phase to zero. Recall that

$$\Delta\mu_s = \mu_s - \mu_s^0 = (\partial F/\partial n_1)_T$$

n_1 representing the number of solvent molecule within the gel, $\mu_s - \mu_s^0$ being respectively the chemical potential of the solvent in the gel and that of the pure solvent. It can easily be checked that $\Delta\mu_s$ can also be calculated from the free energy density per site:

$$\Delta\mu_s = [F' - \phi(\partial F'/\partial\phi)]_{T,\lambda} \tag{A.9}$$

As previously, $\Delta\mu_s$ can be decomposed in a mixing part, $\Delta\mu_{s,\rm mix}$, and an elastic part, $\Delta\mu_{s,\rm el}$, which can be written:

$$\frac{\Delta\mu_{s,\rm mix}}{k_B T} = \ln(1-\phi) + \phi + \chi\phi^2 \tag{A.10a}$$

$$\frac{\Delta\mu_{s,\rm el}}{k_B T} = \frac{A}{N_c}\phi^{1/3}\phi_0^{2/3} - \frac{B}{N_c}\phi \tag{A.10b}$$

Therefore, according to this model, the equation describing the swelling equilibrium in the more general case reads $\Delta/\mu_s = 0$, or:

$$-[\ln(1-\phi) + \phi + \chi\phi^2] + \frac{B}{N_c}\phi = \frac{A}{N_c}\phi^{1/3}\phi_0^{2/3} \tag{A.11}$$

where the terms favouring the swelling (the 'mixing' and the 'crosslinking' terms) have been gathered in the left-hand side, whereas the elastic retraction, which opposes the swelling, appears in the right-hand side.

In the framework of this theory, the equation governing the swelling equilibrium depends therefore on five parameters: A, B, ϕ_0, N_c and χ. It can be made simpler when the solvent is very 'good' and the polymer volume fraction is small enough: the log term in (A.10a) can be expanded to the second order and reduces to

$$\frac{\Delta\mu_{s,\,mix}}{k_B T} = -(1/2-\chi)\phi^2 = -u\phi^2 \qquad (A.12)$$

Thus Eq. A.11 reads:

$$u\phi^2 + \frac{B}{N_c}\phi = \frac{A}{N_c}\phi^{1/3}\phi_0^{2/3} \qquad (A.12)'$$

From this equation, together with A.10, one can easily express the osmotic pressure of the gel Π_{os}^{gel} (or 'swelling pressure' according to Geissler *et al*). Since the osmotic pressure reads $\Pi_{os} = -\Delta\mu_s v_1$, v_1 denoting, as above, the molecular volume of the solvent and $\Delta\mu_s = \Delta\mu_{s,\,mix} + \Delta\mu_{s,\,el}$, one obtains

$$\Pi_{os}^{gel} = \frac{k_B T}{v_1}\left[u\phi^2 + \frac{B}{N_c}\phi - \frac{A}{N_c}\phi^{1/3}\phi_0^{2/3}\right] \qquad (A.12)''$$

A second simplification arises when it can be assumed that the chemical mesh of the gel is large enough and the concentration ϕ_0 of the reference state not too small. Then $(B/N_c)\phi$ can be neglected and one obtains

$$\Pi_{os}^{gel} = \frac{k_B T}{v_1}\left[u\phi^2 - \frac{A}{N_c}\phi^{1/3}\phi_0^{2/3}\right] \qquad (A.13)$$

At swelling equilibrium, $\Delta\mu_s = 0$ or $\Pi_{os}^{gel} = 0$, then

$$u\phi_e^2 = \frac{A}{N_c}\phi_e^{1/3}\phi_0^{2/3}$$

or

$$\phi_e = \left(\frac{A}{u}\right)^{3/5}\phi_0^{2/5}N_c^{-3/5} \qquad (A.13)'$$

with ϕ_e representing the polymer volume fraction of the gel swollen at saturation in pure good solvent. Note that Eq. (A.13) is very similar to the calculation of the chain expansion in Flory's theory (Flory, 1953). This is not surprising since, in Flory's original treatment, the coil is precisely described like a microgel (with one elastic chain!). Here, because of the simplifications which have been introduced, any 'network effect' has vanished and the behaviour of the gel reduces to that of a collection of single chains. The fact that one deals with a network manifests itself only in the presence of the reference state, which is generally identified with the state of preparation, of concentration ϕ_c. If ϕ_c is kept fixed for a series of gels, ϕ_0 should be independent of N_c. This is why $\phi_e \sim N_c^{-3/5}$, and why one does not obtains, by this method, the 'c^* theorem', which would read $\phi_e \sim \phi^* \sim N_c^{-4/5}$ (de Gennes 1979). The latter result can, however, be obtained by describing the gel as a collection of independent chains, but after identifying ϕ_0 with the ϕ^* concentration of the elementary meshes in their Gaussian undeformed state. Then one indeed obtains $\phi_0 \sim N_c^{-1/2}$, thus $\phi_0^{2/5} N_c^{-3/5} \sim N^{-4/5}$. In short, the Flory result is equivalent to the c^* theorem only if the gels are assumed to be synthesized at different concentrations $\phi_c = \phi_0$, varying with the average chain length as $N_c^{-1/2}$. Another option would be to assume that the reference state does not coincide with the preparation state and that, for some reason, $\phi_0 \sim N^{-1/2}$ even if ϕ_c is kept fixed (for more details, see Bastide, Candau and Picot, 1981).

A.1.2 MECHANICAL PROPERTIES: STRESS/STRAIN ISOTHERMS

From the same model, it is easy to calculate the retractive force f_x associated with a deformation, at constant temperature, in the Ox direction, from L_x^i to L_x (i.e. by a deformation ratio λ_x). It is written

$$f_x = (\partial F/\partial L_x)_T \tag{A.14}$$

or

$$f_x = 1/L_x^i(\partial F/\partial \lambda_x)_T$$

As previously, we assume $F = F'.V/v_1$.

Let us consider the 'nominal stress' σ_0, that is, f_x divided by the surface area S_{yz}^i, before deformation, in the direction perpendicular to the x-axis:

$$\sigma_0 = \frac{1}{S_{yz}^i L_x^i}(\partial F/\partial \lambda_x)_T = (1/V).(\partial F/\partial \lambda_x)_T$$

$$= (1/v_1).(\partial F'/\partial \lambda_x)_T \tag{A.15}$$

Usually, a gel deswells under uniaxial compression and swells under uniaxial extension (see Dušek and Prins, 1969, Horkey-Zrinyi 1988). But if the samples are large enough, these changes of volume take time and it is, in general, possible to measure a stress corresponding approximately to a deformation at constant volume. Here, we restrict ourselves to this simple case. Then F_{mix} as well as the controversial 'crosslinking term' $B(\phi/N)k_B T \ln \phi/\phi_0$ can be considered as a constant, and one obtains

$$\sigma_0 = (A/2).(k_B T/v_1)(\phi/N_c).(\phi/\phi_0)^{-2/3}\left(\frac{\partial(\lambda_x^2 + \lambda_y^2 + \lambda_z^2)}{\partial \lambda_x}\right)_T \tag{A.16}$$

where, using Eq. (A.6), the Λ_i terms have been expressed as a function of ϕ, ϕ_0 and the λ_i terms. Note that, by construction, $N_c v_1$ is the molecular volume of the average elementary elastic mesh of the network.

Again because the deformation is assumed to be performed at constant volume, with respect to an 'isotropic state', corresponding to a swelling by $(\phi/\phi_0)^{-1}$ with respect to the reference state, one has $\lambda_y^2 = \lambda_z^2 = 1/\lambda_x$, and therefore

$$\sigma_0 = A\left(\frac{k_B T}{v_1}\right)\frac{\phi^{1/3}}{N_c}\phi_0^{2/3}(\lambda_x - 1/\lambda_x^2) \tag{A.17}$$

In general, for sufficiently swollen gels, the retractive force varies linearly with $(\lambda_x - 1/\lambda_x^2)$. Therefore, one currently defines a modulus of uniaxial deformation G as $\sigma_0/(\lambda_x - 1/\lambda_x^2)$, which is therefore expected to be equal to

$$G = A\left(\frac{k_B T}{v_1}\right)\frac{\phi^{1/3}}{N_c}\phi_0^{2/3} \tag{A.18}$$

It can be easily shown, by an expansion at small λ_x, that G is equal to $E/3$, E being the Young's modulus of the system. If the deformation is truly performed at constant volume, G is then equal to the shear modulus μ, since, in this case, one has also $\mu = E/3$. Here μ should not be confused with the chemical potential of the solvent μ_s.

Equation (A.17) should naturally apply for any ϕ, including $\phi = 1$, that is for bulk networks. A bit surprisingly, deviations from linearity with $(\lambda_x - 1/\lambda_x^2)$ seem to be nearly always observed in the latter case, especially in the case of stretched sample. They seem to be less important in the case of a uniaxial compression. As far as we know, the way the linearity is recovered as the swelling degree increases has not been well studied. This may be partly due to the fact that, in the swollen state, stretching experiments are rarely performed (for experimental convenience, uniaxial compression experiments are in general preferred).

The $\phi^{1/3}$ dependence of the shear modulus has been observed for many systems (see e.g. Horkay, Hecht and Geissler, 1989) and, in some sense, this observation is the main support for the theory. However, some studies have shown a higher dependence of G on ϕ (see Bostiole *et al*, 1984, e.g. Ilavski Bouchal and Dušek, 1989). Recently, modifications of the classical theory were proposed based on assumptions concerning the reference state (Panyukov, 1990a and b; Obukhov, Rubinstein and Colby 1994) and can lead to different values of the modulus exponent (as will be shown below, one could generalize Eq. (A.18) and write $G \sim \phi^l$, which requires also a ϕ^l dependence in the first term of the right-hand side of Eq. A.8′). More specifically, such a change can be achieved within the framework of the Frenkel–Flory–Rehner assumption, by identifying the mean square end-to-end distance in the reference state with that of a free chain in solution at volume fraction ϕ, assuming that the elementary meshes of the network are copiously interspersed in this reference state. It must also be stressed that the $\phi^{1/3}$ dependence is not an unambiguous signature of an affine deformation of the crosslink positions. We return to this point below.

A.1.3 CRITICISM AND EXPERIMENTAL LIMITATIONS OF THE DESCRIPTION

On formal grounds, a number of criticisms has been addressed to this family of theories. We refer especially to those formulated by de Gennes (1979), who pointed out that chains can no longer be Gaussian when the swelling agent is a 'good solvent'. As is now well known, excluded volume interactions provoke a very profound change of the chain statistics, which are liable to have an effect on the elastic properties. Note that this point also makes the definition of the reference state more problematic. Another point is that the Flory–Huggins theory of polymer solutions cannot be considered as fully satisfactory since it underestimates the fluctuations of polymer concentration (or the monomer correlations) in the medium (it is a 'mean field theory'). In practice, it predicts a ϕ^2 dependence of the osmotic pressure of semi-dilute solutions in very good solvent which does not fit the data very satisfactorily. A much better description is given by the des Cloizeaux (1975) law, which predicts $\Pi_{os} \sim \phi^s$, with $s = vd/(v_d - 1) \sim 2.31$, v and d being respectively the excluded volume exponent (~ 0.588) and the dimension of space. In practice, this means that the use of the Flory–Huggins theory for the osmotic pressure of solutions requires a variation of χ with ϕ.

A second type of difficulties is related to the assumptions made for the mechanism of chain deformation in the network. Experimentally, a very small variation for the radius of gyration of a labelled mesh has been found in gels submitted to quite strong changes of volume, produced by osmotic deswelling (Bastide *et al*, 1984). These results, which are also consistent with NMR data (Cohen-Addad, this book and refs therein) do not agree with the assumption of affine displacement of junctions fixed in space (the strict affine hypothesis of Flory–Rehner) and, possibly, not even with the phantom network assumption (affine deformation of mean positions + large fluctuations). Neutron scattering experiments performed on gels containing 'labelled paths' (deuterated chains passing through many crosslinks), deswollen between the state of preparation and the bulk, also indicated a 'loss of affineness' at variance from the expected one (Bastide, Herz and Boue 1985; Bastide, 1985a). More precisely the variation of the deformation along the paths with the length scale of observation was significantly different from those corresponding to the strict affine and the phantom network assumptions. Such type of experiments suggested that the change of volume of a gel (or even its change of shape at constant volume) could be accounted for not only by rearrangements of configurations of the elementary chains but also, and even mainly, by a rearrangement on a larger scale than the mesh size. Through an exchange of the respective positions of the first-neighbour crosslinks, the gel could behave as a sort of tridimensional accordion. A limiting case of such a mechanism can be found in macroscopic networks made by rigid rods connected by flexible crosslinks with $f = 4$, which can be folded and unfolded quite easily (Bastide, Candau and Picot, 1981). The shape of this particular network is changed owing to a highly non-affine process (the length of elementary mesh, the rod, is obviously kept constant). In microscopic systems of this kind, entropic elasticity can, however, exist since the folded conformations are obviously much more probable than the unfolded ones. This led us to propose that, even in networks of flexible chains, the elastic energy could be stored not only in the stretching of the individual chains but also in the configurational entropy of the network on larger length scales.

Theoretical estimates of the elastic moduli of networks of rigid rods have been worked out and, surprisingly, they are not very different, provided the deformation is not too large, from the classical expectations for networks of Gaussian chains (Boué, Edwards and Vilgis, 1988; Vilgis, Boué and Edwards, 1989). Then, in this limiting case, there would be no contribution at all of a single mesh elasticity. It must be stressed that the condition for flexibility of these systems with rigid meshes writes $f \leqslant 2d$ (see e.g. Vilgis, Boué and Edwards, 1989). 'Marginal' systems, for which $f = 2d$, are still floppy, but due only to cooperative movements up to length scales comparable with their size. Thus they presumably exhibit extremely low moduli (Rubinstein, Leibler and Bastide, 1992a). One may then forecast a puzzling consequence: by removing junctions in such a marginal network, one should increase the elastic modulus (Rubinstein *et al*, 1992).

The third type of difficulties encountered when using classical theories of gel swelling and elasticity is that it is not often possible to fit at the same time the swelling and the mechanical data with the same values of the parameters of the theory. When it can be done, the values of these parameters rarely correspond to those expected. For instance, one may be forced to assume that the polymer concentration in the 'reference state' is significantly different from the concentration in the state of preparation (Dušek and Prins, 1969). Besides, the fitted values of χ are often found larger than expected. As a result, the predictions often remain at a qualitative level.

A.2 More general formulation (phenomenological approach + scaling laws)

In short, the classical description, which is very ambitious in its goals, is, at the same time, rather heavy to handle because of the rather large number of parameters involved and is not fully satisfactory in terms of the assumptions regarding the microscopic length scales. This is why it has been proposed to turn to a less ambitious 'phenomenological approach' in which microscopic quantities are eliminated (Bastide, Candau and Leibler, 1980). The key idea is to express, self-consistently, measurable quantities with respect to each other within a model which retains the main features of the Frenkel–Flory–Rehner theory. Such an approach is therefore far less ambitious, since, by construction, it cannot relate macroscopic parameters to chemical parameters as for example the size of the elementary mesh. One expects, however, that the gain in lightness allows us to reach some conclusions more easily.

In such a framework, the first modification is almost obvious. For example, in the case of gels in good solvent, one can conveniently replace the Flory–Huggins free energy by a form consistent with the des Cloizeaux law for the osmotic pressure of semi-dilute solutions. One writes

$$F'_{mix}/kT = u^*\phi^s \tag{A.19}$$

Following the same method as in Eq. (A.10a), one indeed obtains $\Delta\mu_{s,mix}/k_B T = -(s-1)u^*\phi^s$ and then

$$\Pi_{mix} = \frac{k_B T}{v_1}(s-1)u^*\phi^s \tag{A.20}$$

Regarding the elastic contribution to the free energy, a very strong assumption was made. It was postulated that the controversial logarithmic term $B(\phi/N)k_B T \ln \phi/\phi_0$ is absent or can be neglected. The argument in favour of this absence is essentially experimental: this log term would lead to a significant additional positive swelling pressure with respect to Π_{mix} (see Eq. (A.12)''. For gels with sufficiently low ϕ_0 and extensively deswollen with respect to the reference state (in principle identified with the preparation state), Π_{os}^{gel} should even become higher than the osmotic pressure of a semi-dilute solution (identified with Π_{mix}). As far as we know, such a behaviour has not been observed. This is certainly not enough for a definite conclusion. It must therefore be stressed that any deviation from theoretical predictions made with such a simplistic representation should be subject to corrections arising from the presence of a logarithmic term.

The last steps of this phenomenological approach consist of the following:

1. Getting rid of N_c by incorporating this molecular quantity in a unspecified parameter A', whose value necessarily depends on the crosslinking density.
2. Assuming that the factorization appearing in Eq. (A.8)' is conserved, but that the exponent of ϕ is not necessarily equal to 1/3, but to a certain value l. Behind this way of proceeding, one assumes that the non-linearity of F'_{el} with ϕ may not only arise from a competition between dilution of the elastic strands and their extension (for, say, the swelling) but also, possibly, from disinterpenetration effects (i.e. reorganization on length scales larger than the mesh size).

Let us reformulate this idea in a slightly different manner. Consider first a network which can be described as a close-packing of disinterpenetrated meshes (c^* description). From the example of the rigid-rod network it can be conjectured that this expanded system should not be able, under deformation at constant volume, to achieve reorganizations on length scales larger than its mesh. The end-to-end distances would therefore be affinely deformed and the modulus would essentially correspond to the one predicted by the Flory–Rehner theory (if we assume no pre-extension of the meshes). If, as suggested by some neutron scattering experiments (Bastide *et al*, 1984), the gel deswells with respect to this state by folding of the network on length scales larger than the mesh rather than by compression of the latter, one would reach a significant interpenetrated system. The loss of affineness of the meshes under deformation should then increase, because of the enhanced capability, in crumpled networks, to rearrange on length scales larger than the mesh. Then, one may expect the elastic modulus to be lower than the 'affine' value. Following this scheme, one expects, therefore, a non-linear dependence of the modulus with ϕ which does not arise from the change, with swelling degree, of the 'pre-extension' of the elementary meshes. For simplicity, this non-linear dependence was modelled by a scaling law ϕ^l (which, as we have seen above, appears the same way in the free energy density and in the modulus). In sum one writes

$$F'_{\mathrm{el}}/kT = A'\phi^l(\lambda_x^2 + \lambda_y^2 + \lambda_z^2) \tag{A.21}$$

or, in the case when the gel remains isotropic,

$$F'_{\mathrm{el}}/kT = 3A'\phi^l \tag{A.21}$$

where, as stated above, A' is a parameter depending on the structural features of the network and l an exponent smaller than unity ($0.3 < l < 0.6$).

In order to eliminate the indetermination on A', one can express it from an experimentally measured quantity, the polymer concentration at maximum swelling ϕ_e. After incorporation of the above modifications, Eq. (A.12)″ becomes

$$\Pi_{\mathrm{os}}^{\mathrm{gel}} = \frac{k_B T}{v_1}[(s-1)u^*\phi^s - 3A'(1-l)\phi^l] \tag{A.22}$$

As previously, the swelling equilibrium condition is given by $\Pi_{\mathrm{os}}^{\mathrm{gel}} = 0$ and reads as follows:

$$(s-1)u^*\phi_e^s = 3A'(1-l)\phi_e^l \tag{A.23}$$

and

$$\phi_e = \left[\frac{3(1-l)A'}{(s-1)u^*}\right]^{1/(s-l)}$$

thus

$$A' = \frac{(s-1)u^*}{3(1-l)}\phi_e^{s-l} \tag{A.24}$$

where u^* can be taken from osmotic pressure data of semi-dilute solution (if we first ignore what we call the Geissler–Hecht–Horkay correction in section 5.3). Obviously, A', as given by Eq. (A.24) can now be reincorporated in Eqs (A.21) and (A.22):

$$F'/k_B T = u^*\phi^s + \frac{(s-1)u^*}{3(1-l)}\phi_e^{s-l}\phi^l(\lambda_x^2 + \lambda_y^2 + \lambda_z^2) \tag{A.25}$$

$$\Pi_{\mathrm{os}}^{\mathrm{gel}} = \frac{k_B T}{v_1}\{(s-1)u^*\}[\phi_e^s - \phi_e^s(\phi/\phi_e)^l] \tag{A.26}$$

As above, derivation of the elastic modulus is performed easily and one obtains

$$G = \frac{k_B T}{v_1}\frac{2(s-1)u^*}{3(1-l)}\phi_e^{s-l}\phi^l \tag{A.27}$$

This relation gives therefore:

1. The variation of G for a given gel at different swelling degrees ($\sim \phi^l$), which is obtained in a sense by construction.
2. The variation of G for a series of gels of different crosslinking densities swollen at saturation ($\phi = \phi_e$):

$$G = \frac{k_B T}{v_1} \frac{2(s-1)u^*}{3(1-l)} \phi_e^s \tag{A.28}$$

and for $1 = 1/3$

$$G = \frac{k_B T}{v_1}(s-1)u^* \phi_e^s = \Pi_{os}^{mix} \; (= \Pi_{os}^{sol} \text{ in the framework of the strong additivity assumption}). \tag{A.29}$$

For $s = 2.31$ (or 9/4, depending on the value taken for v) one obtains the result originally predicted in the framework of the 'c^* theorem' (de Gennes, 1979 and refs therein, see also Munch *et al*, 1977). But it must be stressed that this result is obtained here without ϕ_e being necessarily proportional to ϕ^* of the average mesh. A more restrictive set of assumptions was used in the original version of this phenomenological approach (Bastide, Candau and Leibler, 1980). The present derivation is taken from Bastide (1985c), where a discussion of different possible levels of assumptions is given.

Again the result of Eq. (A.27) can be reinjected in the free energy and in the osmotic pressure:

$$F' = u^* k_B T \phi^s + \frac{v_1 G}{2}(\lambda_x^2 + \lambda_y^2 + \lambda_z^2) \tag{A.29}$$

$$\Pi_{os}^{gel} = \left[\frac{k_B T}{v_1}(s-1)u^* \phi^s - \frac{k_B T}{v_1}(s-1)u^* \phi_e^s (\phi/\phi_e)^l \right]$$

$$\Pi_{os}^{gel} = \Pi_{os}^{sol} - \frac{(1-1)}{2/3} G(\phi) \tag{A.30}$$

which reduces to the following very simple expression if l takes the classical value 1/3.

$$\Pi_{os}^{gel} = \Pi_{os}^{sol} - G(\phi) \tag{A.31}$$

INDEX

CPSIA information can be obtained
at www.ICGtesting.com
Printed in the USA
BVHW072011180919
558841BV00006B/16/P

9 780471 939719